PRINTREADING

Part 1

For RESIDENTIAL CONSTRUCTION

Fourth Edition

Thomas E. Proctor
Leonard P. Toenjes

AMERICAN TECHNICAL PUBLISHERS, INC.
HOMEWOOD, ILLINOIS 60430-4600

American Technical Publishers, Inc. Editorial Staff

Editor in Chief:
Jonathan F. Gosse
Production Manager:
Peter A. Zurlis
Technical Editor:
Michael B. Kopf
Copy Editor:
Catherine A. Mini
Illustration/Layout:
James M. Clarke
Erin E. Clifford
William J. Sinclair
Ellen E. Pinneo
Aimée M. Brucks
Maria R. Aviles
CD-ROM Development:
Carl R. Hansen

4 5 6 7 8 9 – 04 – 9 8 7 6 5 4 3

Printed in the United States of America

ISBN 978-0-8269-0409-6

This book is printed on 30% recycled paper.

Acknowledgments

The authors and publisher are grateful to the following companies for providing technical information and assistance.

American Hardwood Export Council
APA—The Engineered Wood Association
ASME International
ASTM International
California Redwood Association
CertainTeed Corporation
David White Instruments
The Garlinghouse Company
Hulen & Hulen Designs
ITW Ramset/Redhead
James Hardie Building Products
Kohler Co.
Kolbe & Kolbe Milwork Co., Inc.
Leica Geosystems
Plan Ahead, Inc.
Rodger A. Brooks, Architect
Sioux Chief Manufacturing Company, Inc.
Southern Forest Products Association
Topcon Laser Systems, Inc.
Trus Joist, A Weyerhaeuser Business
Vanguard Piping Systems, Inc.
Victaulic Company of America
William Brazley & Associates
Wire Reinforcement Institute
Wood Truss Council of America

Contents

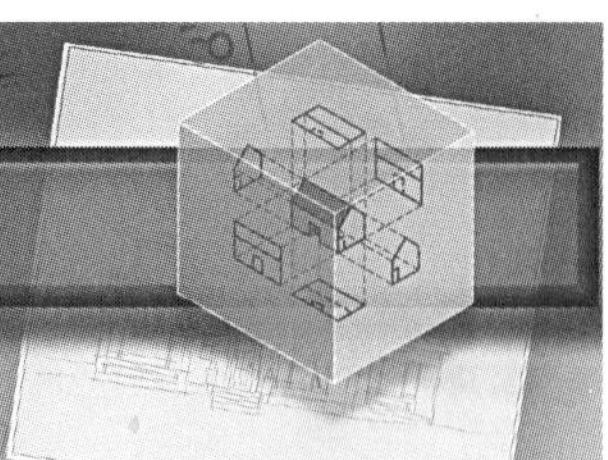

CD-ROM Contents

- Using This CD-ROM
- Quick Quizzes™
- Illustrated Glossary
- Stewart Residence – Prints
- Stewart Residence – 3-D Model
- Master Math™ Problems
- Media Clips
- Reference Material

Introduction

Printreading for Residential Construction–Part 1, 4th Edition, presents printreading fundamentals and provides printreading activities related to residential construction. Topics presented include conventional drafting, computer-aided design (CAD), symbols and abbreviations, plot plans, floor plans, elevations, sections, and details. Trade information related to printreading for residential construction is also presented.

Two sets of plans are included in the vinyl sleeve at the back of the book. Plans for the Wayne Residence are covered throughout the book. Plans for the Stewart Residence, used in Chapter 11 for the final exam, are included in the vinyl sleeve and also on the *Printreading for Residential Construction–Part 1* CD-ROM. In addition, the Campbell Residence is included for additional printreading activities. A list of the prints is included on the last page of the book.

The CD-ROM in the back of the book contains a wealth of information to supplement the book including the following:

- Quick Quizzes™ provide an interactive review of topics covered in chapters 1–10 of the book. Each chapter Quick Quiz™ includes 10 questions. Each question screen includes textbook reference and glossary icons for obtaining related information about the question. A video icon appears on selected question screens and provides access to a media clip related to the question.
- An Illustrated Glossary provides a helpful reference to key terms included in the text. Selected terms are linked to illustrations and media clips that augment the definition provided.
- The Stewart Residence Prints allow interactive navigation of the electronic version. The electronic prints can be viewed on screen.
- The Stewart Residence 3-D Model provides a three-dimensional representation of the Stewart Residence prints.
- Media Clips provide a convenient link to selected video and animation clips.
- Master Math™ Problems provide formulas for interactive calculation and application of math-related concepts.
- The Reference Material button accesses links to web sites that contain useful related manufacturer and reference information.

Information about using the *Printreading for Residential Construction–Part 1* CD-ROM is included on the last page of the book. To obtain information about related training products, visit the American Tech web site at www.go2atp.com.

The Publisher

ANSWERING QUESTIONS

Chapters 1 through 10 conclude with a variety of sketching exercises, review questions, and trade competency tests. Chapter 11 contains a final review exam, six print-specific exams dealing with different aspects of the prints, and a final printreading exam. Specific instructions are given to complete each sketching exercise. Review questions and trade competency tests include Multiple Choice, Completion, Identification, Matching, True-False, and Printreading questions. Record your answer in the space(s) provided.

Multiple Choice

Select the response that correctly completes the statement. Write the appropriate letter in the space provided.

D **10.** Dimension lines may be terminated by ___.
A. arrowheads
B. slashes
C. dots
D. all of the above

Completion

Determine the response that correctly completes the statement. Write the appropriate response in the space provided.

R-19 **18.** Batt insulation on the rear wall provides an insulation value of ___.

Identification

Select the response that correctly matches the given word(s). Write the appropriate letter in the space provided.

D **1.** Ceiling outlet
B **2.** Bifold doors
A **3.** Double-hung window
C **4.** 240 V receptacle

(A) (B) (C) (D)

Matching

Select the response that correctly matches the given word(s). Write the appropriate letter in the space provided.

Answer	Item	Response
D	**1.** Gable roof	A. Double slope in two directions
E	**2.** Hip roof	B. Double slope in four directions
C	**3.** Flat roof	C. Minimum slope of ⅛″
A	**4.** Gambrel roof	D. Single slope in two directions
B	**5.** Mansard roof	E. Single slope in four directions

True-False

Circle T if the statement is true. Circle F is the statement is false.

(T) F **18.** Hidden lines on floor plans show features above the cutting plane.

Printreading

Study the plan being referred to. Questions may be True-False, Multiple Choice, Completion, Matching, and so forth. Write the answer in the space(s) provided.

Refer to Torrance Residence on page 139.

C **1.** A ___ house is shown.

A. one-story
B. one-story with basement
C. one-and-one-half story
D. two-story with basement

11'-4" **2.** The living room measures ___ × ___.

dashed **3.** The arch between the entry and the living room is shown with ___ lines.

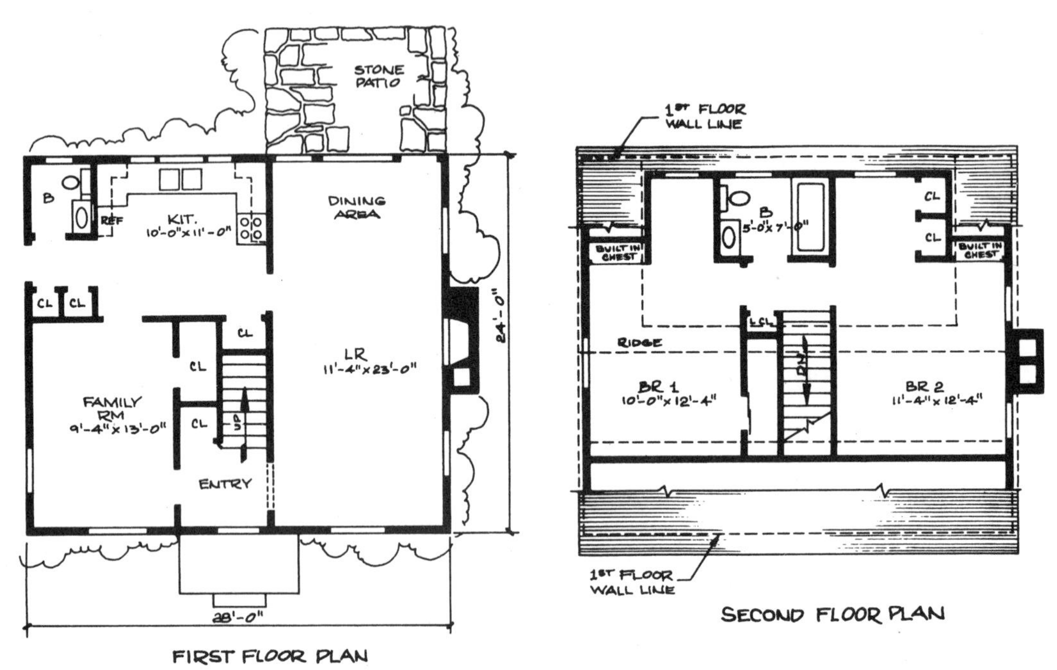

TORRANCE RESIDENCE

Working Drawings and Prints

1

Working drawings contain the graphic information necessary to complete a construction job. Plot plans, floor plans, elevations, sections, and details comprise working drawings. Working drawings are made by the conventional method using T-squares, triangles, and other drafting instruments or by the computer-aided design (CAD) method using computers, plotters, printers, and other computer hardware and software. Prints are reproductions of working drawings. Tradesworkers read and follow prints to complete a construction job.

WORKING DRAWINGS

A *working drawing* is a drawing that contains the graphic information necessary to complete a construction job. *Specifications* are written supplements to working drawings that provide additional building information. Working drawings include plot plans, floor plans, elevations, sections, and details. A title block identifies each sheet.

A builder or prospective owner (client) meets an architect to discuss the planning of a house. The architect determines the parameters of the available building space based on items such as lot size, lot shape, and available project budget. House styles are discussed and client needs are determined including intended usage, desired number of rooms, and approximate sizes of rooms. Materials, equipment, finishes, and fixtures are also discussed. The approximate price range of the house is determined, zoning requirements are discussed, and building codes are reviewed.

Compliance with various building codes and legislative requirements is necessary to obtain a building permit. For example, the intended building occupancy and usage could require compliance with the Americans with Disabilities Act (ADA). Variations in door opening sizes and countertop heights may be required in certain circumstances. Other access and egress accommodations may also be required. With this information, the architect prepares a series of rough sketches for client review. The client makes final decisions and the architect proceeds to complete the working drawings.

Southern Forest Products Association

Working drawings convey graphic information about a building from an architect to the tradesworkers.

Plot Plans

A *plot plan* is a scaled drawing that shows the shape and size of the building lot; location, shape, and overall size of a house on the lot; and the finish floor elevation. Plot plans are drawn to smaller scales than floor plans. Solid contour lines on a plot plan show the finish grade contour of the building lot. Dashed contour lines, when included on a plot plan, show existing grade contours. Other information shown on a plot plan includes a symbol designating true North and the point of beginning from which building corners and heights, location of streets, easements, and utilities are established. See Figure 1-1.

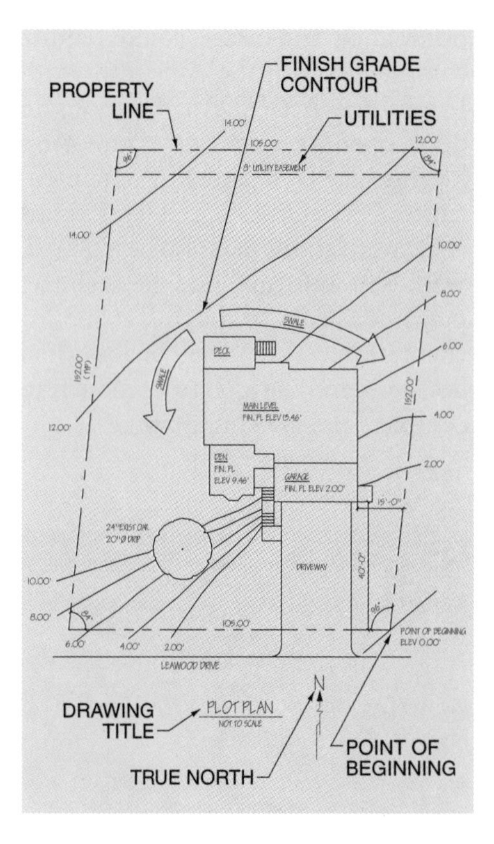

Figure 1-1. Plot plans show the shape and size of the building lot and location and shape of the house.

Floor Plans

A *floor plan* is a scaled view of the various floors in a house looking directly down from a horizontal cutting plane taken 5′-0″ above each finished floor. Floor plans show the layout of rooms and give information about windows, doors, cabinets, fixtures, and other features. The shapes, sizes, and relationships of rooms are also shown on floor plans. Dimensions show sizes of rooms, hallways, wall thicknesses, and other measurements. Symbols and abbreviations provide additional information about the rooms. See Figure 1-2. A separate floor plan is required for each story of the house. Floor plans are generally the first drawings of a set of plans to be drawn.

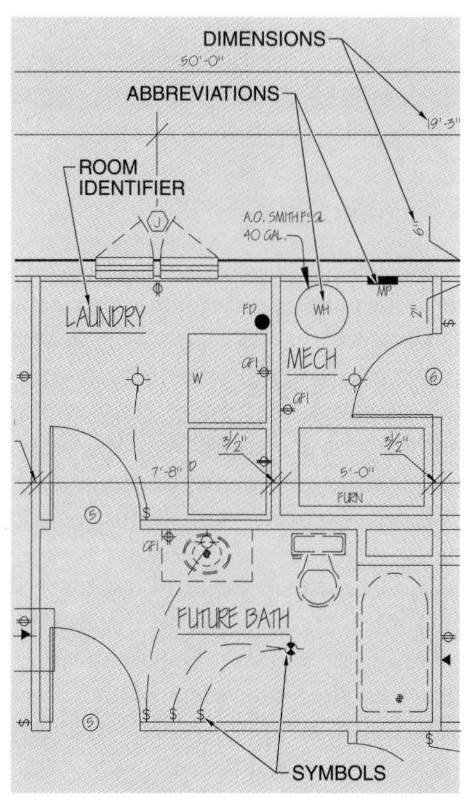

Figure 1-2. Floor plans show the shapes, sizes, and relationships of rooms.

Elevations

An *elevation* is a scaled view looking directly at the walls. Elevations show the true shape of the walls without any allowances for perspective in the drawings. Dimensions include floor-to-floor heights and heights of windows above finished floors. The two types of elevations are exterior and interior elevations. See Figure 1-3.

Exterior Elevations. An *exterior elevation* is a scaled view that shows the shape and size of the outside walls of the house and the roof. At a minimum, four exterior elevations are required to show exterior walls. Exterior elevations are generally drawn to the same scale as the floor plan. Building materials such as brick, stone, fiber-cement siding, vinyl siding, exterior insulation and finish systems (EIFS), and various wood finish materials are specified for the exterior walls through the use of symbols and notations. Door and window openings and their types and sizes are shown in their proper locations. Roof styles, slopes, and types of roof coverings are also shown.

> Elevations contain very few dimensions other than vertical dimensions.

Interior Elevations. An *interior elevation* is a scaled view that shows the shapes, sizes, and finishes of interior walls and partitions of the house. Interior elevations may be drawn to the same scale as the floor plan or to a larger scale when showing more detail. Interior elevations commonly show details of kitchen base and wall cabinets, sinks, dishwashers, ranges, and other built-in appliances. In addition, interior elevations show details of bathroom fixtures such as water closets, bathtubs, showers, and vanities.

Interior elevations also show special wall treatments in rooms. For example, details of interior elevations may show fireplaces, bookcases, and other built-in features. Symbols and notations are used to specify materials and specific installation instructions may be given.

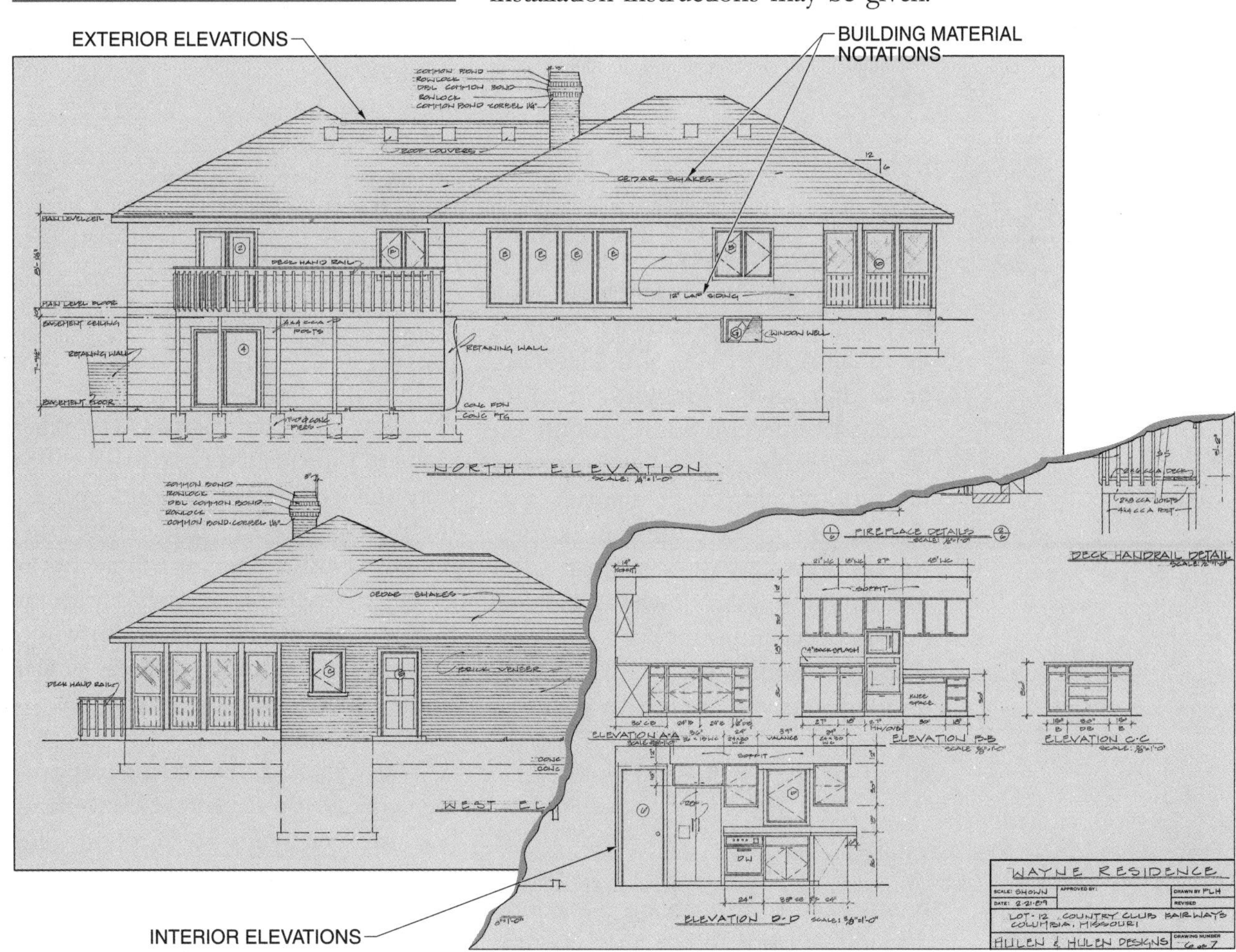

Figure 1-3. Elevations show the sizes and shapes of exterior walls, interior walls, and partitions.

Sections

A *section* is a scaled view created by passing a cutting plane through a portion of a building. See Figure 1-4. Common sections are taken through outside walls to show information about foundation footings and walls, wall and floor framing, height of windows above floors, and eaves and roof construction.

Cutting planes for sections are shown on floor plans. Direction arrows on the cutting plane line show the line of sight. Coded references shown with the cutting plane line give the sheet number of the section.

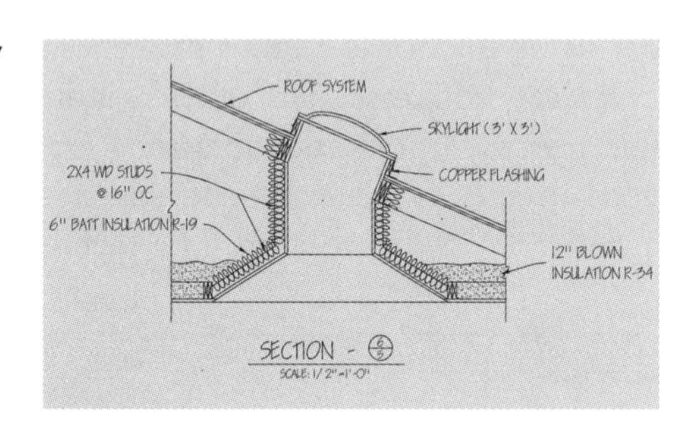

Figure 1-4. Sections show features revealed by a cutting plane.

Details

A *detail* is a scaled plan, elevation, or section drawn to a larger scale to show special features. Whenever a part of a building cannot be shown clearly at the small scale of the plan, elevation, or sections, it is redrawn at a larger scale so that necessary information can be shown more clearly. For example, a fireplace, framing for a stairwell, special interior or exterior trim, an entrance doorway, or a section of a foundation may be drawn as a detail. Section details show the cross-sectional shape of features such as foundation footings and windows. The scale for the detail is determined by the complexity of the detail. See Figure 1-5.

Title Blocks

A *title block* is an area on a working drawing or print that is used to provide written information about the drawing or print. See Figure 1-6. The title block is located along the right side or bottom of the sheet. Title blocks give the number of the sheet and total number of sheets in the set of plans. For example, 1 OF 7 denotes the first sheet of a set of plans containing seven sheets. In a larger set of plans, initials may precede the sheet number to indicate a particular trade area. For example, E1 OF 3 denotes the first electrical sheet of three electrical sheets. Other letters commonly used to denote specific trade area prints are P for plumbing and M for mechanical including heating, ventilating, and air conditioning (HVAC).

In addition to the sheet number, the title block may include the following information:

- design firm name
- architect seal
- project or building number
- date
- revisions
- drafter initials
- checker initials
- owner or client name
- building site address
- scale of drawings

PRINTS

A *print* is a reproduction of a working drawing. Originally, these reproductions were referred to as blueprints because the process used to make them produced a white line on a blue background. Any number of copies or blueprints could be made from working drawings by using a process similar to the process used for making photographic prints.

Today, diazo and electrostatic prints are generally preferred over blueprints because their white backgrounds and dark lines make them more legible. Electrostatic prints are popular because they can be easily enlarged or reduced. See Figure 1-7.

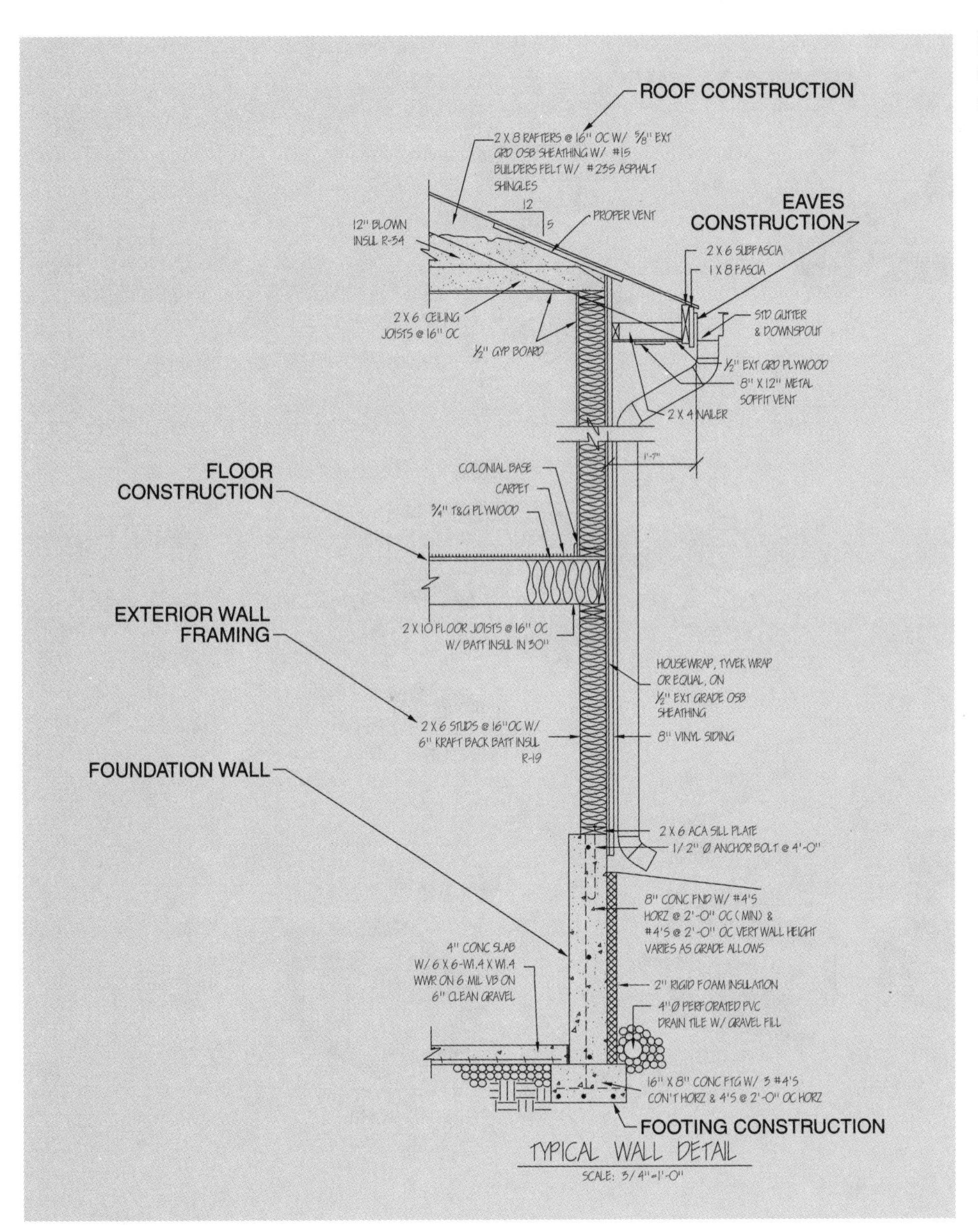

Figure 1-5. Details show special features of plans as elevation or section views.

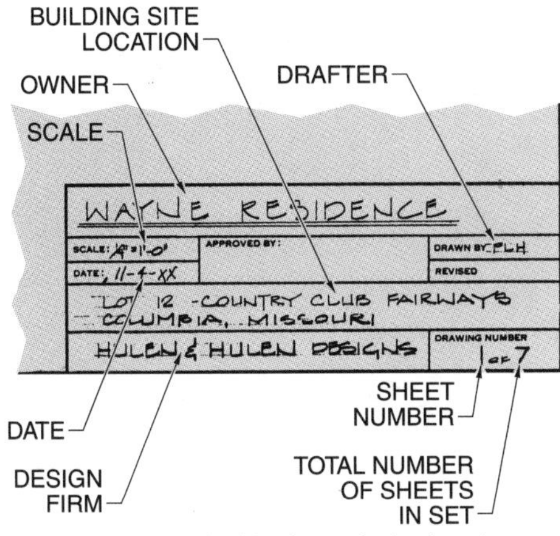

Figure 1-6. Title blocks include the sheet number, architect or design firm name, owner, and other pertinent information.

Blueprints

The use of blueprints began in 1840 when a method of producing paper sensitized with iron salts was developed. The paper underwent a chemical change when exposed to light. Original drawings made on translucent paper (paper that allowed light to pass through) were placed over the sensitized paper in a glass frame that held the paper tightly. The drawings were then exposed to sunlight. A chemical reaction occurred wherever the light struck the sensitized paper.

Blueprints are rarely used today and fade when continuously exposed to sunlight.

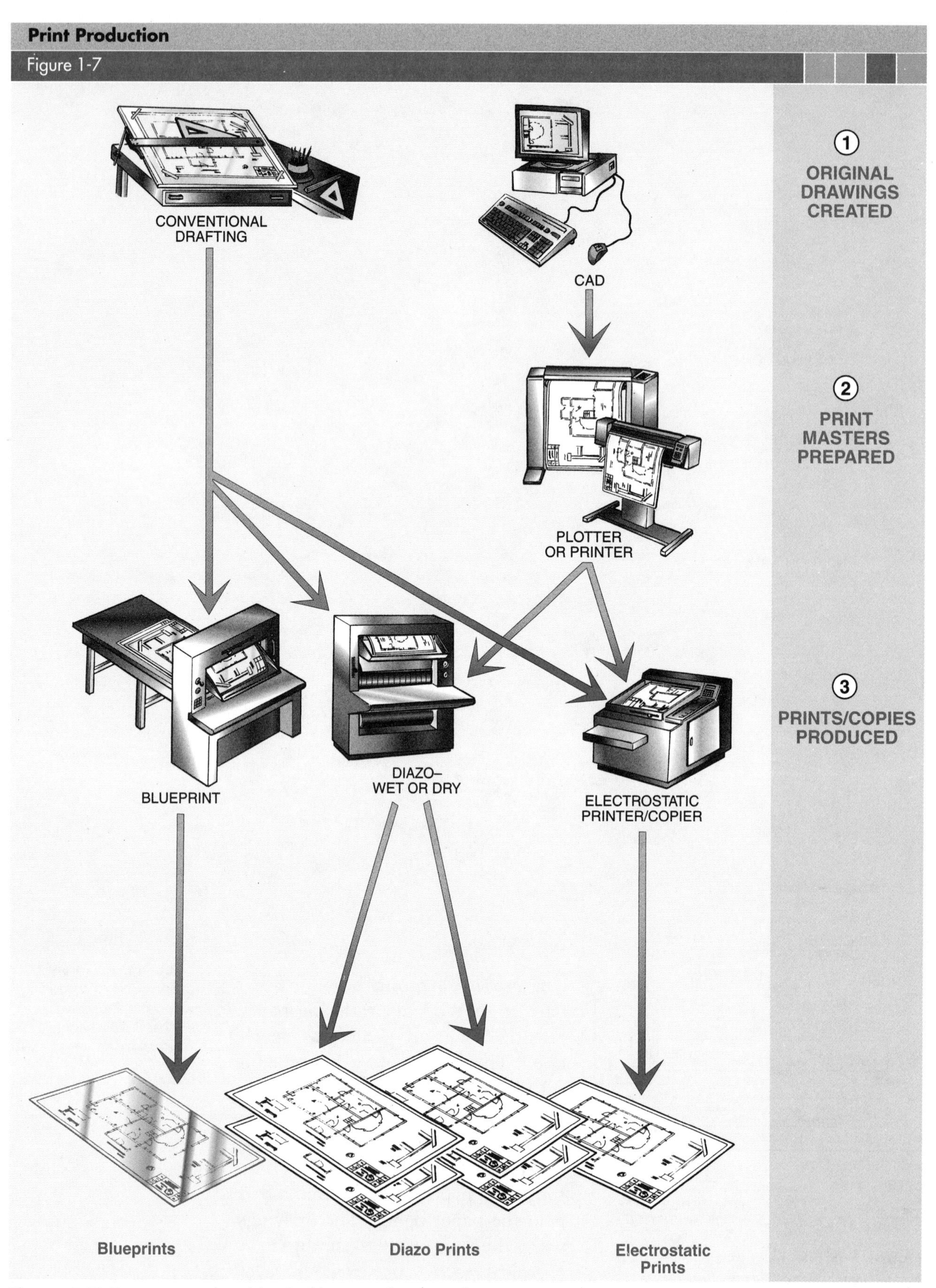

Figure 1-7. Prints are produced by the blueprint, diazo, or electrostatic process.

When the sensitized paper was washed in water, the part protected by the pencil or ink lines on the original drawing would show as white lines on a blue background. A fixing bath of potassium dichromate, a second rinse with water, and print drying completed the process. Blueprints are not common today since other methods have been developed to replace them.

Diazo Prints

Many prints used today have blue or black lines on a white background and are made by the diazo process. The diazo process has the advantage of providing excellent reproductions with very good accuracy because the paper is not soaked with water and then dried as blueprints are. Diazo prints, with their white background, are easier to read than blueprints. Additionally, the white background provides a convenient area for writing field notes or making required changes.

Two types of sensitized paper are used in the diazo process, one for each development method. These sensitized papers are coated with a chemical that, when exposed to ultraviolet light, becomes a part of a dye complex. An original drawing or a copy, on translucent material, is placed over a sheet of the sensitized paper, with the yellow side of the paper held against the original, and is fed by a belt conveyor into a print machine. The two sheets revolve around a glass cylinder containing an ultraviolet lamp and are exposed to the light. The sensitized paper is exposed through the translucent original in the clear areas but not where lines or images block the light. The sheets are separated, the original is returned to the operator, and the sensitized paper is transported through the developing area. The sensitized paper is then developed by either a wet diazo or dry diazo method.

Wet Method. In the wet development method, the sensitized paper passes under a roller that moistens the exposed top surface, completing the chemical reaction to bring out the image. Prints made by this method have black or blue lines on a white background.

Dry Method. In the dry development method, the sensitized paper is passed through a heated chamber where its surface is exposed to ammonia vapor. The ammonia vapor precipitates the dye to bring out the image. Prints made using the dry diazo method have black or blue lines on a white background. The dry diazo method is used today to provide high-quality reproductions on Mylar® (plastic film) or sepia (brown line) copies.

Electrostatic Prints

Electrostatic prints are produced using the same process that office copiers use. Full-size working drawings are exposed to light and projected directly through a lens onto a negatively charged drum.

The drum is discharged by the projected light from the nonimage areas but retains the negative charge in the unexposed areas. The drum then turns past a roller where black toner particles are attracted to the negatively charged image areas on the drum surface. As the drum continues to turn in synchronization with the positively charged copy paper, toner particles are attracted to the paper and fused to it by heat and pressure. See Figure 1-8.

Prints made using the electrostatic method have black lines on a white background. The advantages of the electrostatic process include easy enlargement and reduction of drawings, small storage size, quick retrieval and duplication, and reduced shipping costs.

An engineering copier, which uses the electrostatic method, can produce up to 1200 D-size prints per hour.

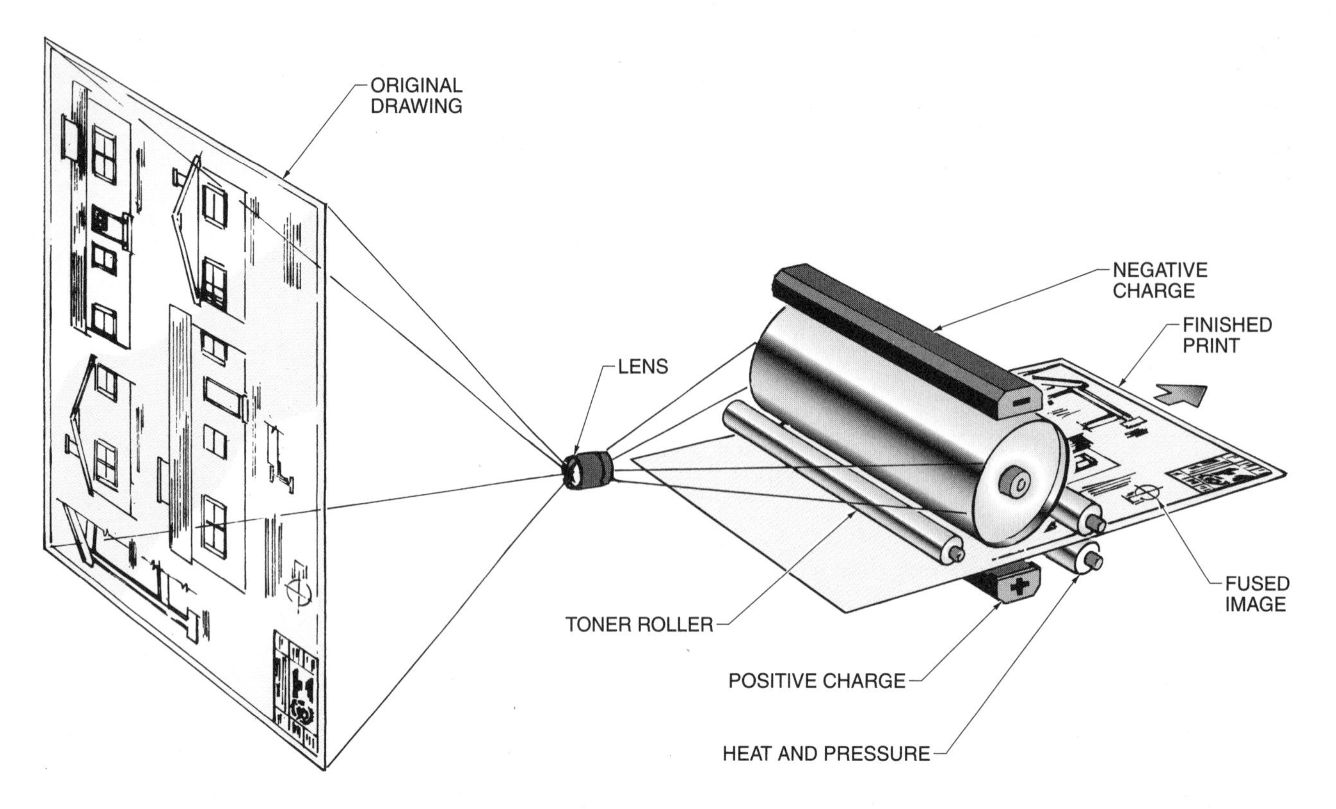

Figure 1-8. Electrostatic prints are produced as light is projected through a lens onto a negatively charged drum that offsets the image to positively charged paper.

The major disadvantage of electrostatic prints is the potential for distortion by projection through a lens.

DRAFTING METHODS

Working drawings for prints may be made using conventional drafting practices or computer-aided design software. A "language" of standard lines, symbols, and abbreviations is used in conjunction with drafting principles so that drawings are consistent and easy to read. The American National Standards Institute (ANSI) in conjunction with ASME International (formerly the American Society of Mechanical Engineers) publishes standards related to drafting conventions in their Y series standards.

> Linetypes on working drawings are established in ASME Y14.2M, *Line Conventions and Lettering.*

Conventional Drafting

Basic tools such as T-squares, triangles, scales, and pencils are used to produce working drawings by the conventional method. Other drafting instruments such as dividers and compasses are also used to produce working drawings. See Figure 1-9. Drafting machines (com-

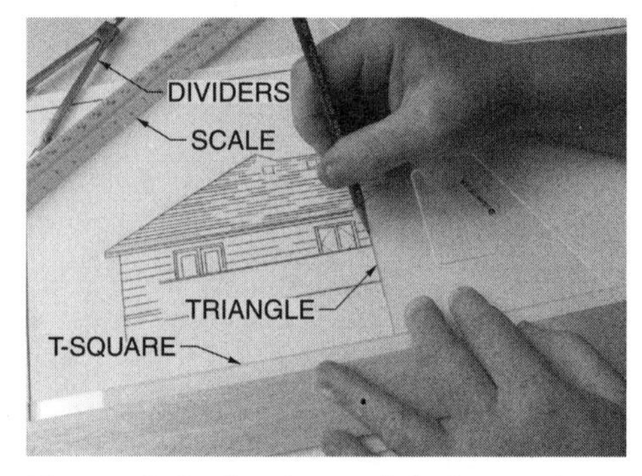

Figure 1-9. Conventional drafting tools include T-squares, triangles, scales, and pencils. Drafting instruments, such as dividers, may also be used to create drawings.

bination T-square, scale, and triangles) and parallel straightedges (combination drafting board and modified T-square) are commonly used for production as well. Drawings are started after properly taping the drafting paper to the drafting board. Initial linework is done using light construction lines, which are later darkened to produce the final drawing.

T-Squares. A *T-square* is a drafting tool used to draw horizontal lines and as a reference base for positioning triangles. The head of a T-square is held firmly against one edge of the board to ensure accuracy. T-squares are made of wood, plastic, or aluminum and are available in various lengths. Common T-squares are 24″ to 36″ in length.

Triangles. A *triangle* is a drafting tool used to draw vertical and inclined lines. The base of a triangle is held firmly against the blade of the T-square to ensure accuracy. Two standard triangles–30°-60° and 45°–are available in a variety of sizes. A 30°-60° triangle is used to produce vertical lines and inclined lines of 30° or 60° sloping to the left or right. A 45° triangle is used to produce vertical lines and inclined lines of 45° sloping to the left or right. The triangles may be used together to produce inclined lines every 15°. Triangles are made of clear plastic. See Figure 1-10.

Drafting Instruments. Although a wide variety of precision drafting instruments is available, the compass and dividers are the most commonly used instruments. Compasses and dividers are available in varying sizes. See Figure 1-11.

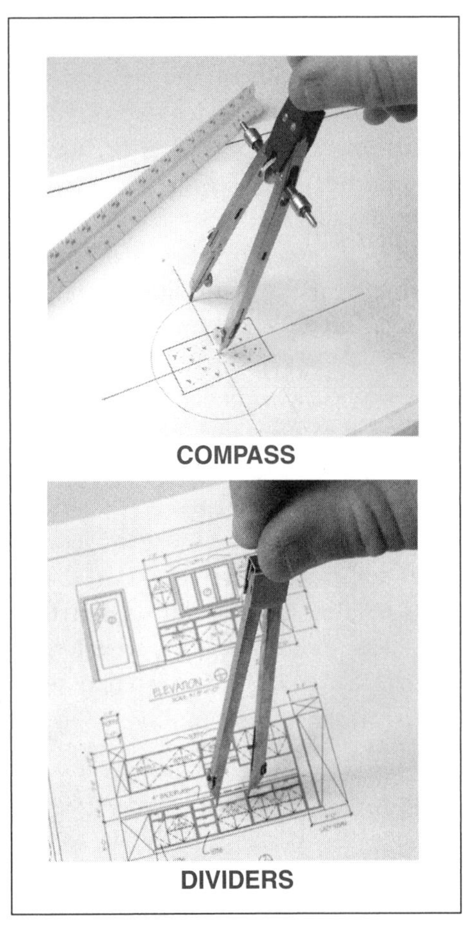

Figure 1-11. A compass is used to draw arcs and circles. Dividers are used to transfer measurements.

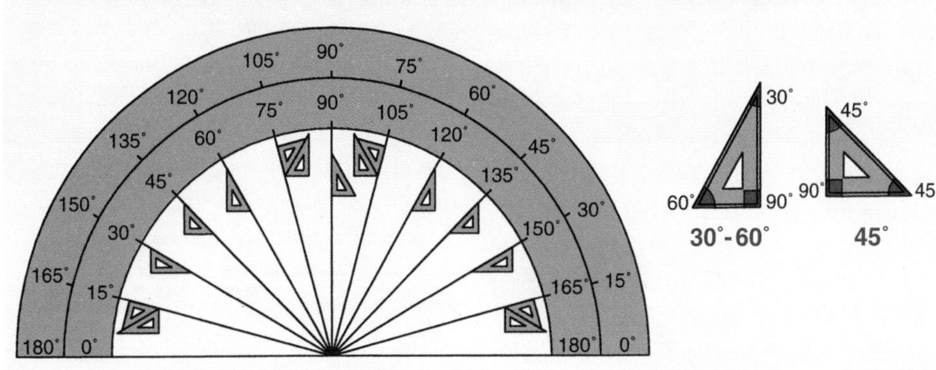

Figure 1-10. The 30°-60° and 45° triangles are used to draw lines 15° apart.

A *compass* is a drafting instrument used to draw arcs and circles. One leg of a compass contains a needlepoint that is positioned on the centerpoint of the arc or circle to be drawn. The other leg contains the pencil lead used to draw the line. Two types of compasses are center-wheel and friction compasses. On center-wheel compasses, the center wheel is adjusted to change the radius of the arc. On friction compasses, arcs of various radii are obtained by opening or closing the legs. Center-wheel compasses are the most popular and most accurate.

Dividers are a drafting instrument used to transfer dimensions. Each leg of a set of dividers contains a needlepoint. Two types of dividers are center-wheel and friction dividers. Friction dividers are more useful for general work.

Other drafting instruments include irregular (French) curves and architectural templates. Irregular curves are used to draw curves that do not have consistent radii. Architectural templates are used to save time when drawing standard items such as doors, windows, and cabinets.

Architect's scales are based on 12 units to the foot while engineer's scales are based on 10 units to the foot.

Scales. A *scale* is a drafting tool used to measure lines and reduce or enlarge them proportionally. Three types of scales are the architect's scale, civil engineer's scale, and mechanical engineer's scale. See Figure 1-12.

An *architect's scale* is a scale used when producing drawings of buildings and structural parts. The most common type of architect's scale is triangular in shape. One edge of the scale is a standard ruler divided into inches and sixteenths of an inch. The other edges contain 10 scales that are labeled 3, 1½, 1, ¾, ½, ⅜, ¼, 3/16, 3/32, and ⅛. For example, the ¼ scale means that ¼″ = 1′-0″. For larger scale drawings, the 1½″ = 1′-0″ or the 3″ = 1′-0″ scales are used.

Scales

Figure 1-12

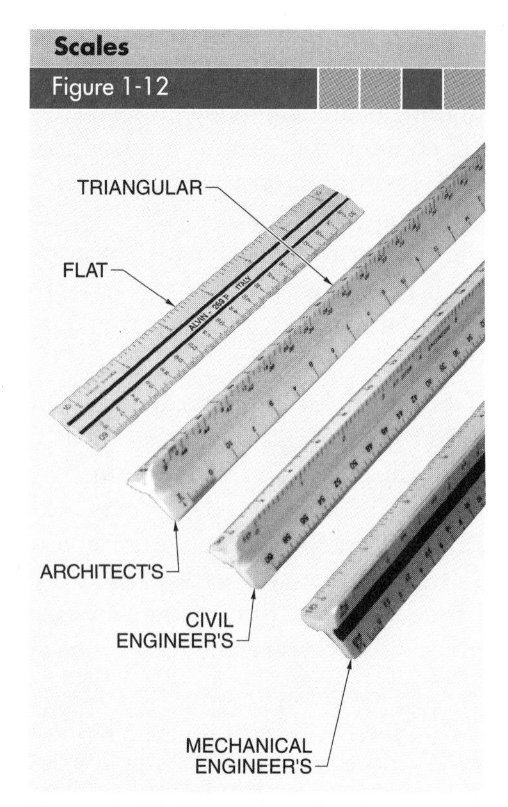

Figure 1-12. Three types of scales are used to produce scaled drawings. The architect's scale is typically used for building trades plans.

A *civil engineer's scale* is a scale used when creating maps and survey drawings. Plot plans may also be drawn using a civil engineer's scale. A civil engineer's scale is graduated in decimal units. One-inch units on the scale are divided into 10, 20, 30, 40, 50, or 60 parts. These units are used to represent the desired measuring unit such as inches, feet, or miles. For example, a property line that is 100′-0″ long drawn with the 20 scale (1″ = 20′-0″) measures 5″ on the drawing.

A *mechanical engineer's scale* is a scale used when drawing machines and machine parts. A mechanical engineer's scale is similar to an architect's scale except that the edges are limited to fractional scales of ⅛, ¼, ½, and 1. Decimal scales are also available.

Architect's and engineer's scales are commonly made of plastic or wood.

Pencils. Wood or mechanical pencils are used to draw lines. Wood pencils have a stamp near one end indicating the lead hardness. When using a mechanical pencil, a lead of the desired hardness is inserted into the jaws of the pencil.

Grades of lead range from 6B (extremely soft) to 9H (exceptionally hard). Hard leads are used to draw fine, precise lines. Medium leads are used to draw object lines. Soft leads are used primarily for sketching. The architect's range is HB, F, H, and 2H. F and H are most commonly used for producing drawings. See Figure 1-13.

Soft grades of drafting leads are thicker in diameter and produce broader lines. Hard grades are thinner and produce fine lines. In addition to sketching, soft grades are used to draw object and border lines on working drawings. Hard grades are used to draw centerlines and section lines.

Computer-Aided Design (CAD)

Computer-aided design (CAD) is also known as computer-aided drafting or computer-aided drafting and design (CADD). Advantages of CAD include speed, accuracy, consistency, changeability, duplication, and storage. A major additional benefit of a CAD system is its information management capabilities. As drawings are made or changed, information about the materials used in the drawings is stored and updated, allowing the operator to extract it for estimating or pricing purposes. Advantages of CAD for tradesworkers include legibility and consistency of line weights, notes, and symbols. In addition, drawings may be printed in colors for greater clarity.

CAD systems use hardware (computer equipment) and software (computer programs) to input (generate), manage (duplicate, file, and retrieve), and output (print or plot) drawings. A wide variety of hardware and software is available for producing CAD-generated drawings. See Figure 1-14.

Figure 1-14. A CAD system includes a keyboard, central processing unit (CPU), monitor, and input devices.

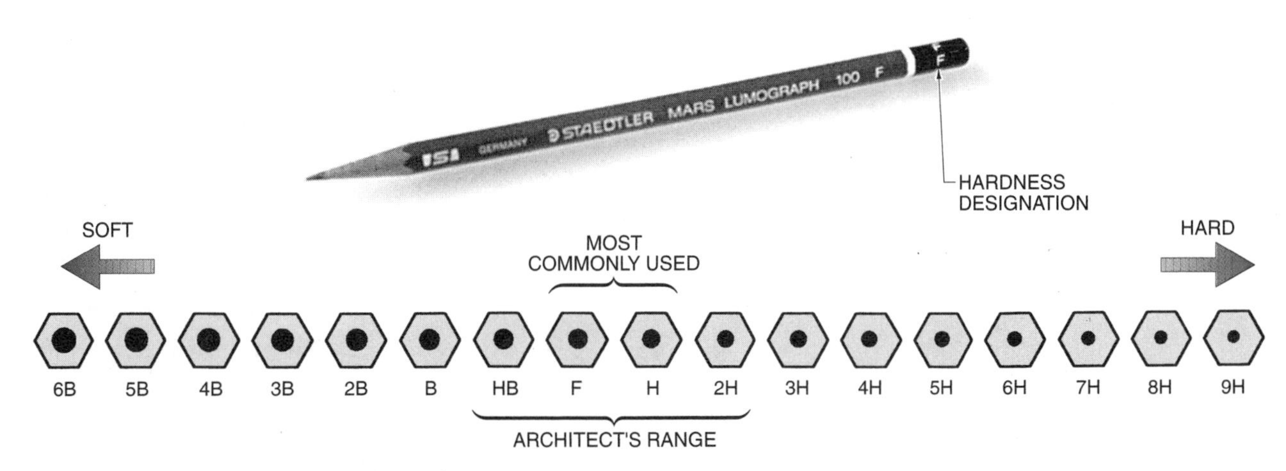

Figure 1-13. Pencil leads range from extremely soft to exceptionally hard.

Drawings produced using conventional drafting methods can be scanned and converted to a CAD format to allow the drawings to be easily updated.

Input Systems. Input systems are composed of a software such as AutoCAD® and hardware such as a keyboard, computer, and monitor. Other input devices such as a mouse, trackball, or other pointing device may be required.

A *keyboard* is an input device containing standard alpha and numeric keys as well as function keys. A *function key* is a key that performs a special task. A keyboard may be used to initiate a command, enter drawing coordinates, or enter text that will appear on the drawings.

A *computer* is a device that receives information from an input device, processes it, and displays the results on a monitor. A hard drive and a large amount of random access memory (RAM) are required to efficiently run CAD programs.

A *monitor* is a high-resolution color display screen that allows drawings, data, and text to be displayed. Components in a CAD drawing are organized by layers, which are often drawn using different colors. For example, a floor plan for a building may be drawn in blue, the electrical plan drawn in red, and the mechanical plan drawn in yellow. Color allows features on the individual layers to be easily distinguished from one another. Layers can be turned on or off to view only the desired information.

A mouse and trackball are input devices used to implement drawing commands that create and modify drawings. A mouse is moved on a tabletop, affecting the movement and location of the cursor on the monitor. A trackball consists of a large ball fitted into a housing. As the ball is rotated, the cursor on the monitor also moves.

A complete set of prints is usually stored in the general contractor's job trailer or at a central location within the residence.

CAD drawings can be easily modified. Three-dimensional CAD drawings can be rotated, revolved, or viewed from other angles. Drawing files are easily copied and stored. Compact discs (CDs) containing CAD drawing files may be mailed, or CAD drawing files may be shared via e-mail or the Internet.

Output Systems. Output systems can display drawings on a monitor, print drawings on a printer, or plot drawings using an electrostatic printer or inkjet plotter. See Figure 1-15. One or two monitors may be used to display drawings. A 21″ or larger monitor is recommended for displaying architectural drawings.

Printers output drawings by transferring the image to paper after receiving electronic information about the drawings from the computer. Prints can be made on opaque paper, tracing vellum, or film in one or several colors.

A laser printer used to output CAD drawings functions the same as a laser printer used for a word-processing program. Information is electronically transmitted to print paper that attracts toner particles, forming the image of the drawing. The particles are then permanently fused to the paper by heat.

Inkjet plotters generate drawings by accurately spraying fine droplets of ink on paper. The paper in an inkjet plotter moves in one direction. As the paper passes below the spray head, ink is deposited on the paper to achieve the desired design.

CAD drawings can be used to create realistic renderings of buildings.

CAD Drawings. CAD drawings are more consistent and standardized than drawings completed using conventional drafting methods. Line weights, symbols,

PRINTER

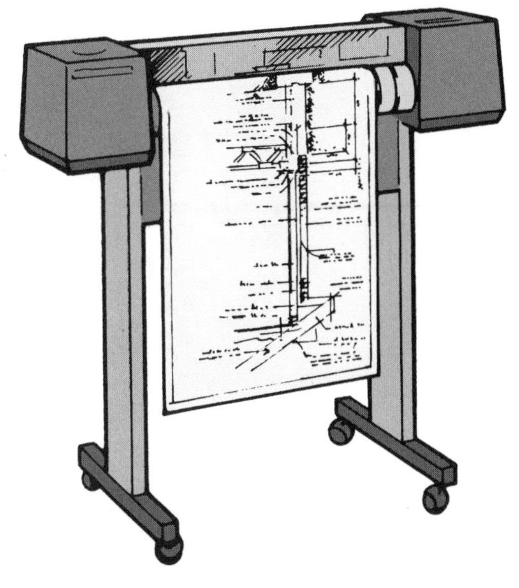

PLOTTER

Figure 1-15. CAD output systems may include a printer or plotter.

and lettering on CAD drawings are precise and easy to read. The same elements on conventional drawings reflect the individual technique of the architect.

CAD drawings are becoming more prevalent due to lower prices of computer systems equipped with sufficient memory and storage capacity for CAD software. Architects are using CAD more readily because of the variety of software programs available and their ability to perform complicated tasks quickly. See Figure 1-16.

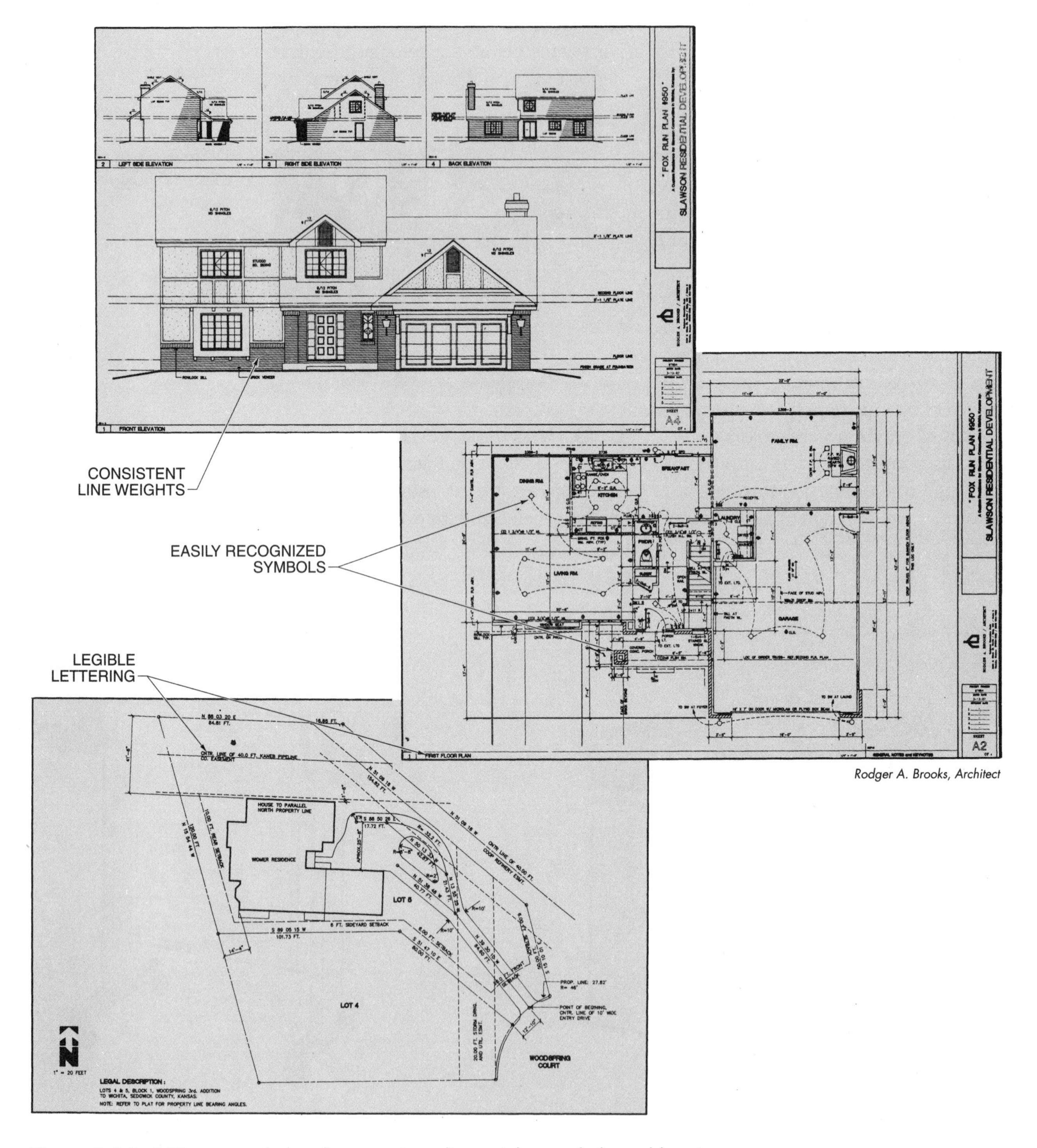

Figure 1-16. CAD-generated plans have consistent line weights, symbols, and lettering.

Working Drawings and Prints 1

Review Questions

Name ______________________________ Date ______________

Multiple Choice

_______________ **1.** A blueprint has ___ lines on a ___ background.
- A. white; blue
- B. white; black
- C. blue; black
- D. black; white

_______________ **2.** The major disadvantage of electrostatic prints is their ___.
- A. sensitivity to light
- B. difficulty of duplication
- C. large storage space requirements
- D. potential for distortion by projection through the lens

_______________ **3.** Shapes, sizes, and relationships of rooms are most clearly shown on ___.
- A. plot plans
- B. floor plans
- C. elevations
- D. details

_______________ **4.** Translucent paper ___.
- A. absorbs light
- B. allows light to pass through it
- C. reflects light
- D. prevents light from passing through it

_______________ **5.** T-squares are made of ___.
- A. wood
- B. aluminum
- C. plastic
- D. all of the above

_______________ **6.** The dry diazo process utilizes ___ to produce prints.
- A. silicates
- B. peroxide
- C. chlorine
- D. ammonia

_______________ **7.** The primary use of compasses on drawings is to draw ___.
- A. horizontal and inclined lines
- B. vertical and inclined lines
- C. arcs and curves
- D. none of the above

_______________ **8.** The ___ scale is preferred when drawing buildings and other structural parts.

A. architect's
B. civil engineer's
C. mechanical engineer's
D. drafter's

_______________ **9.** An F pencil lead is harder than a ___ pencil lead.

A. 2H
B. 4H
C. 6B
D. none of the above

_______________ **10.** Coded references on a cutting plane line show the ___.

A. scale of the details
B. sheet number of the section view
C. architect's initials
D. all of the above

_______________ **11.** Two basic types of elevations are ___.

A. rough and finished
B. top and bottom
C. interior and exterior
D. scaled and freehand

_______________ **12.** Section details show the ___ shape of features.

A. reduced
B. perspective
C. cross-sectional
D. all of the above

_______________ **13.** The first drawing(s) of a set of plans to be drawn is(are) generally the ___.

A. elevations
B. floor plans
C. details
D. plot plan

_______________ **14.** Cutting planes for floor plans are taken ___ above the finished floor.

A. 3′-0″
B. 4′-0″
C. 5′-0″
D. 6′-0″

_______________ **15.** The name and seal of the ___ are commonly found in the title block.

A. architect
B. drafter
C. contractor
D. owner

_______________ **16.** Regarding floor plans, ___.

A. the layout of rooms is shown
B. symbols and abbreviations give additional information
C. windows, doors, cabinets, fixtures, and other features are shown
D. all of the above

_______________ **17.** Regarding title blocks, ___.

A. one title block is completed for each set of plans
B. initials representing trade areas may precede sheet numbers
C. an inspection checklist is a part of title blocks
D. none of the above

_______________ **18.** Regarding prints, ___.

A. blue line prints may be made by the blueprint method
B. blue line prints may be made by the electrostatic method
C. blue or black line prints may be made with the dry diazo method
D. all of the above

_______________ **19.** Regarding scales, ___.

A. an architect's scale is graduated in decimal units
B. an architect's scale contains a ruler and 10 scales
C. a civil engineer's scale is graduated in fractional units
D. none of the above

_______________ **20.** Regarding CAD, ___.

A. the mouse and trackball are output systems
B. a laser printer is part of an input system
C. an ink-jet plotter sprays tiny ink droplets on paper
D. prints are initially produced on carbon paper

True-False

T F **1.** The Y series of ANSI standards includes drafting conventions.

T F **2.** Electrostatic prints are produced using the same process used by office copiers.

T F **3.** Tradesworkers prepare working drawings of the final building design for approval by the owner.

T F **4.** Working drawings contain all graphic information necessary to complete a job.

T F **5.** Section views can show information about foundation footings and walls.

T F **6.** Elevations are generally drawn to the same scale as floor plans.

T F **7.** Each sheet of a set of working drawings contains a title block.

T F **8.** True South is designated on the plot plan to show building orientation on a lot.

T F **9.** Plot plans are drawn to a smaller scale than floor plans.

T F **10.** A separate floor plan is required for each story of a house.

T F **11.** Floor plans are generally the first drawings of a set of plans to be drawn.

T F **12.** The cutting plane for a section view is shown on the floor plan.

T F **13.** The scale of a print may be given in the title block.

Completion 1-1

_______________ **1.** ___ are reproductions of working drawings.

_______________ **2.** The mechanical engineer's scale is used when drawing ___ parts.

_______________ **3.** The location of streets, easements, and utilities is shown on the ___ plan.

_______________ **4.** The ___ of the T-square should be held firmly against the edge of the drawing board when in use.

_______________ **5.** ___ are used to transfer dimensions on drawings.

_______________ **6.** ___ on plans give the size of an object.

_______________ **7.** ___ are written documents giving additional information about the plans.

_______________ **8.** The slope of the roof is shown on the ___ elevations.

_______________ **9.** Solid lines on plot plans show ___ contours.

_______________ **10.** Line weights, symbols, and lettering on ___ drawings are precise and easy to read.

Completion 1-2

_______________ **1.** The angle at A contains ___°.

_______________ **2.** The angle at B contains ___°.

_______________ **3.** The angle at C contains___°.

_______________ **4.** The angle at D contains ___°.

_______________ **5.** The angle at E contains ___°.

_______________ **6.** The angle at F contains ___°.

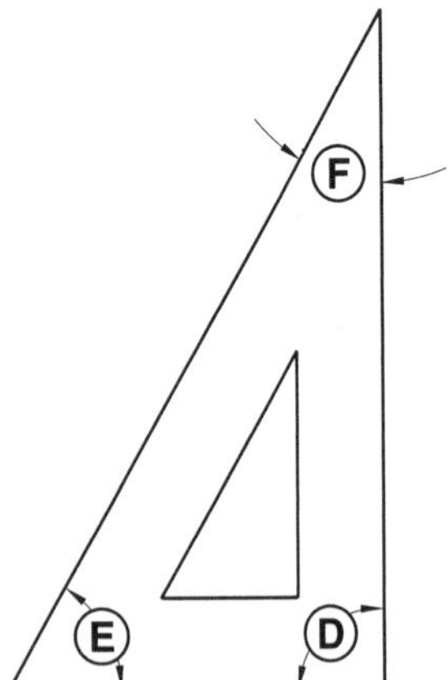

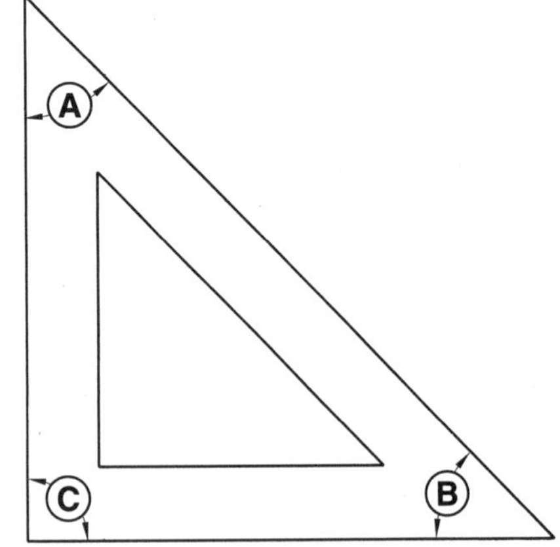

Working Drawings and Prints 1

Trade Competency Test

Name ______________________________ Date ______________

Identification 1-1

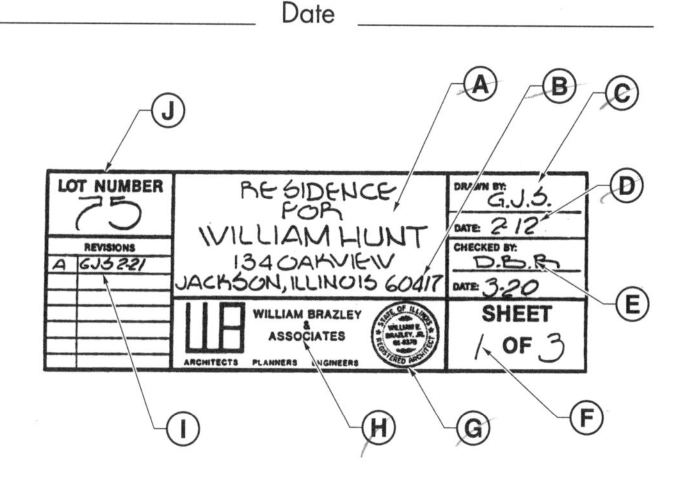

H 1. Architect name

G 2. Architect seal

A 3. Owner name

D 4. Date

F 5. Sheet number

B 6. Owner address

C 7. Drafter initials

I 8. Revisions

E 9. Checker initials

J 10. Lot number

Identification 1-2

F 1. 15°

C 2. 30°

D 3. 45°

E 4. 75°

B 5. 120°

A 6. 135°

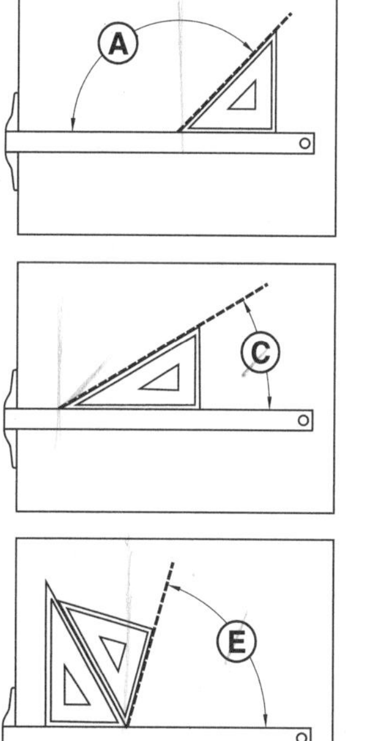

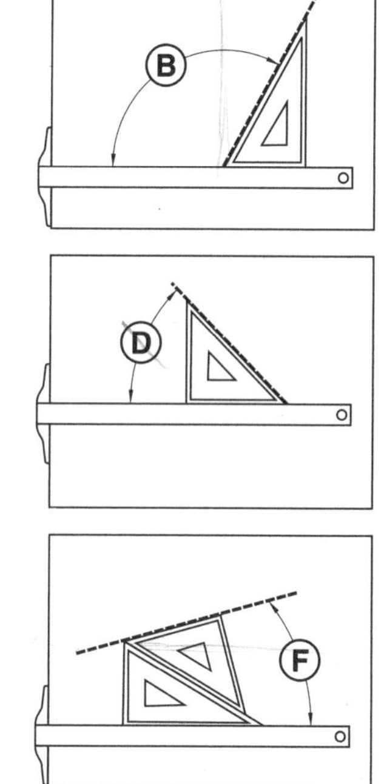

Identification 1-3

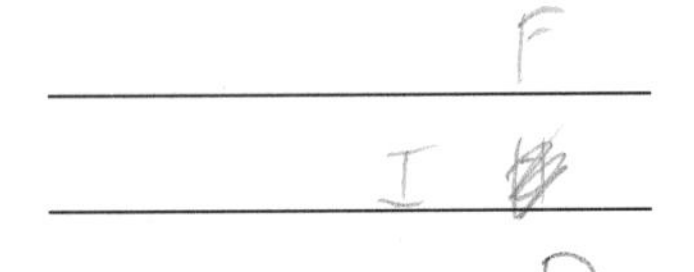

_______________ **1.** Plot plan

_______________ **2.** Floor plan

_______________ **3.** Details

_______________ **4.** Exterior elevation

_______________ **5.** Interior elevation

_______________ **6.** North

_______________ **7.** Title block

_______________ **8.** Cutting plane

_______________ **9.** Symbol

_______________ **10.** Scale

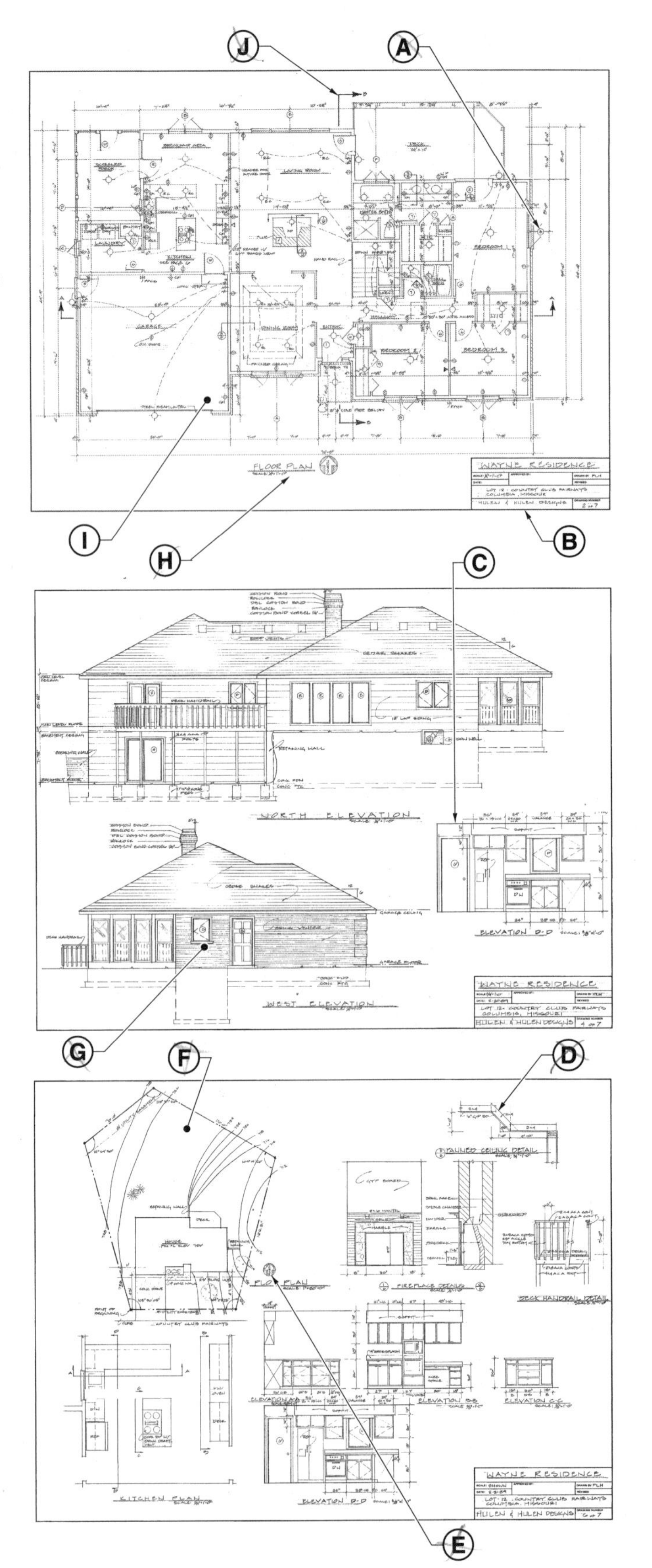

Working Drawing Concepts 2

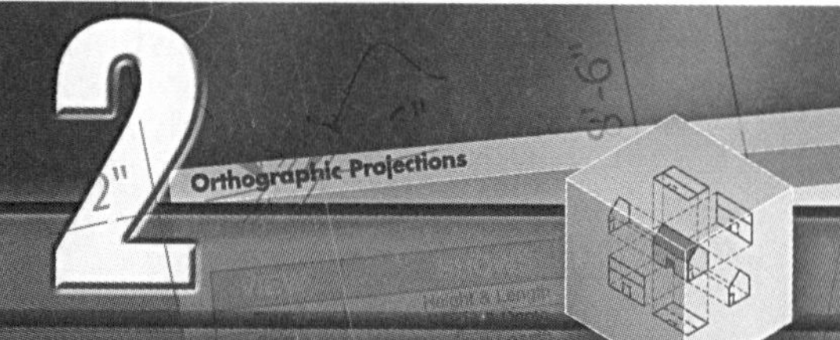

Working drawings drawn as orthographic projections are based on concepts developed from sketches. Plan, elevation, and section views are drawn to scale and dimensioned to show actual sizes.

Sketches provide a quick, concise method of communicating ideas. Pictorial sketches provide architects, clients, and builders with general information about the appearance of a finished dwelling. Three basic types of pictorial drawings are perspective, isometric, and oblique.

SKETCHING

Sketching is the process of drawing without instruments. Sketches are made by the freehand method. The only tools required are pencil, paper, and eraser. See Figure 2-1.

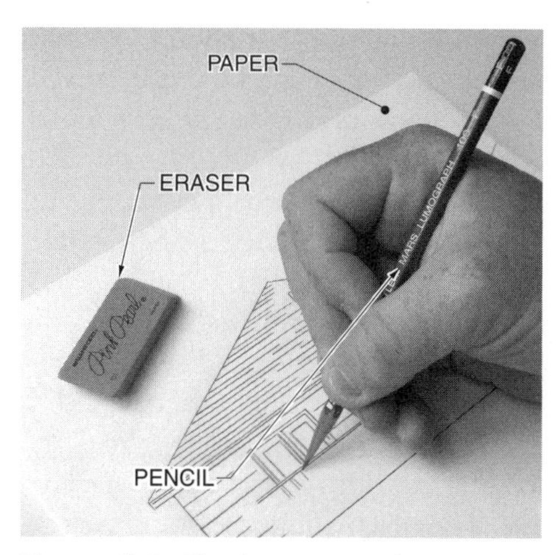

Figure 2-1. Sketches convey ideas graphically and are drawn using basic tools such as pencils, paper, and erasers.

Sketching pencils are either wood or mechanical. Wood pencils must be sharpened, and the lead must be pointed. One type of mechanical pencil contains a thick lead that is pointed with a lead pointer, file, or sandpaper. Another type of mechanical pencil contains a thin lead that does not require pointing. Softer leads, such as HB, F, and H, are commonly used for sketching.

Paper selected for sketching depends upon the end use of the sketch. Plain paper is commonly used. Tracing vellum is used if the sketch is to be duplicated on a diazo printer. Papers and tracing vellums are available in pads or sheets in standard sizes designated A, B, or C. Size A is 8½″ × 11″; size B is 11″ × 17″; size C is 17″ × 22″. The paper is either plain or preprinted with grids to facilitate sketching. Preprinted paper is available in a variety of grid sizes for orthographic and pictorial sketches. A grid size of ¼″ is common. Grids are commonly printed in light-blue, nonreproducing inks.

Erasers are designed for use with specific papers and leads. The eraser selected should be soft enough to remove pencil lines without smearing them or damaging the paper. Pink Pearl® and white vinyl erasers are commonly used when sketching.

> Erasing shields are used to protect a portion of a sketch when erasing.

Sketching Techniques

When sketching, the pencil point should be pulled across the paper. Pushing the pencil point can tear the paper. While pulling the pencil, it should be slowly rotated to produce lines of consistent width.

Horizontal, vertical, inclined, and curved lines are used to produce three-view and pictorial drawings. Shading techniques are not used with three-view drawings; however, pictorial drawings may be shaded.

Computer software such as AutoSketch® and Microsoft® Paint can also be used for sketching with a personal computer. Sketching is done on the monitor with input from a mouse or other input device. Lines and commands are selected from a menu and moved about the screen to create the sketch. The sketches are printed on paper using a standard printer or plotter.

Long horizontal, vertical, or slanted lines can be sketched using a series of short lines, which are then connected to complete the long lines.

Horizontal Lines. A *horizontal line* is a line that is level or parallel to the horizon. When sketching a horizontal line, the end points are located with dots to indicate the position and length of the line. For short lines, the end dots are connected with a smooth wrist movement from left to right (for a right-handed person). Long lines may require intermediate dots. If grid paper is used, intermediate dots are not required. For long lines, a full arm movement may be required to avoid making an arc.

The top or bottom edges of the paper or pad may be used as a guide when sketching horizontal lines. Light construction lines are drawn first to establish the straightness of the line. The line is then darkened. Construction lines may be omitted when more sketching experience is gained.

Vertical and Slanted Lines. A *vertical line* is a line that is plumb or upright. When sketching a vertical line, the end dots are located and the line is drawn from the top to the bottom. The side edges of the paper or pad may be used as a guide when sketching vertical lines.

A *slanted line* is an inclined line that is neither horizontal nor vertical. When sketching a slanted line, the end dots are located and the line is drawn from left to right (for a right-handed person). To facilitate sketching, the paper is rotated so that the inclined lines are in either a horizontal or vertical position.

Plane Figures. A *plane figure* is a geometric shape with a flat surface. Common plane figures include circles, triangles, quadrilaterals, and polygons. See Figure 2-2.

A *circle* is a plane figure generated around a centerpoint. All circles contain 360°. The *circumference* is the outside edge of a circle. The *diameter* is the distance from circumference to circumference passing through the centerpoint. The *radius* is one-half the diameter. A *chord* is a line from circumference to circumference that does not pass through the centerpoint. An *arc* is a portion of the circumference. A *quadrant* is one-fourth of a circle. A quadrant has a 90° angle. A *sector* is a pie-shaped segment of a circle. A *semicircle* is one-half of a circle. Semicircles always contain 180°.

When sketching a circle, the centerpoint is located and several intersecting diameter lines are drawn. The radius is marked off on these lines and connected with a series of arcs. The lines are darkened to produce a smooth circle.

A *triangle* is a three-sided plane figure that contains 180°. A *right triangle* is a triangle that contains one 90° angle. An *obtuse triangle* is a triangle that includes one angle greater than 90°. An *acute triangle* is a triangle in which all angles are less than 90°. Equilateral triangles are acute triangles. An *isosceles triangle* is a triangle that contains two equal sides and two equal angles.

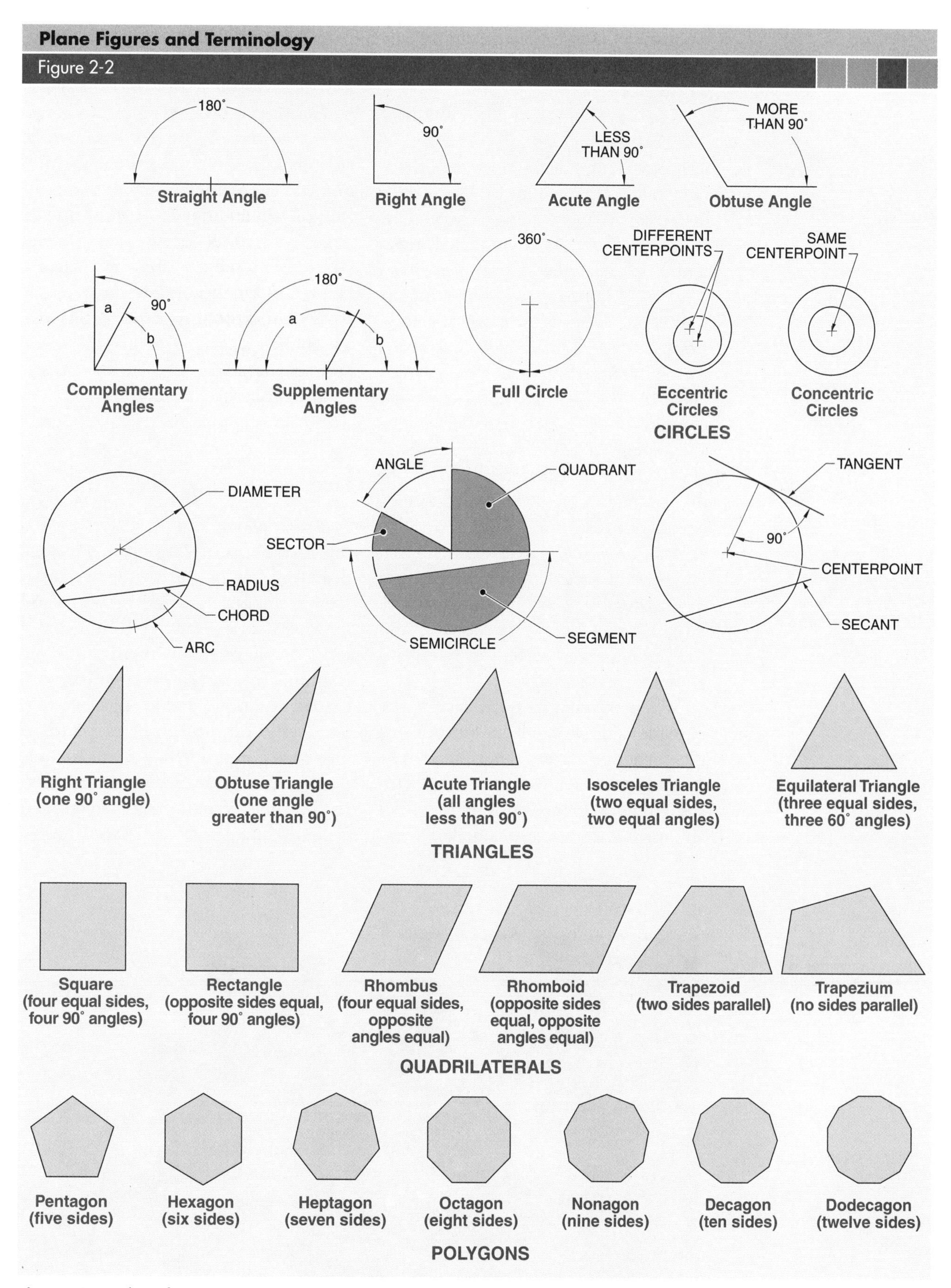

Figure 2-2. Plane figures are geometric shapes with flat surfaces.

An *equilateral triangle* is an acute triangle that contains three equal sides and three 60° angles.

When sketching a triangle, the base is drawn, the angles of the sides are determined, and straight lines are drawn. Generally, one or more of the sides is dimensioned and the angles noted.

A *quadrilateral* is a four-sided plane figure. All quadrilaterals contain 360°. A *square* is a quadrilateral that contains four equal sides and four 90° angles. A *rectangle* is a quadrilateral that contains four 90° angles with opposite sides equal. A *rhombus* is a quadrilateral that contains four equal sides with opposite angles equal and no 90° angles. A *rhomboid* is a quadrilateral that has opposite sides equal, opposite angles equal, but does not contain any 90° angles. A *trapezoid* is a quadrilateral that has two sides parallel. A *trapezium* is a quadrilateral that has no sides parallel. Squares, rectangles, rhombuses, and rhomboids are classified as parallelograms in mathematics.

When sketching a quadrilateral, the baseline is drawn and the locations of the corner points are determined. The corner points are connected with straight lines. Dimensions of two sides and angles are often included.

A *polygon* is a many-sided plane figure that is bounded by straight lines. A *regular polygon* is a polygon that has equal sides and equal angles. An *irregular polygon* is a polygon that has unequal sides and unequal angles. Regular polygons are named according to the number of their sides. For example, a triangle has three sides, a quadrilateral has four sides, a pentagon has five sides, and a hexagon has six sides.

When sketching a polygon, the length of each side is marked off at the appropriate angle. The lines are darkened to complete the polygon.

Pictorial Drawings

A *pictorial drawing* is a three-dimensional representation of an object. Pictorial drawings show the three principal measurements of height, length, and depth on one drawing. Three basic types of pictorial drawings used in the building trades are perspective, isometric, and oblique drawings. See Figure 2-3. Common uses of pictorial drawings for residential work include exterior perspective drawings, isometric drawings of plumbing systems, and oblique drawings of cabinets.

The Garlinghouse Company

An exterior perspective drawing provides a realistic view of a building.

Pictorials
Figure 2-3

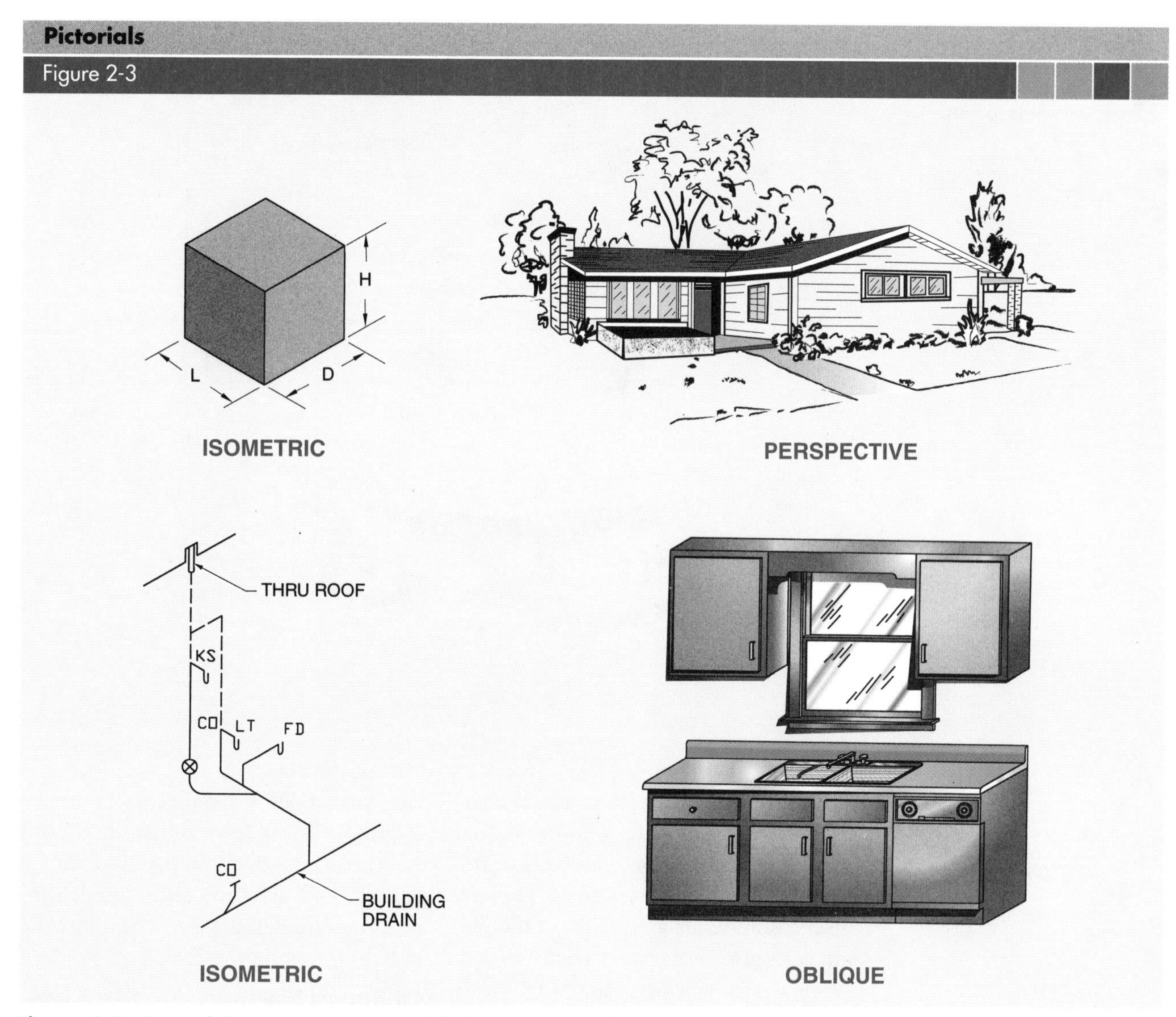

Figure 2-3. Pictorial drawings show height (H), length (L), and depth (D) in one drawing.

Perspective Drawings. A *perspective drawing* is a pictorial drawing with all receding lines converging to vanishing points. Perspective drawings resemble photographs. Perspective drawings may be made with one, two, or three vanishing points. The object may be located above, on, or below the horizon to produce a worm's-eye, eye-level, or bird's-eye view, respectively. The number of vanishing points used and location of the object in relation to the horizon determine the type of perspective drawing. Two vanishing points and an eye-level view are commonly used for perspective drawings of houses. See Figure 2-4.

When sketching a perspective drawing, the type of drawing is determined and the appropriate number of vanishing points are located. The height of the object is established and receding lines that converge at the vanishing points are drawn. Object lines are darkened to complete the drawing. The proper location of the initial height line, its relationship to the horizon, and the distance between the vanishing points are critical in producing a realistic perspective drawing.

On isometric piping drawings, all horizontal pipes are drawn as 30° lines.

Figure 2-4. Receding lines of perspective drawings converge to one, two, or three vanishing points.

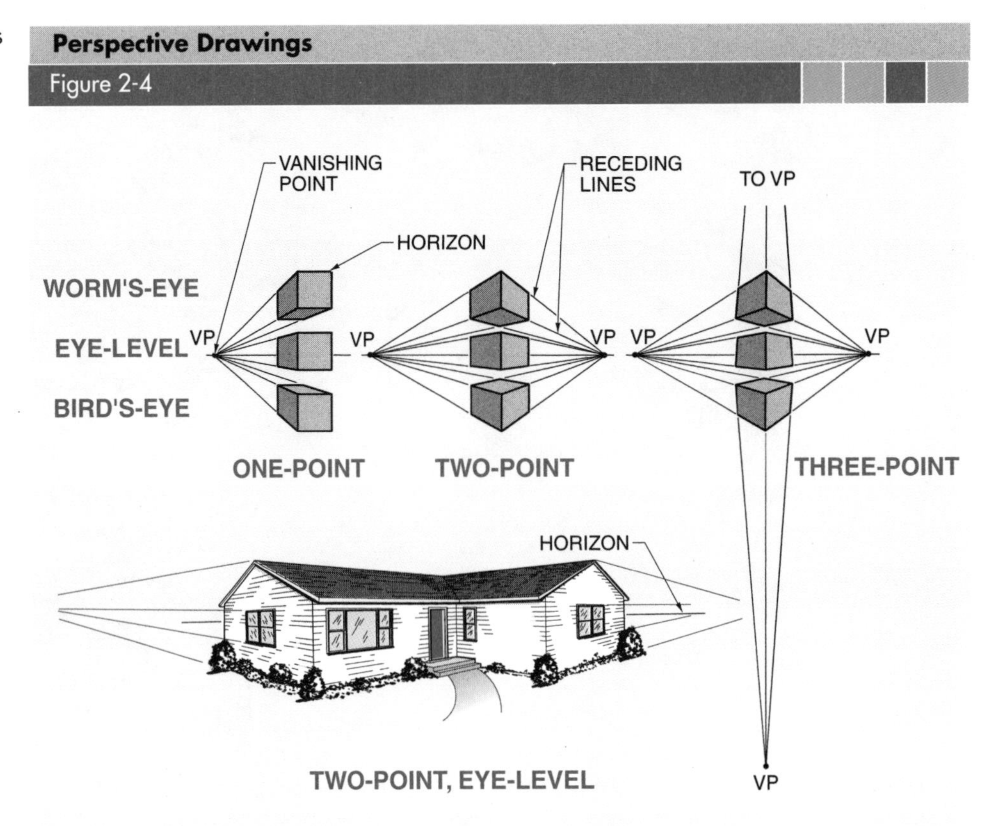

Isometric Drawings. An *isometric drawing* is a pictorial drawing with horizontal lines drawn 30° above (or below) the horizon. Vertical lines remain vertical. All measurements are made on the 30° and vertical axes or lines parallel to them. Nonisometric lines are drawn by locating and connecting their end points on isometric lines.

Circles on isometric drawings appear as ellipses. An *ellipse* is a plane figure generated by the sum of the distances from two fixed points. These centerpoints are located to draw an ellipse.

When sketching an isometric drawing, the isometric axes are drawn and end points for the principal measurements of height, length, and depth are located. Vertical lines are projected from the length and depth end points, and receding lines are projected from the height end point to construct two isometric surfaces. The top surface of the cube is formed by projecting receding lines at the corners. The end points for other lines on the isometric axes or lines parallel to them are located. These end points are connected and all nonisometric lines, circles, and arcs are completed. Object lines are darkened to complete the drawing. See Figure 2-5.

Oblique Drawings. An *oblique drawing* is a pictorial drawing with one surface drawn in true shape and receding lines projecting back from the face. Receding lines are commonly drawn at 30° or 45° angles. In oblique cavalier drawings, receding lines are drawn full-scale. In oblique cabinet drawings, receding lines are drawn at one-half scale. Oblique cabinet drawings are the most common type of oblique drawing. Circles appear in true shape on the front surface of oblique drawings and as ellipses on receding surfaces.

Pictorial drawings are also referred to as presentation drawings since they are commonly used to present design ideas to potential owners.

Isometric Drawings
Figure 2-5

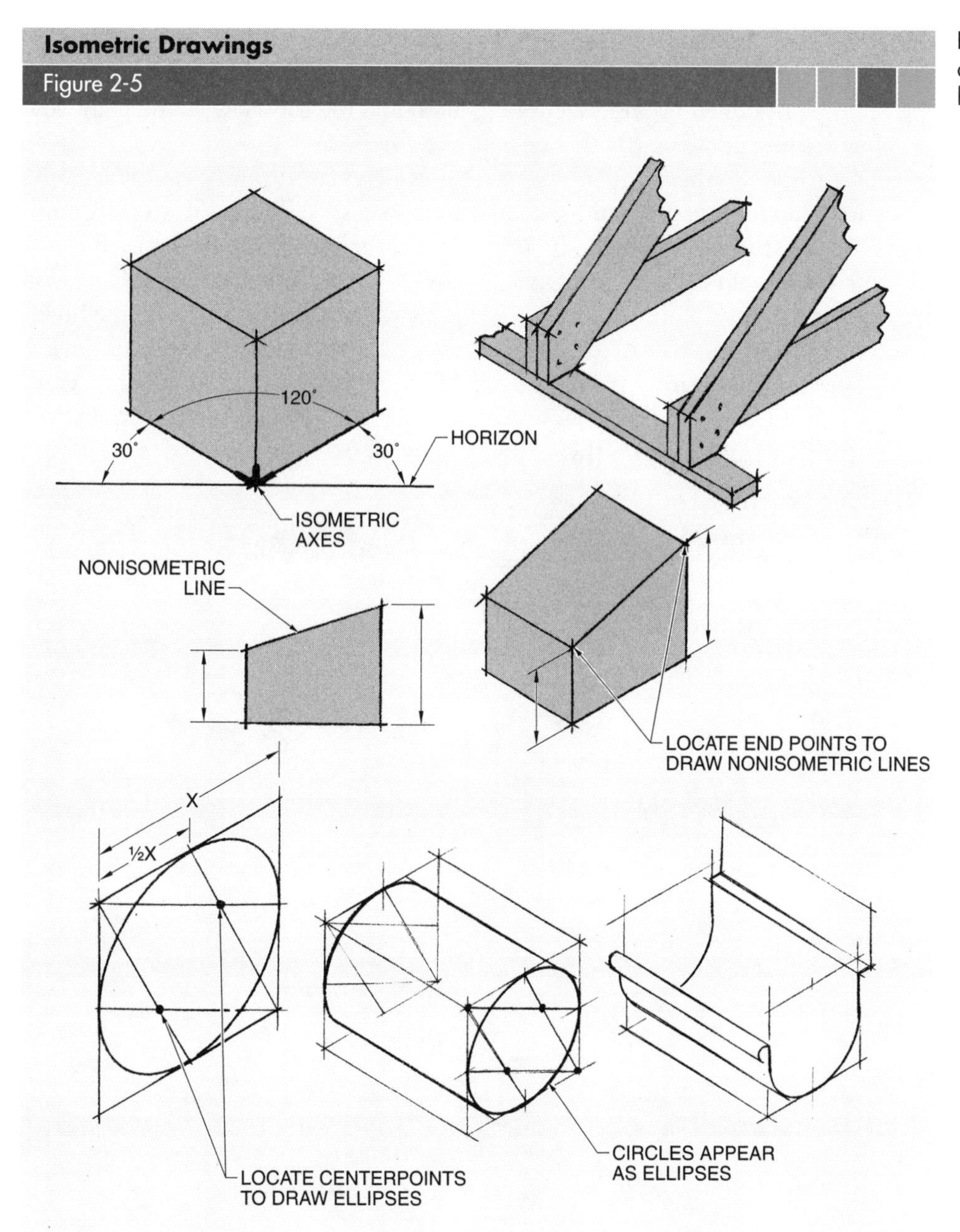

Figure 2-5. The isometric axes are drawn 30° above horizontal.

When sketching an oblique cabinet drawing, the true shape of the front surface is drawn to scale showing the height and length of the object. Receding lines are drawn to one-half scale. All measurements along the oblique axis or lines parallel to them are made. Nonoblique lines are drawn by locating their end points on oblique lines. All nonoblique lines, circles, and arcs are completed. Object lines are darkened to complete the drawing. See Figure 2-6.

Orthographic Projections

An *orthographic projection* is a drawing in which each face of an object is projected onto flat planes, generally at 90° to one another. Each view is a two-dimensional drawing showing two principal measurements. The most common orthographic projection is a three-view, or multiview, drawing. The front view shows height and length; the side view shows height and depth; the top view shows depth and length. The concept of three-view drawing is used in all fields of architecture, engineering, and the building trades to provide accurate graphic representations of the construction to be completed.

In an orthographic projection, lines are projected from every corner of the object to be drawn onto an imaginary transparent plane. Three planes are usually sufficient to show all details of most objects. These three planes produce the front, top, and side view of the object. In printreading, the front view is referred to as the front elevation, the side view as the side elevation, and the top view as the plan view. See Figure 2-7.

When describing a building to be built on a specific plot of ground, compass directions are used for each elevation. For example, the North Elevation is a view of the north side of the building and the South Elevation is a view of the south side of the building. Generally, four elevations are required to show all exterior views of a building.

Figure 2-6. The oblique axis of an oblique drawing is drawn 30° or 45° above horizontal.

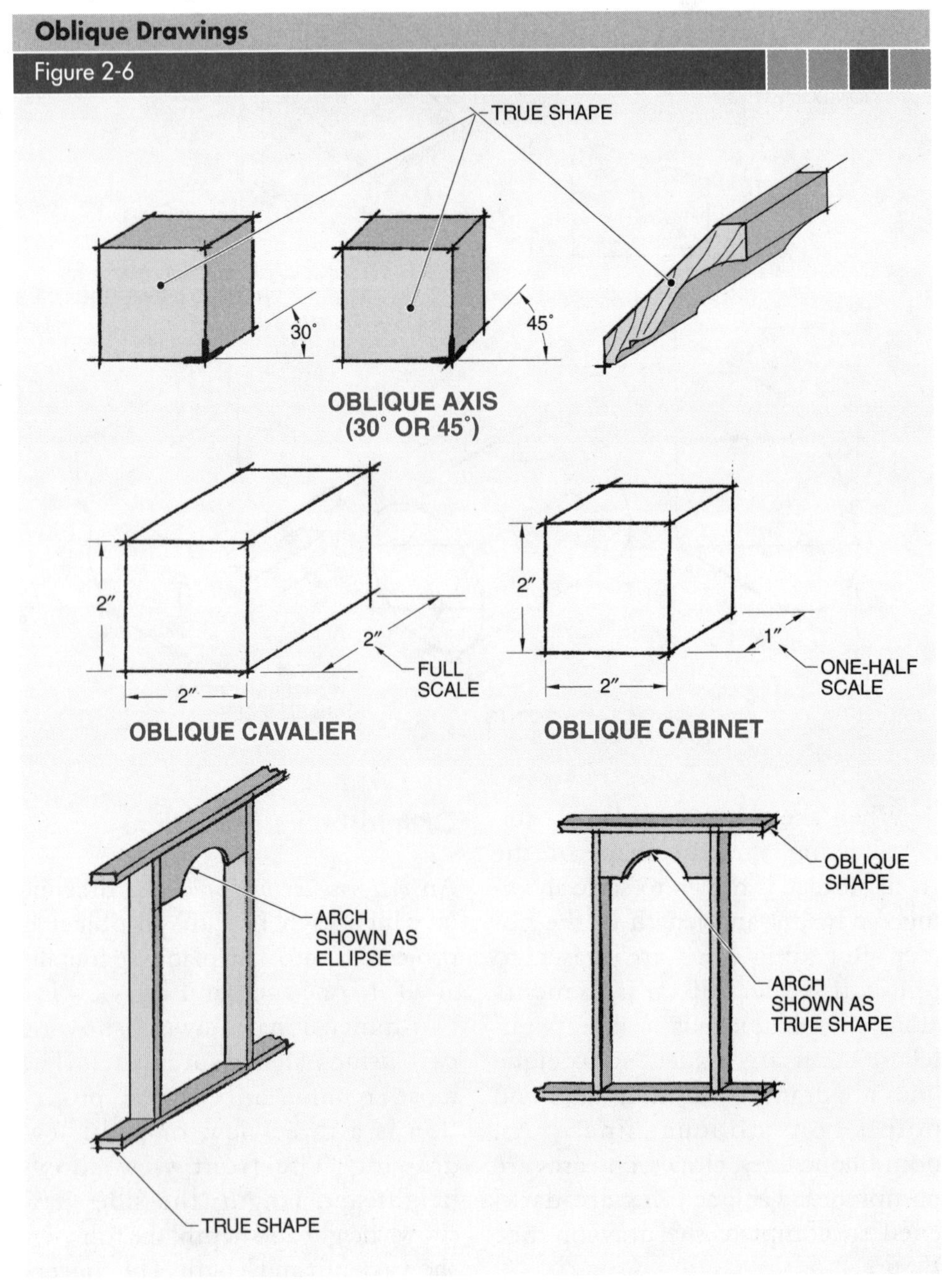

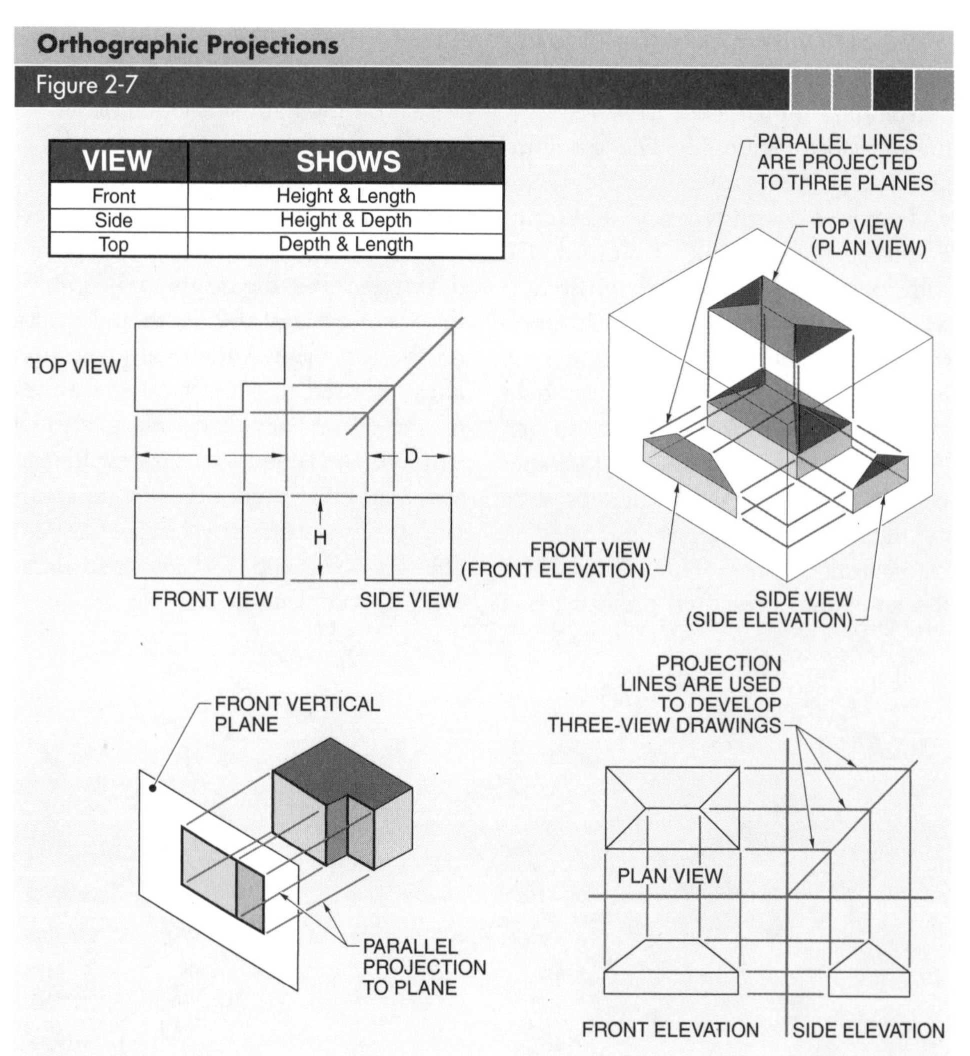

Figure 2-7. The three-view drawing is the most common type of orthographic projection.

The top view of a three-view drawing shows the roof plan of a building. In order to see the floor plan, which is generally considered the most important single drawing of a set of plans, an imaginary horizontal cutting plane is passed through the building 5′-0″ above the finished floor. Floor plans for basements and other floor levels of a multistory house are drawn by using additional imaginary cutting planes at 5′-0″ above each finished floor level.

Projection lines that connect the parts of one view to another show the relationship of the elevations to the plan view. A transparent isometric box shows the house and indicates the various elevations and plan views. The box is then unfolded to show the relationship of views. Projection lines show that the principal measurements of height, length, and depth are consistent throughout the views. See Figure 2-8.

Southern Forest Products Association

A set of working drawings includes several orthographic projections that relate to one another.

Visualizing the relationship of points from one elevation to another aids in understanding the concept of three-view drawings. See Figure 2-9. The simplified pictorial of the tri-level house contains wall surfaces designated A, B, C, and D. Points on these surfaces are designated with lowercase letters a through h. Surfaces A and B are seen on the front elevation in their true shape and size. Surfaces C and D are seen on the right side elevation in their true shape and size. Points on one view are lines on the other view. For example, the point representing the main roof ridge is designated ab on the front elevation. The order in which the letters are presented indicates that point a is closer to the observer than point b. The roof ridge is shown as horizontal line ab on the right side elevation.

When sketching a three-view drawing, the true shape of the front view showing height and length of the object is drawn to scale. Projection lines are drawn to define the height and depth of the side view and the depth and length of the top view. Additional projection lines are used to project other features of the object including offsets, slanted surfaces, and centerpoints for circles and arcs. Lines that cannot be seen on a particular view are drawn as hidden (dashed) lines. All object and hidden lines are darkened to complete the drawing.

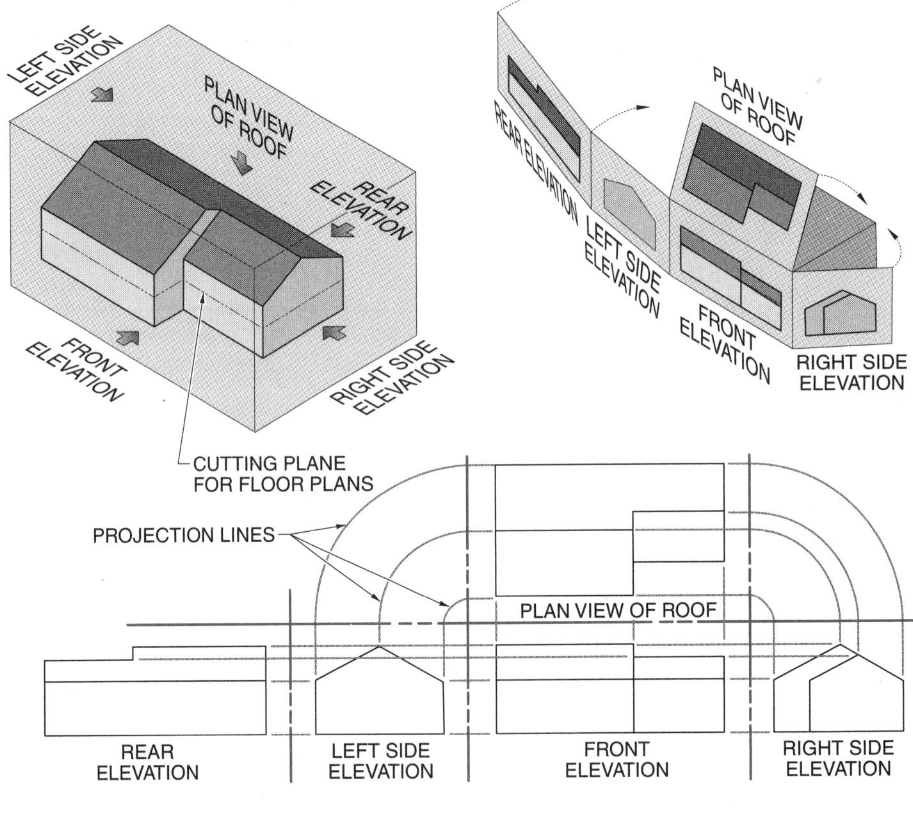

Figure 2-8. Elevation views are related by projection of features from one view to another.

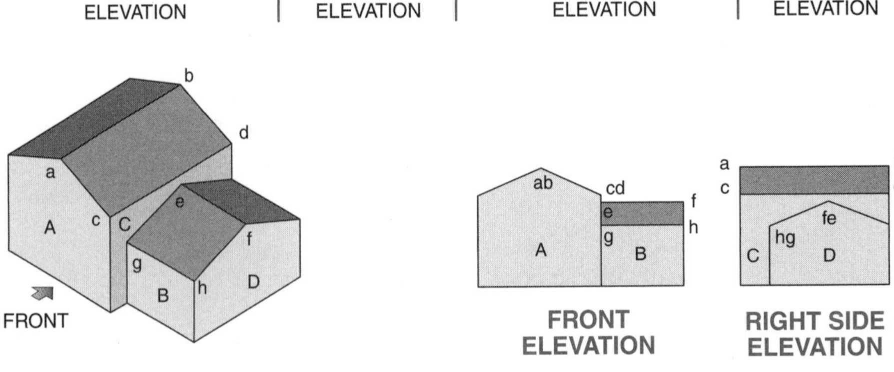

Figure 2-9. A point on an elevation is shown as a line on the adjacent elevation and vice versa.

DIMENSIONING

Prints used on a construction project are reproductions of architectural working drawings drawn to scale. A road map is a common example of a drawing made to scale. An area of several thousand square miles is shown on a small piece of paper by using a scale of a certain number of miles per inch.

Prints are drawn small enough so they can be handled easily, yet large enough to show necessary information clearly. Common drawing and print paper sizes in the United States are shown in the following table:

Letter Designation	Size
A	8½″ × 11″ (sheet)
B	11″ × 17″ (sheet)
C	17″ × 22″ (sheet)
D	22″ × 34″ (sheet)
E	34″ × 44″ (sheet)
E+	34″ × 44″ + (roll)

The length of each line on a print is reduced, or scaled, to a constant fraction of its true length so that all parts of the building are in exact relationship to each other. The scale most commonly used for floor plans and elevations is ¼″ = 1′-0″. For example, the floor plan of a 36′-0″ × 60′-0″ house drawn to the scale of ¼″ = 1′-0″ is drawn as a 9″ × 15″ rectangle (36′ × ¼ in./ft = 9″; 60′ × ¼ in./ft = 15″). For details of doors, windows, or other features, a larger scale, such as 1½″ = 1′-0″, may be used.

A complete set of plans is seldom drawn to the same scale because of the need to show details at a larger scale or the need to show the plot plan at a smaller scale. The scale for a sheet may be shown in the title block and/or below each plan, elevation, detail, or section view.

Architect's Scale

A triangular architect's scale has six ruled faces designed to measure in ten different scales. One of the edges is identical to a 12″ ruler divided into ¹⁄₁₆″ increments. The following are the ten scales on the triangular architect's scale:

- 3″ = 1′-0″
- 1½″ = 1′-0″
- 1″ = 1′-0″
- ¾″ = 1′-0″
- ½″ = 1′-0″
- ⅜″ = 1′-0″
- ¼″ = 1′-0″
- ³⁄₁₆″ = 1′-0″
- ⅛″ = 1′-0″
- ³⁄₃₂″ = 1′-0″

Architect's scales are read from left to right or right to left depending on the scale. See Figure 2-10. A ¼″ = 1′-0″ scale is read from right to left beginning at the 0 on the right end of the scale. The same set of markings is used for both the ¼″ = 1′-0″ scale and the ⅛″ = 1′-0″ scale. The correct line in relation to the scale used must be read. For example, 18′-0″ on the ¼″ = 1′-0″ scale is on the line representing 57′-0″ on the ⅛″ = 1′-0″ scale. Inches are read between the 0 and the end of the architect's scale.

Metric scales are available when creating drawings in the SI metric system. Common scales on a metric scale are 1:1, 1:2, 1:2.5, 1:5, 1:10, 2:1, 5:1, and 10:1.

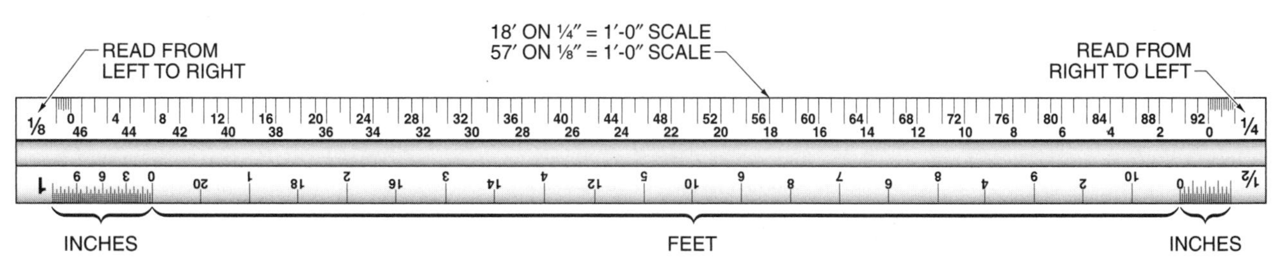

Figure 2-10. An architect's scale is used to produce scaled drawings.

Linear Measurement

Accurate readings of feet, inches, and fractions of an inch from a ruler or tape measure are essential in building construction. In addition, a tape measure can be used to scale measurements from a print.

When reading a tape measure or ruler, the number of feet is determined first. Many tape measures have highlighted areas indicating feet. See Figure 2-11. Next, the number of full inches is determined by reading the last number on the rule prior to the measurement.

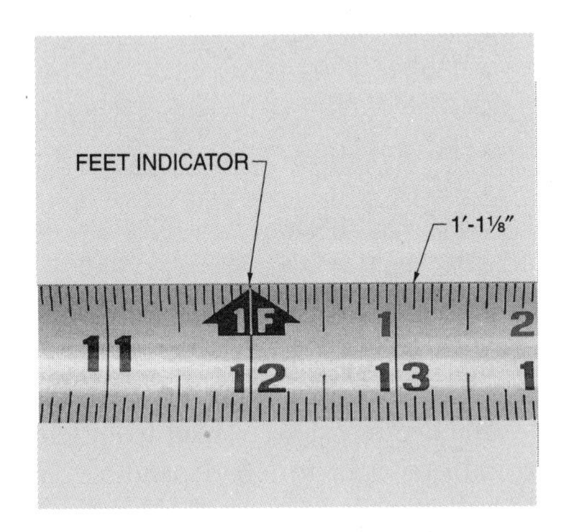

Figure 2-11. Most tape measures have highlighted portions indicating feet.

Next, the fractional part of the inch must be determined. Rulers divide an inch into equal fractional parts with longer lines denoting larger fractions and progressively shorter lines indicating smaller divisions.

Most tape measures and rulers used in the building trades are divided into 1/16″ increments. See Figure 2-12. There are 16 equal increments in each inch. Each inch on a tape measure or ruler is divided into 1/16″ increments that are indicated with marks of varying lengths. The ½″ mark is the longest mark between the two full inch marks. The ¼″ marks are slightly shorter than the ½″ marks and are placed one-half the distance between the ½″ mark and the full inch mark on either side. The ⅛″ marks are slightly shorter than the ¼″ marks and are placed half the distance between the ¼″ marks and the full inch marks and the ¼″ marks and the ½″ mark. The 1/16″ marks are slightly shorter than the ⅛″ marks and placed half the distance between the adjoining ⅛″ marks.

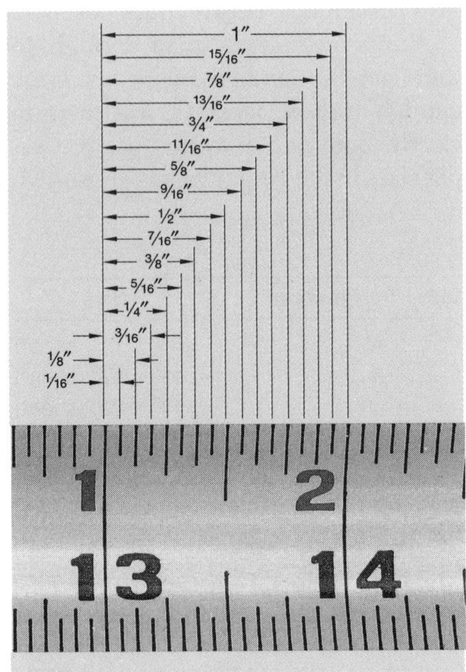

Figure 2-12. Tape measures and other rulers used in the building trades are often divided into 1/16″ increments.

Some rulers are divided into 32nds or 64ths of an inch using progressively shorter lines. With practice, reading the fractional marks based on their length and position within the adjoining inch and fractional marks can become quick and accurate.

Scaling Prints with a Tape Measure. A tape measure or ruler divided into inches and 1/16″ increments may be used to make and read drawings at the ¼″ = 1′-0″ scale. The results, however, may not be as accurate because the small divisions on the architect's scale are not available. Prints should not be measured for scale at a construction site to obtain a dimension unless the dimension is verified by the architect or engineer.

Tradesworkers may use a tape measure on a job site to determine a specific detail or to transmit information back to the architect. Each $\frac{1}{4}''$ space on a tape measure is equal to 1′-0″ on a print with a $\frac{1}{4}'' = 1'\text{-}0''$ scale. Each $\frac{1}{16}''$ space on the tape measure is equal to 3″. Therefore, a distance of $1\frac{3}{16}''$ on a print represents 4′-9″ at the scale of $\frac{1}{4}'' = 1'\text{-}0''$. The number of $\frac{1}{4}''$ spaces must be counted to find the number of feet. This number is added to the number of $\frac{1}{16}''$ spaces, which represent 3″ each to obtain the total of 4′-9″. See Figure 2-13.

Some dimensions on a print are found by simple mathematics. For example, if an overall dimension of 26′-0″ is shown for a wall having window openings 6′-6″ from each corner, the distance from the center of one window opening to the center of the other window opening is found by adding the given measurements for each window opening and subtracting them from the total wall length (6′-6″ + 6′-6″ = 13′-0″; 26′-0″ - 13′-0″ = 13′-0″).

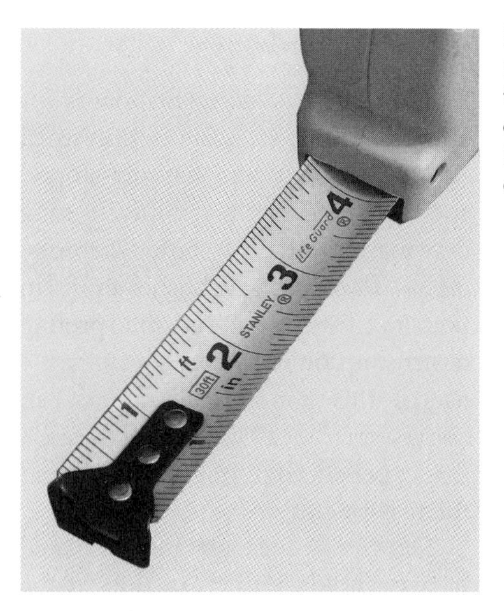

Figure 2-13. A tape measure can be used by a tradesworker to measure a component and transmit information back to the architect.

SYMBOLS AND CONVENTIONS

Symbols are drawn to scale to indicate their relative sizes. A *symbol* is a pictorial representation of a structural or material component used on prints. Walls, windows, doors, plumbing and fixtures, footings, partitions, chimneys, roofs, and other features are drawn in proportion to their size. See Figure 2-14.

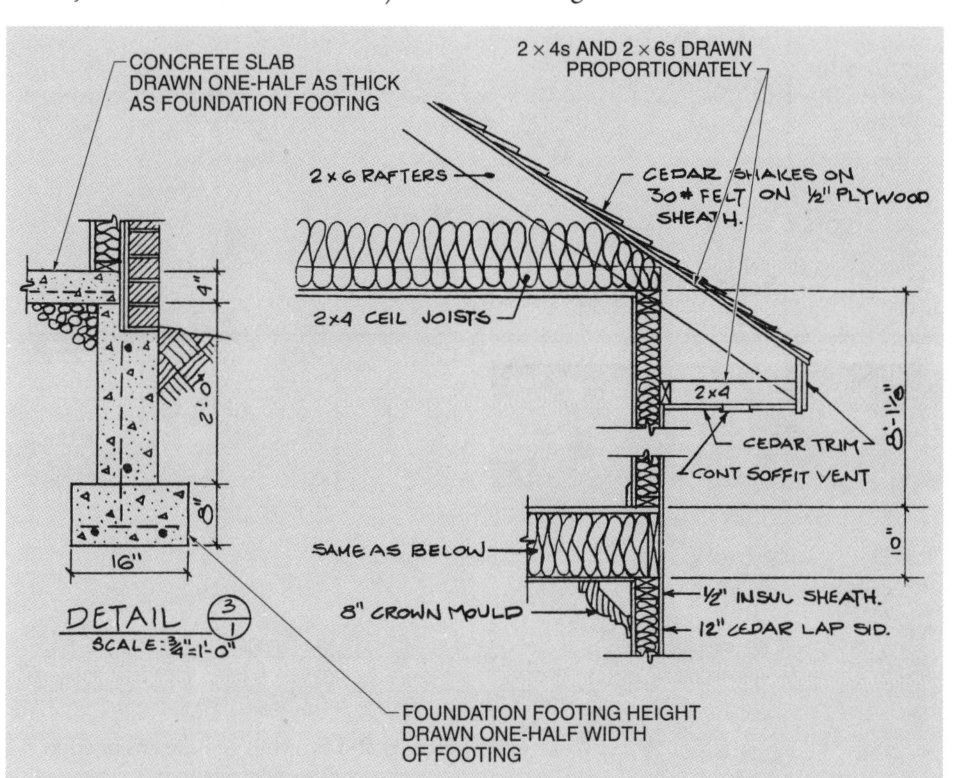

Figure 2-14. Symbols are drawn in proportion to other construction materials.

Dimensioning Conventions

The American National Standards Institute (ANSI) in conjunction with ASME International has developed specific standards for dimensioning drawings. ASME Y14.5M, *Dimensioning and Tolerancing,* establishes uniform practices for stating and interpreting dimensions on drawings. Through the use of this standard, architects can convey their ideas with the assurance that experienced printreaders can read them correctly.

Dimension lines may be terminated by arrowheads, slashes, or dots at the points on a drawing where they meet extension lines. The dimension is placed above the dimension line unless the space is too small. If the space is too small, the dimension is then placed in the nearest convenient space and related to the space by a leader line. See Figure 2-15.

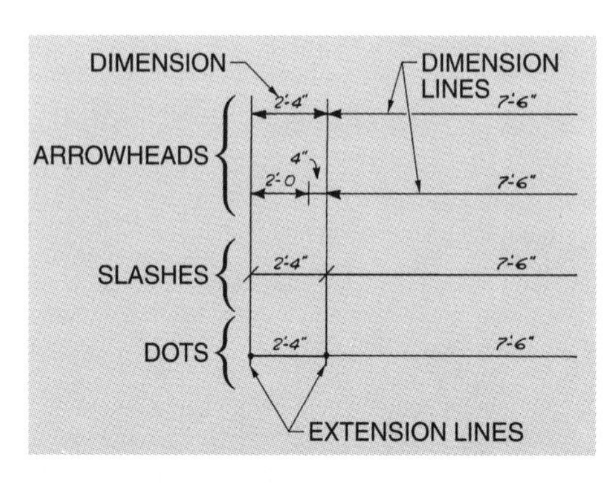

Figure 2-15. Arrowheads, slashes, or dots may be used to terminate dimension lines.

Solid masonry walls are dimensioned to their outside faces.

Exterior Walls. Framed, masonry, and masonry veneer walls are dimensioned according to dimensioning standards. See Figure 2-16. The preferred method for framed walls is to dimension to the outside faces of stud corner posts. Most of the remaining structural members can be located in relation to these rough framing members. It is easier to locate openings for doors and windows if the dimensions start from the outside faces of stud corner posts than if the dimensions are given from the outside face of sheathing since sheathing is applied after the rough openings are located.

Exterior Wall Dimensioning

Figure 2-16

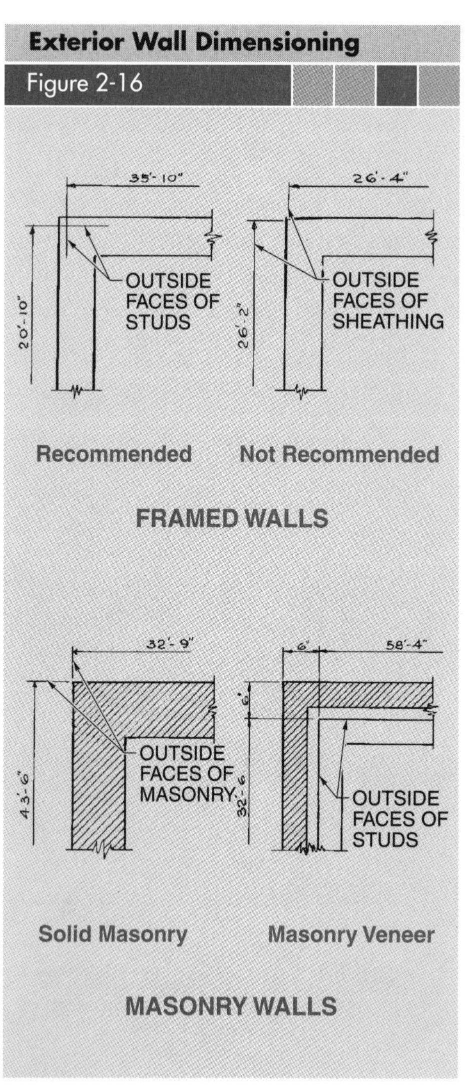

Figure 2-16. Walls are dimensioned to facilitate construction methods.

Solid masonry walls are dimensioned to their outside faces. Masonry veneer walls are dimensioned to the outside faces of the studs and to the outside faces of the masonry veneer wall. The outside dimensions of the foundation are usually the same as the outside dimensions of the masonry veneer wall. The faces of the studs must be located in relationship to the face of the masonry veneer wall because the frame structure is built before bricks for the veneer wall are laid.

Interior Partitions. Interior partitions vary in thickness depending on the finish. Dimensions are drawn to the center or the face of a stud partition. See Figure 2-17. Stud partitions have a nominal thickness of 4″, not including the finish wall covering. Because of variations that may occur on the job site, the 4″ nominal thickness aids in setting partitions accurately. Actual dimensions of a 2 × 4 stud are 1½″ × 3½″. Thus, a partition with ½″ drywall on both sides is 4½″ thick (½″ + 3½″ + ½″ = 4½″). Partitions in bathrooms and kitchens may be thicker to accommodate plumbing soil stacks.

Concrete masonry units, tile, or other materials of standard width are generally dimensioned to the face of the finish material.

> Interior partition thickness may be increased for rooms such as in-home theaters and game rooms.

Windows and Doors. Locations of openings for windows and doors on the floor plans of frame and brick veneer houses are dimensioned to the center of openings. Locations of openings for windows and doors on floor plans of masonry houses are dimensioned to the center of the openings or to the finish masonry abutting the window or door. Dimensions to the centers of the openings are preferred. See Figure 2-18.

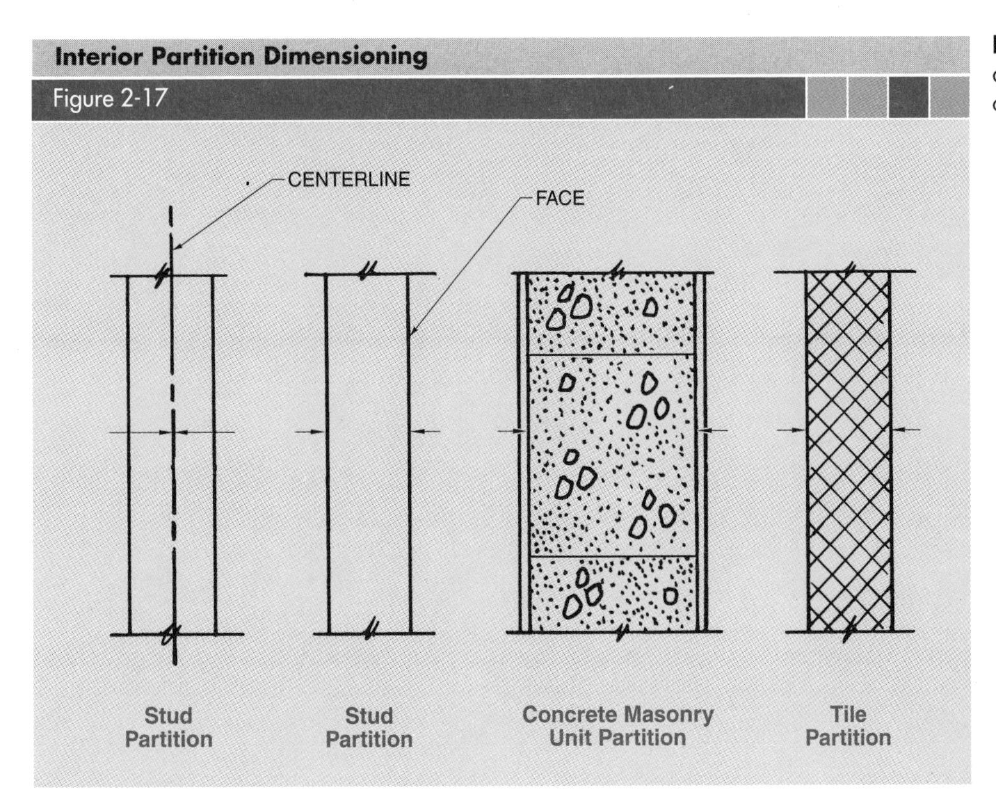

Figure 2-17. Partitions are dimensioned to their centerlines or faces.

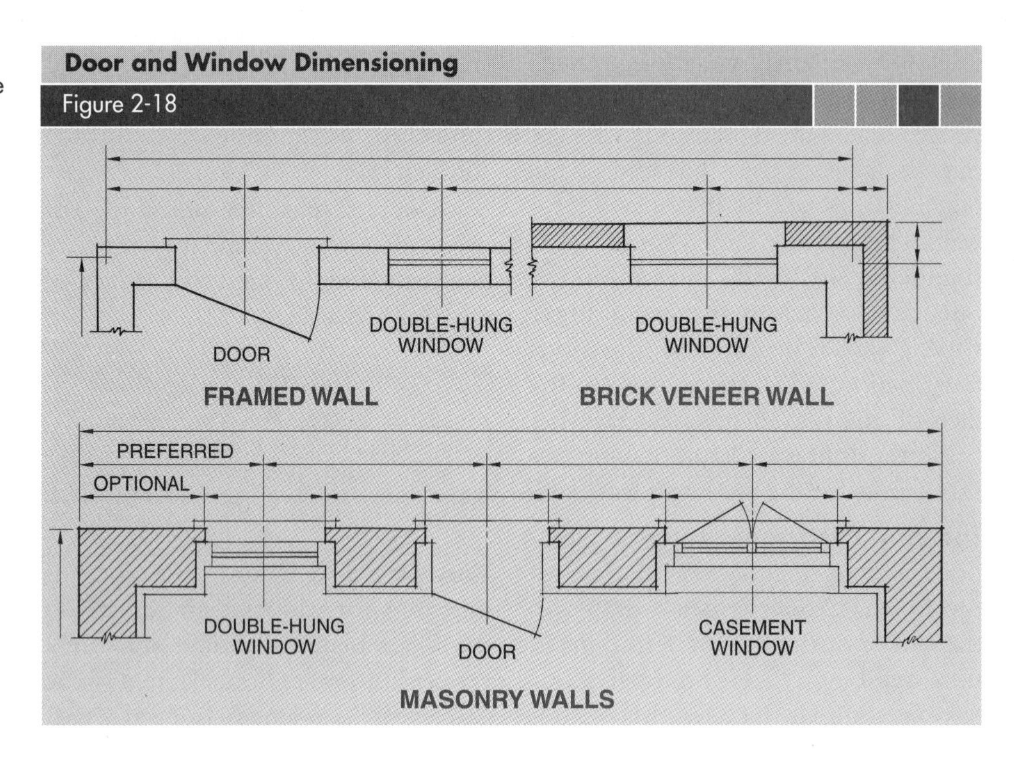

Figure 2-18. Openings for windows and doors are dimensioned to their centerlines.

Sketching

Name ______________________ Date ____________

Sketching 2-1

Sketch the missing view of each three-view drawing.

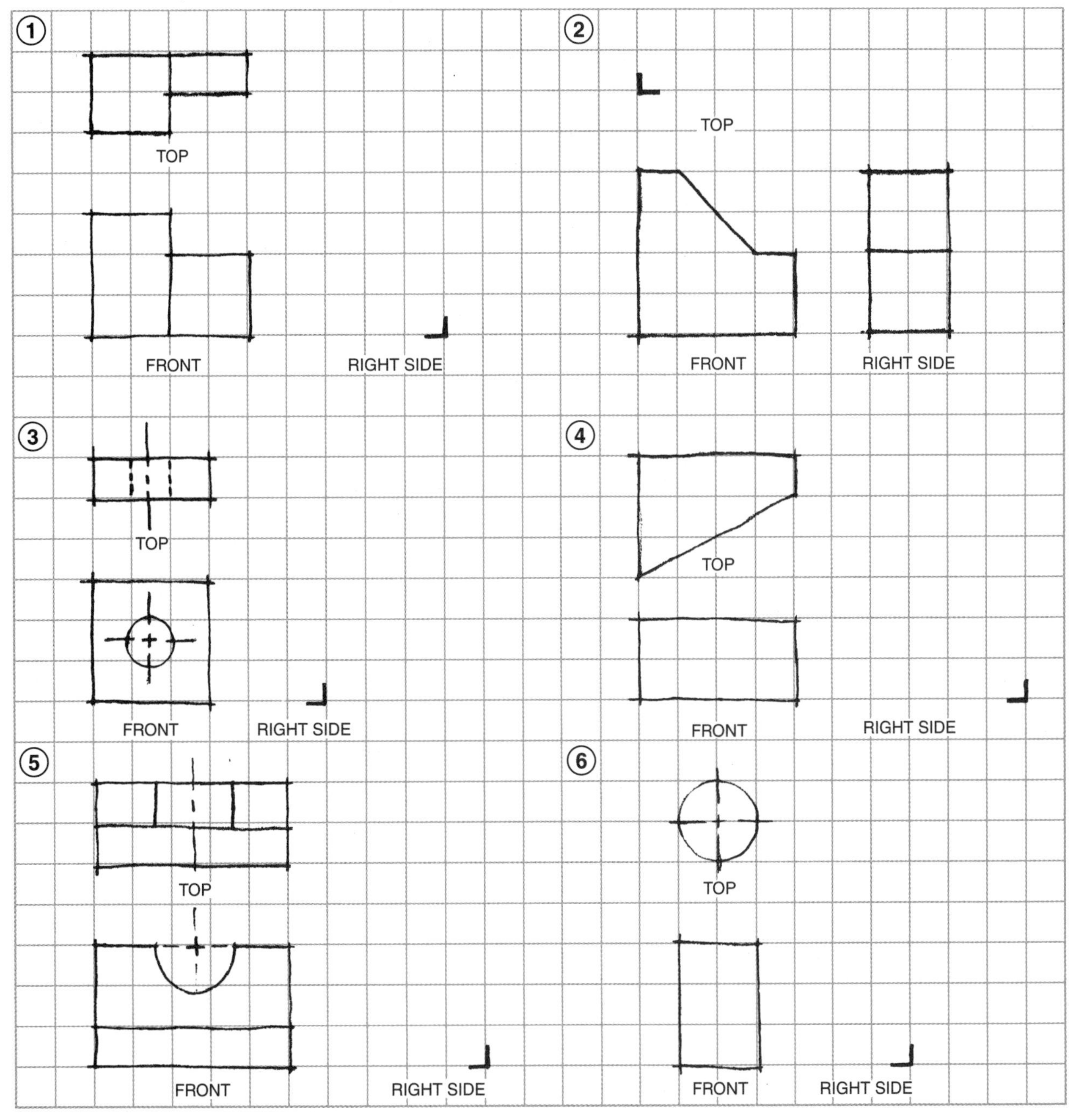

Sketching 2-2

Sketch oblique cabinet drawings of the multiview drawings on a separate sheet of paper.

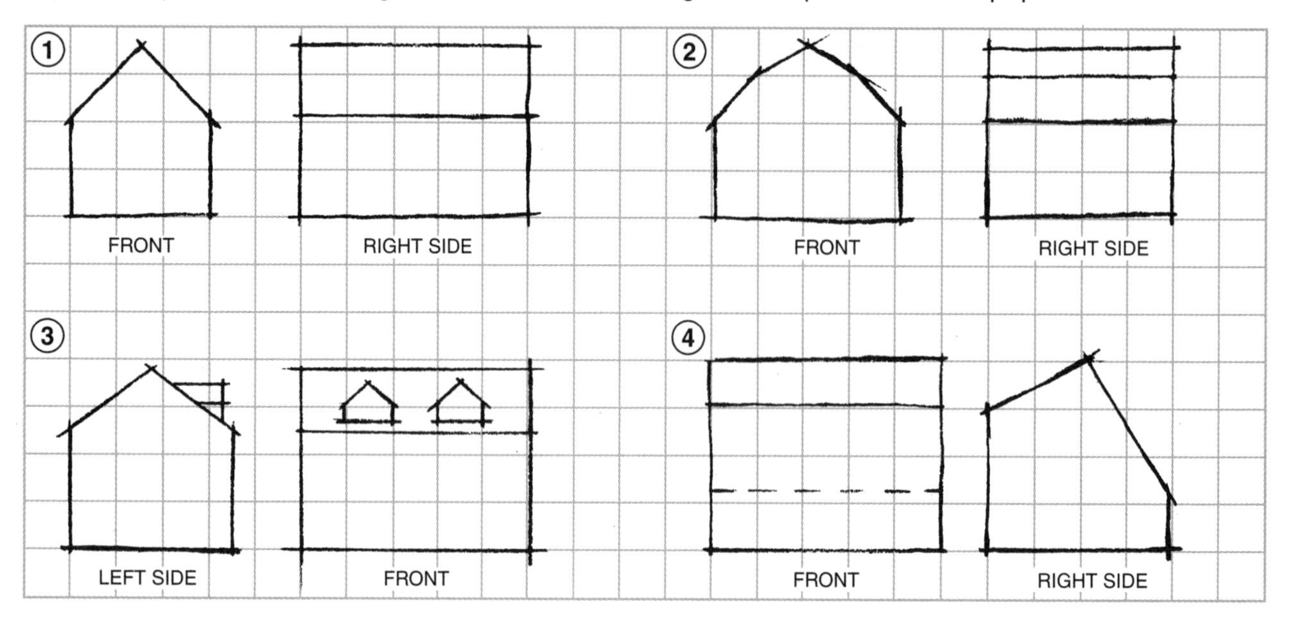

Sketching 2-3

Sketch front and right side views of the oblique drawings on a separate sheet of paper.

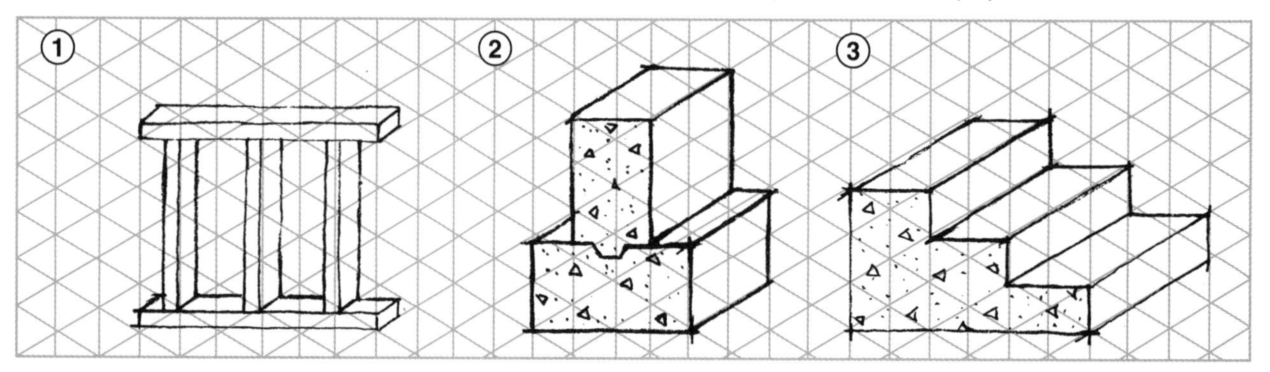

Sketching 2-4

Sketch front and right side views of the isometric drawings on a separate sheet of paper.

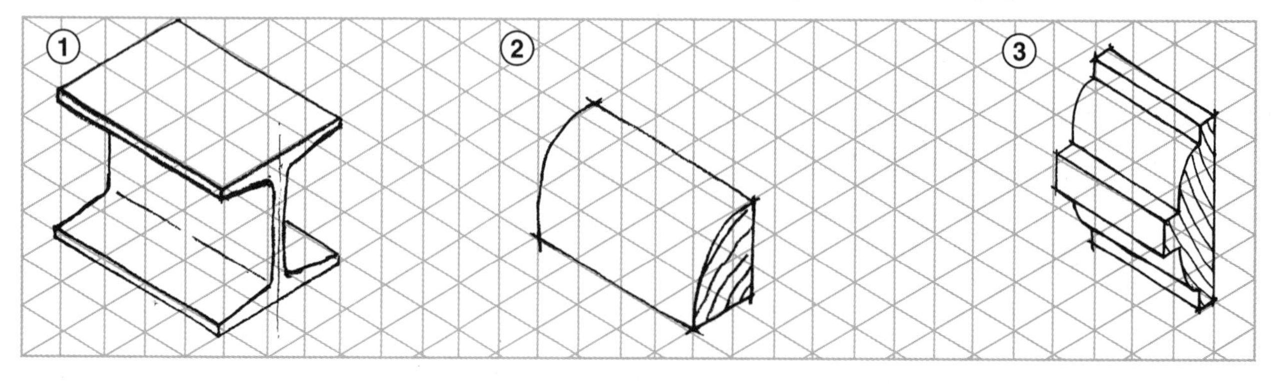

Sketching 2-5

Sketch oblique drawings of the section views of the moldings. Use 30° receding lines to the right.

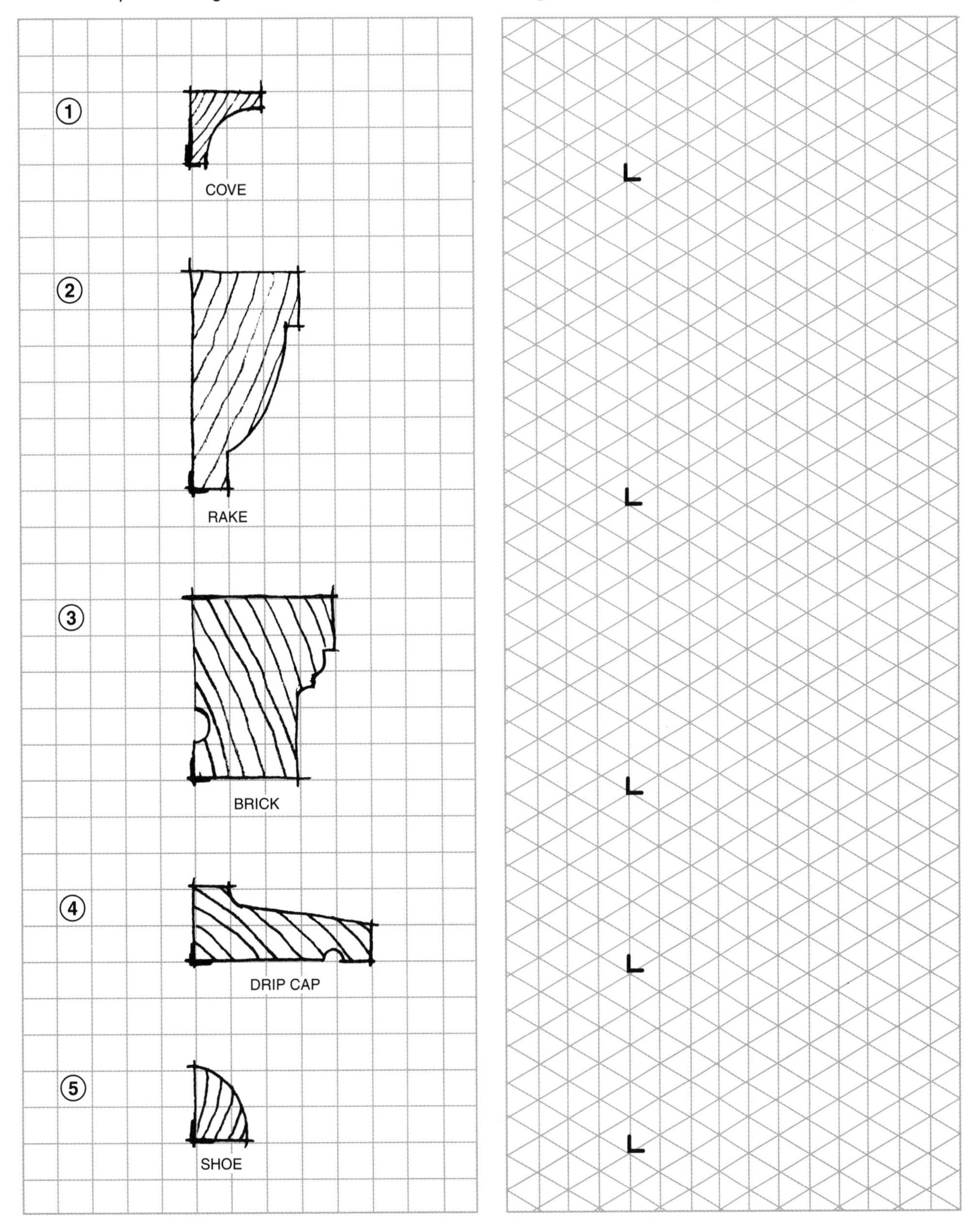

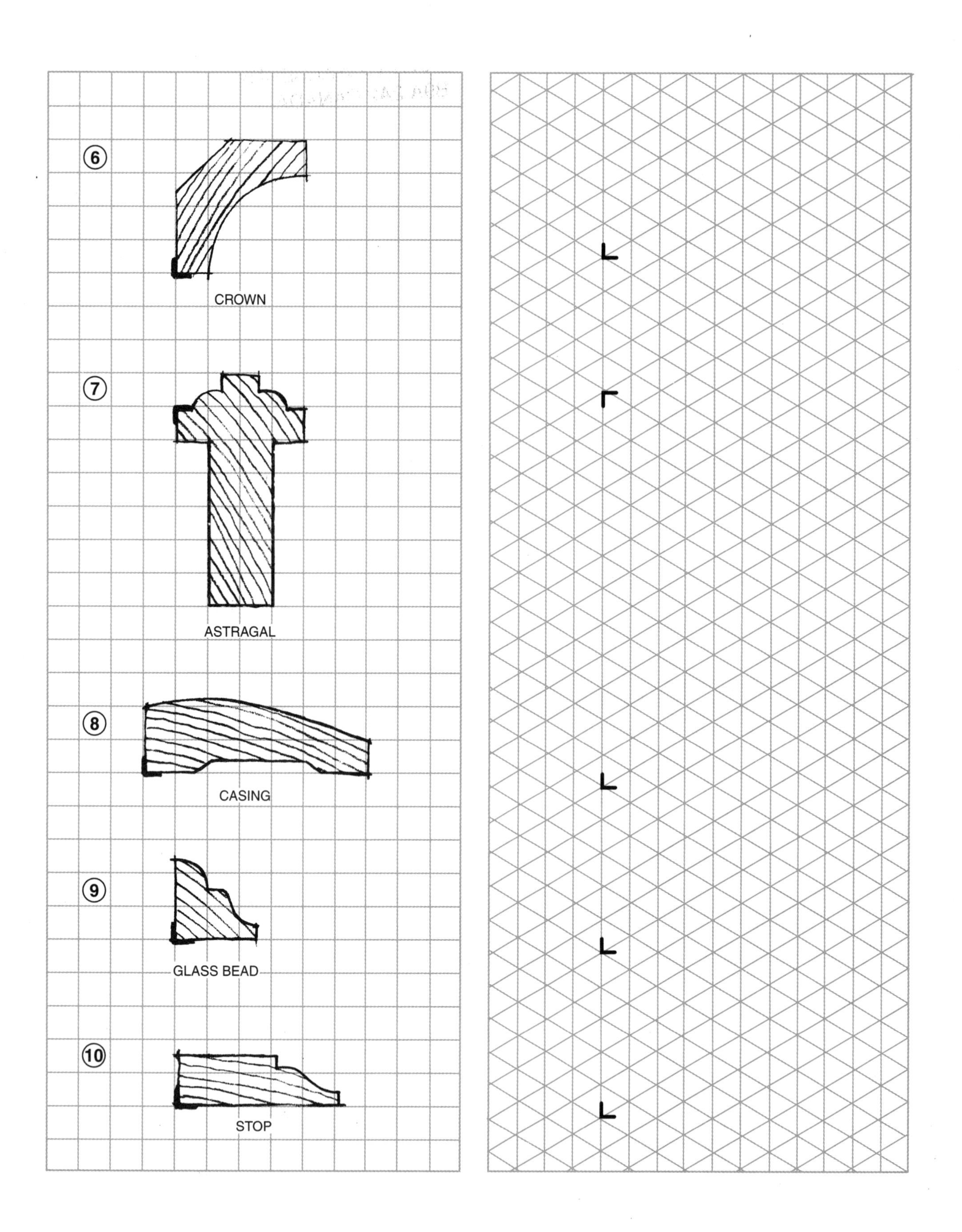

6
CROWN
7
ASTRAGAL
8
CASING
9
GLASS BEAD
10
STOP

Working Drawing Concepts 2

Review Questions

Name ______________________ Date ______________

Multiple Choice

_______________ **1.** A basic type of pictorial drawings is ___.
- A. perspective
- B. isometric
- C. oblique
- D. all of the above

_______________ **2.** C size paper is ___.
- A. 8½″ × 11″
- B. 11″ × 17″
- C. 17″ × 22″
- D. 18″ × 24″

_______________ **3.** Isosceles and equilateral triangles are ___ triangles.
- A. acute
- B. obtuse
- C. right
- D. none of the above

_______________ **4.** Quadrilaterals are ___ figures.
- A. three-sided plane
- B. three-sided solid
- C. four-sided plane
- D. four-sided solid

_______________ **5.** In an oblique cabinet drawing, receding lines are drawn at ___ scale.
- A. one-fourth
- B. one-half
- C. three-fourths
- D. full

_______________ **6.** The radius of a circle is ___ the diameter.
- A. the same length as
- B. one-half
- C. twice
- D. two and one-half times

_______________ **7.** The top view of a three-view drawing of a house shows the ___ elevation.
- A. front
- B. rear
- C. side
- D. none of the above

_______________ **8.** The preferred method of dimensioning framed walls is to dimension to the ___.
- A. centerlines of stud corner posts
- B. inside face of stud corner posts
- C. outside face of stud corner posts
- D. most convenient location

_______________ **9.** A ___ movement is used when sketching long lines.
- A. smooth wrist
- B. full arm
- C. rigid
- D. none of the above

_______________ **10.** Dimension lines may be terminated by ___.
- A. arrowheads
- B. slashes
- C. dots
- D. all of the above

_______________ **11.** The diameter of a circle is a ___.
- A. straight line passing anywhere through the circle
- B. curved line passing anywhere through the circle
- C. straight line passing through the centerpoint
- D. curved line passing through the centerpoint

_______________ **12.** Each view of a three-view drawing gives ___ principal measurement(s).
- A. one
- B. two
- C. three
- D. four

_______________ **13.** A trapezoid has ___.
- A. opposite sides parallel
- B. opposite sides parallel and equal
- C. two sides parallel
- D. two sides parallel and two sides equal

_______________ **14.** A wood stud partition with ⅜″ drywall on each side of 2 × 4 studs is ___″ thick.
- A. 3½
- B. 4⅛
- C. 4¼
- D. 4½

_______________ **15.** A bird's-eye perspective is drawn ___ the horizon line.
- A. above
- B. on
- C. below
- D. none of the above

True-False

T F **1.** Shading techniques are generally not used when producing three-view drawings.

T F **2.** A quadrant is one-half of a circle.

T F **3.** A right triangle contains one 90° angle.

T F **4.** Compass directions are commonly used to name exterior elevation drawings.

T F **5.** An arc is a portion of the circumference of a circle.

T F **6.** Oblique drawings may have one, two, or three vanishing points.

T F **7.** All circles contain 360°.

T F **8.** A complete set of plans is seldom drawn to the same scale.

T F **9.** A rhombus contains one 90° angle.

T F **10.** Circles on isometric drawings appear as ellipses.

T F **11.** Architect's scales are always read from left to right.

T F **12.** Solid masonry walls are dimensioned to their outside faces.

T F **13.** Rough openings for doors and windows are generally dimensioned to their centerlines.

T F **14.** The scale most commonly used for floor plans is ¼″ = 1′-0″.

T F **15.** Stud partitions vary in thickness depending on the finish.

Completion 2-1

______________ **1.** ___ is the process of drawing without instruments.

______________ **2.** ___ are plane figures with four 90° angles and four equal sides.

______________ **3.** Sketching pencils are either ___ or mechanical.

______________ **4.** Tracing ___ should be used when a sketch is to be duplicated on a diazo printer.

______________ **5.** Semicircles always contain ___°.

______________ **6.** ___ different scales are on the triangular architect's scale.

______________ **7.** AutoSketch® is a(n) ___ software program used for sketching.

______________ **8.** ___ lines are drawn level or parallel with the horizon.

______________ **9.** ___ polygons contain sides of equal length and equal angles.

______________ **10.** Pictorial drawings show the three principal measurements of ___, length, and depth.

______________ **11.** Perspective drawings may have one, two, or three ___ points.

______________ **12.** Concrete block partitions are generally dimensioned to the ___ of the finish material.

_______________ **13.** The ___ view of a three-view drawing shows the height and depth of an object.

_______________ **14.** The sum of two supplementary angles always equals ___°.

_______________ **15.** ___ lines are plumb or upright.

_______________ **16.** Plane figures have ___ surfaces.

_______________ **17.** Circles on receding surfaces of oblique drawings appear as ___.

_______________ **18.** Receding lines of oblique drawings may be 30° or ___°.

_______________ **19.** An acute angle contains less than ___°.

_______________ **20.** A(n) ___ of a circle is pie shaped.

_______________ **21.** ___ are three-sided plane figures.

_______________ **22.** A straight line always contains ___°.

_______________ **23.** Right triangles always contain one ___° angle.

_______________ **24.** ___ lines are straight lines that are neither horizontal nor vertical.

_______________ **25.** Horizontal lines of isometric drawings are drawn ___° above the horizon.

Completion 2-2

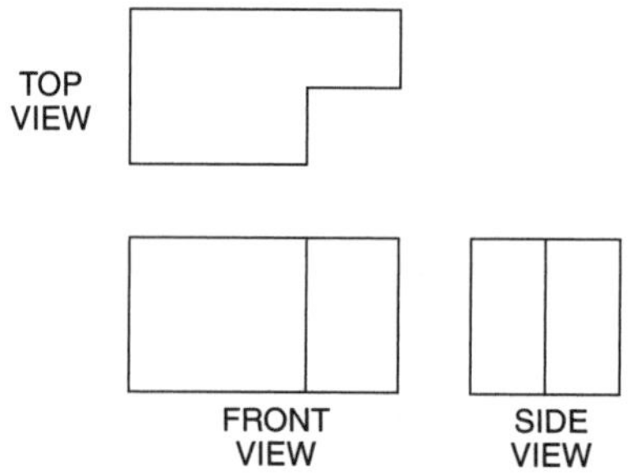

_______________ **1.** The ___ view shows height and length.

_______________ **2.** The ___ view shows height and depth.

_______________ **3.** The ___ view shows depth and length.

TOP VIEW

FRONT VIEW

SIDE VIEW

Working Drawing Concepts 2

Trade Competency Test

Name ______________________ Date ____________

Identification 2-1

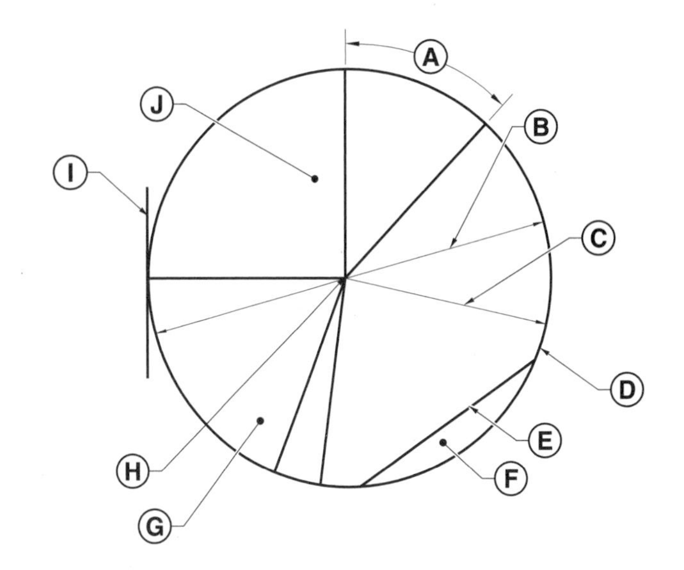

______ 1. Tangent

______ 2. Chord

______ 3. Radius

______ 4. Centerpoint

______ 5. Angle

______ 6. Segment

______ 7. Quadrant

______ 8. Diameter

______ 9. Sector

______ 10. Circumference

Identification 2-2

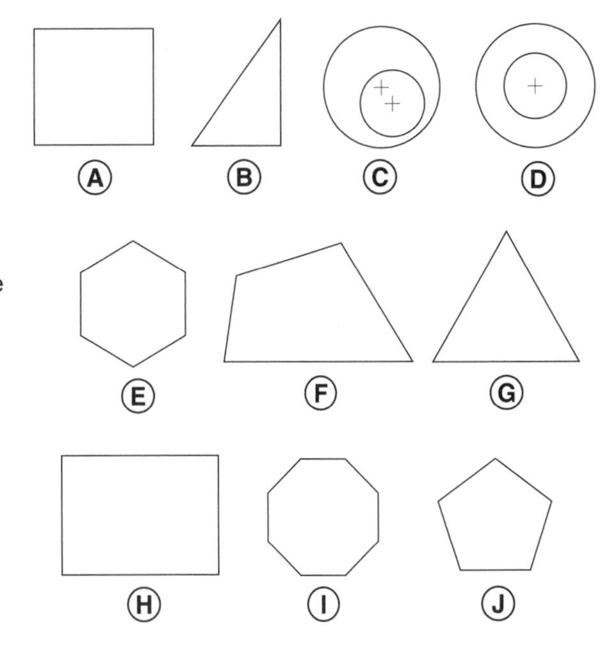

______ 1. Eccentric circles

______ 2. Square

______ 3. Hexagon

______ 4. Equilateral triangle

______ 5. Pentagon

______ 6. Right triangle

______ 7. Octagon

______ 8. Concentric circles

______ 9. Trapezium

______ 10. Rectangle

Identification 2-3

Refer to Randall Residence.

_______________ **1.** West Elevation (A) (B) (C) (D)

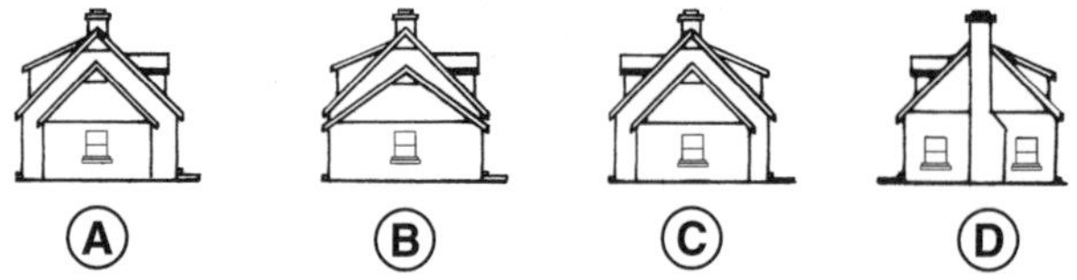

_______________ **2.** South Elevation (A) (B) (C)

_______________ **3.** East Elevation (A) (B) (C) (D)

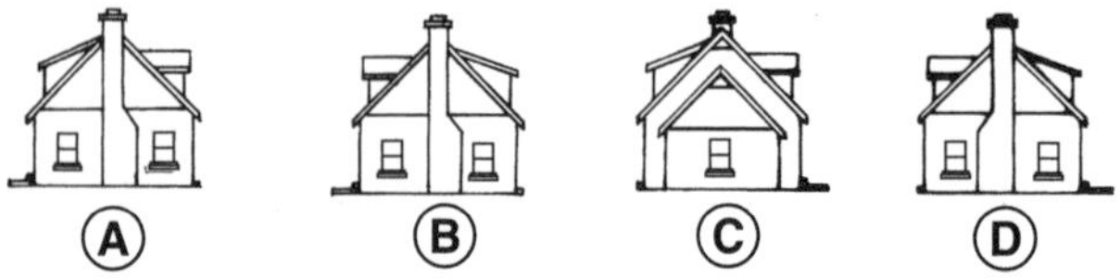

_______________ **4.** North Elevation (A) (B) (C)

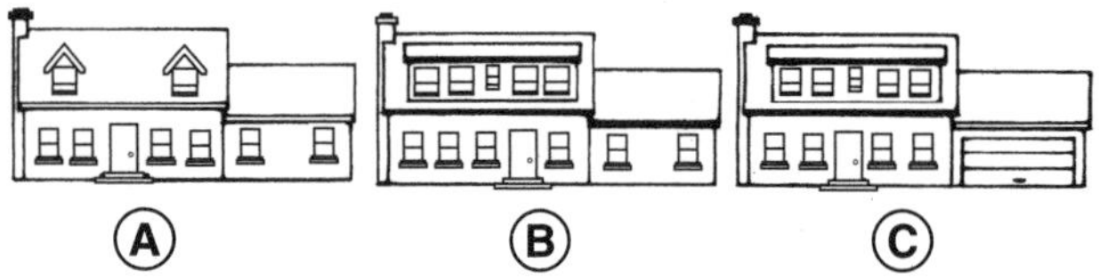

RANDALL RESIDENCE

Identification 2-4

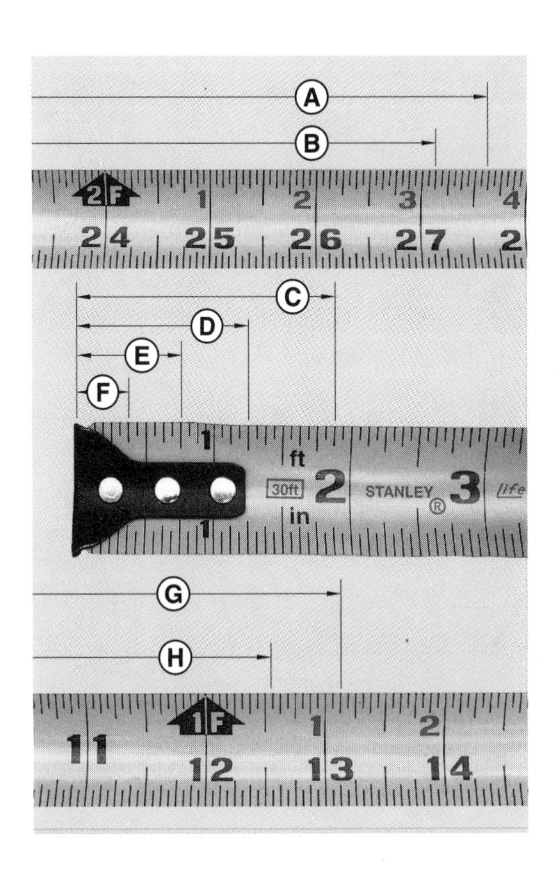

_______________ **1.** ⅜″

_______________ **2.** ¾″

_______________ **3.** 1¼″

_______________ **4.** 1⅞″

_______________ **5.** 2′-3⅛″

_______________ **6.** 2′-3⅝″

_______________ **7.** 1′-0 9/16″

_______________ **8.** 1′-1⅛″

Identification 2-5

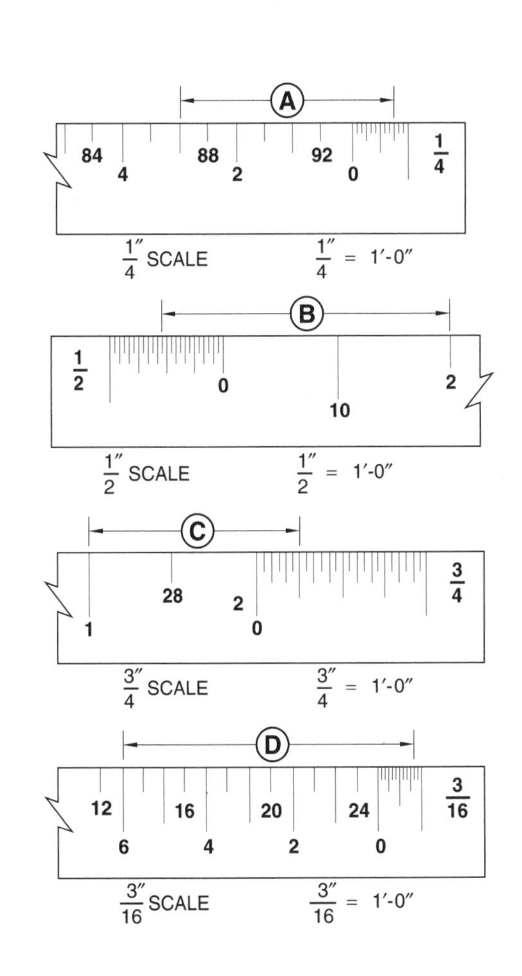

_______________ **1.** 2′-6½″

_______________ **2.** 1′-3″

_______________ **3.** 3′-9″

_______________ **4.** 6′-10″

Completion 2-1

1. __________ ¼″ = 1′-0″

2. __________ ¼″ = 1′-0″

3. __________ ¼″ = 1′-0″

4. __________ ⅜″ = 1′-0″

5. __________ ¾″ = 1′-0″

6. __________ ½″ = 1′-0″

7. __________ 1″ = 1′-0″

8. __________ 1½″ = 1′-0″

9. __________ 1½″ = 1′-0″

10. __________ 3″ = 1′-0″

Completion 2-2

Determine the missing dimensions.

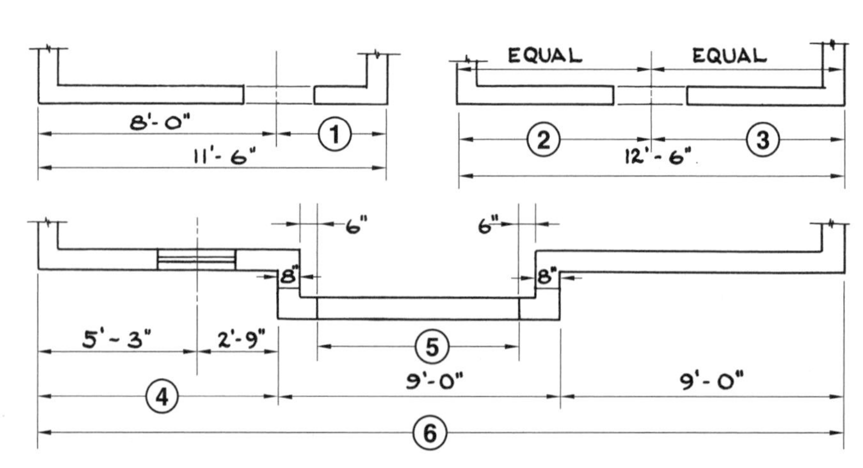

1. ____________________

2. ____________________

3. ____________________

4. ____________________

5. ____________________

6. ____________________

Multiple Choice

Refer to Joseph Residence on page 51.

_______________ **1.** A(n) ___ drawing of the house is shown.
- A. perspective
- B. isometric
- C. oblique
- D. none of the above

_______________ **2.** A person standing at position X would see the ___ Elevation.
- A. North
- B. South
- C. East
- D. West

_______________ **3.** A person standing at position Y would see the ___ Elevation.
- A. North
- B. South
- C. East
- D. West

_______________ **4.** The kitchen entrance will be shown on the ___ Elevation.
- A. North
- B. South
- C. East
- D. West

_______________ **5.** ___ door(s) will be shown on the South Elevation.
- A. No
- B. One
- C. Two
- D. Three

_______________ **6.** ___ window(s) will be shown on the East Elevation.
- A. No
- B. One
- C. Two
- D. Three

_______________ **7.** ___ window(s) will be shown on the West Elevation.
- A. No
- B. One
- C. Two
- D. Three

_______________ **8.** The chimney will be shown head-on on the ___ Elevation.
- A. North
- B. South
- C. East
- D. West

_______________ **9.** Wall surfaces ___ and ___ will be shown on the North Elevation.

A. A; B
B. C; D
C. A; C
D. B; D

_______________ **10.** The front entrance will be shown on the ___ Elevation.

A. North
B. South
C. East
D. West

_______________ **11.** A total of ___ door(s) and ___ windows will be shown on the South Elevation.

A. 1; 2
B. 2; 2
C. 2; 3
D. 2; 4

_______________ **12.** A total of ___ door(s) and ___ window(s) will be shown on the North Elevation.

A. 1; 1
B. 1; 3
C. 2; 2
D. 3; 3

_______________ **13.** Wall surfaces ___ will be shown on the East Elevation.

A. A and C
B. B and D
C. F, E, and G
D. none of the above

_______________ **14.** Wall surface F will be shown on the ___ Elevation.

A. North
B. South
C. East
D. West

_______________ **15.** Wall surface A will be shown on the ___ Elevation.

A. North
B. South
C. East
D. West

Refer to Multiple Choice on page 49.

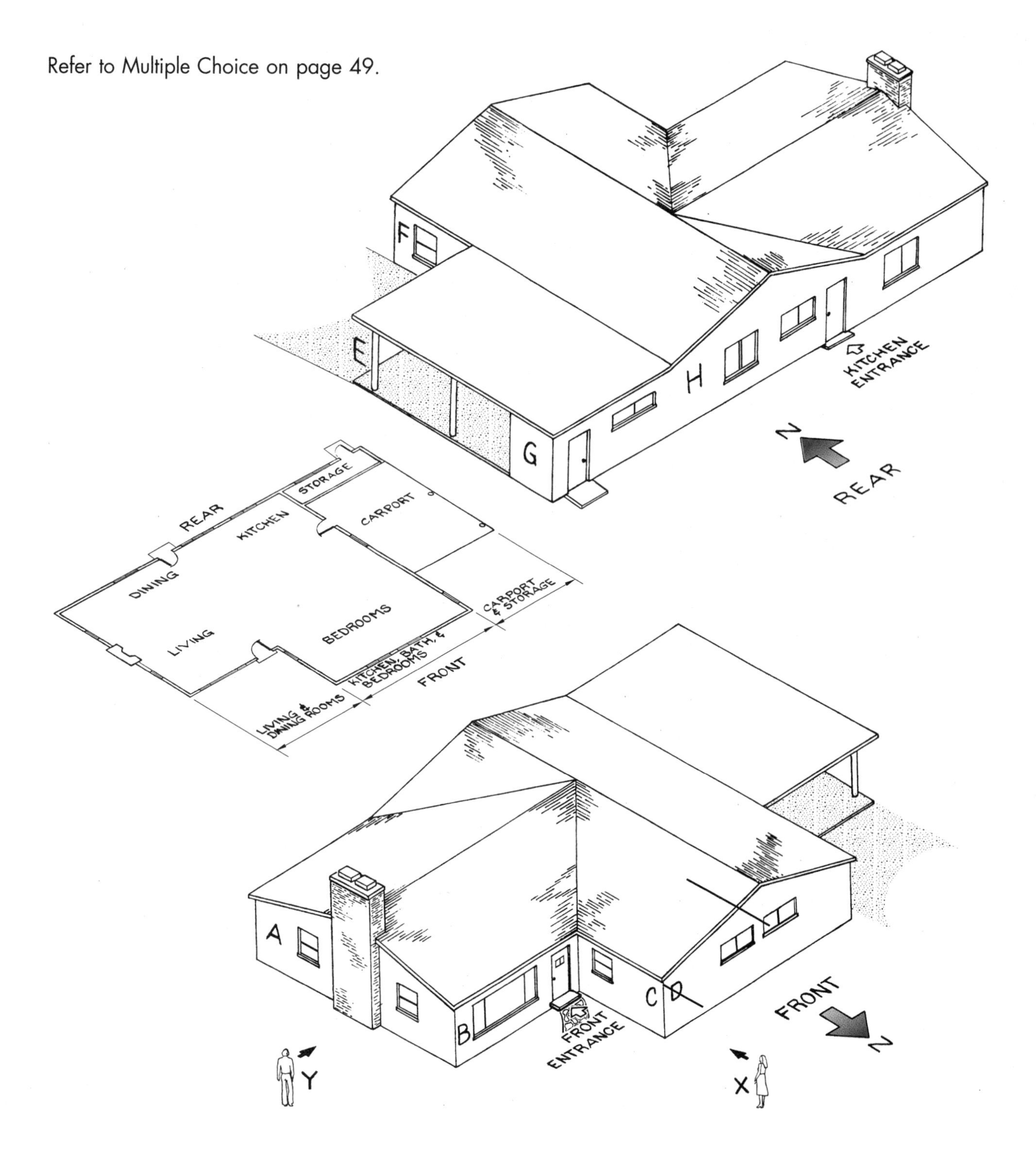

JOSEPH RESIDENCE

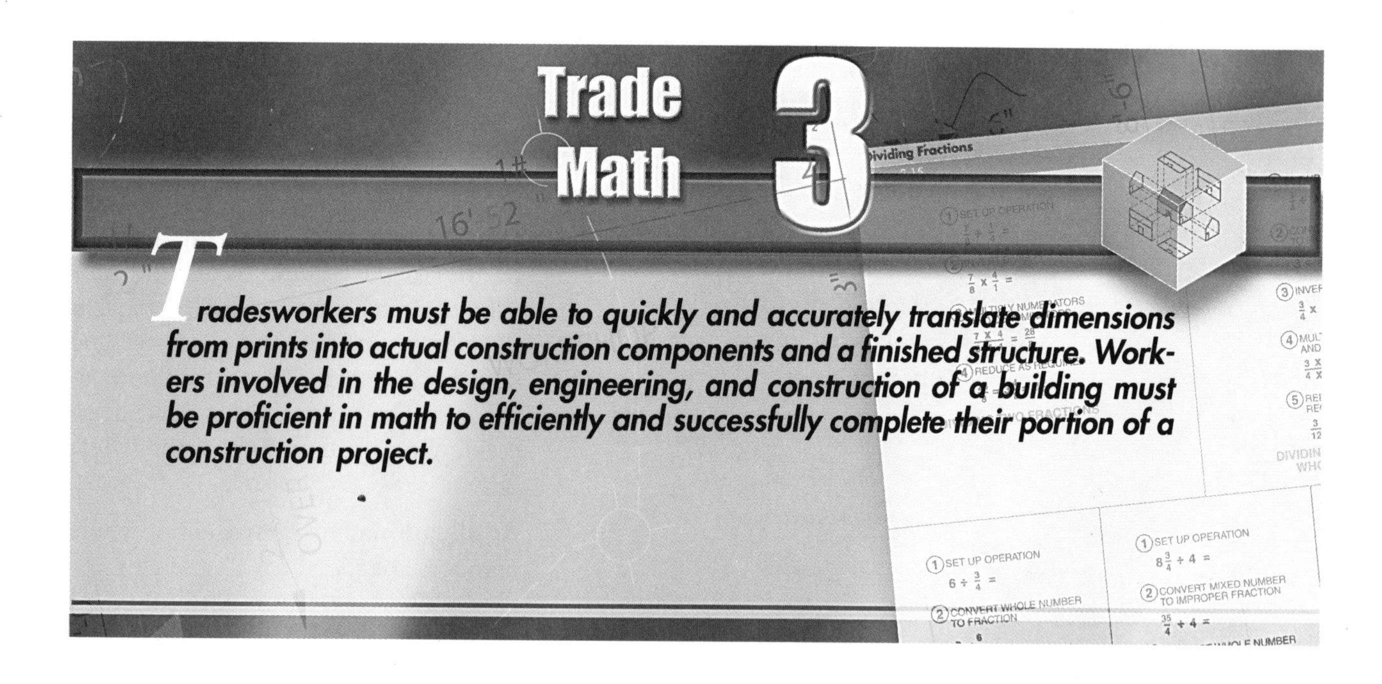

Trade Math 3

Tradesworkers must be able to quickly and accurately translate dimensions from prints into actual construction components and a finished structure. Workers involved in the design, engineering, and construction of a building must be proficient in math to efficiently and successfully complete their portion of a construction project.

WHOLE NUMBERS

A *whole number* is a number that does not have a fractional or decimal part. Whole numbers are commonly used for counting individual components and other items such as doors, windows, fixtures, bolts, or pipes. Numbers such as 1, 2, 3, 5, 12, 135, 1345, and 10,367 are whole numbers.

Whole numbers are either even or odd numbers. An *even number* is a number that can be divided by two without a remainder or decimal occurring. An *odd number* is a number that cannot be divided by two without a remainder or decimal occurring. See Figure 3-1. A *prime number* is an odd number that can only be divided by 1 or itself without a remainder or decimal occurring. Examples of prime numbers are 1, 3, 5, 7, 11, 17, and 23.

A *place* is the position that a digit occupies; it represents the value of the digit. See Figure 3-2. A digit has different values according to the place is occupies. For example, the digits 15 represent fifteen items, while the digits 51 represent fifty-one items.

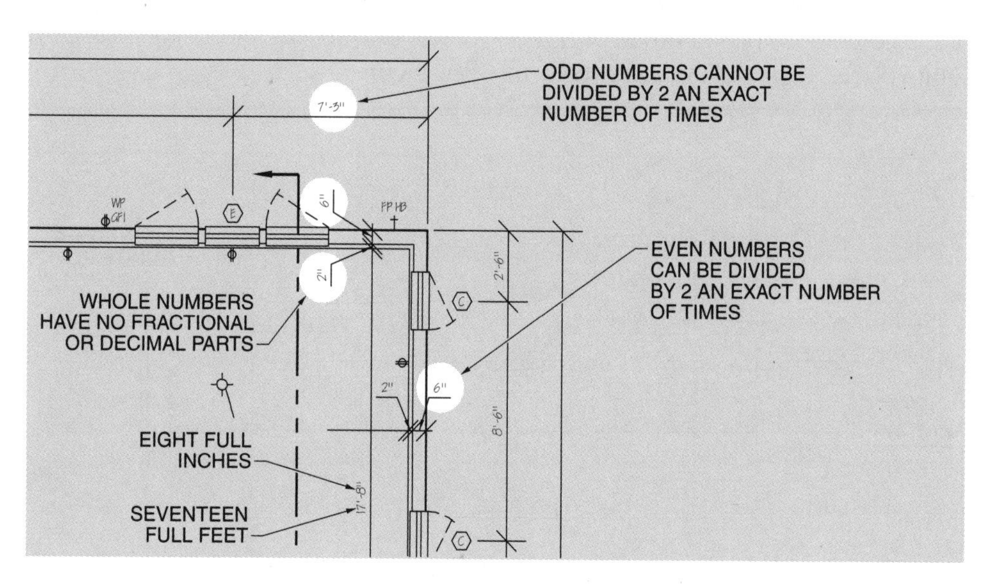

Figure 3-1. Whole numbers represent full units such as inches and feet.

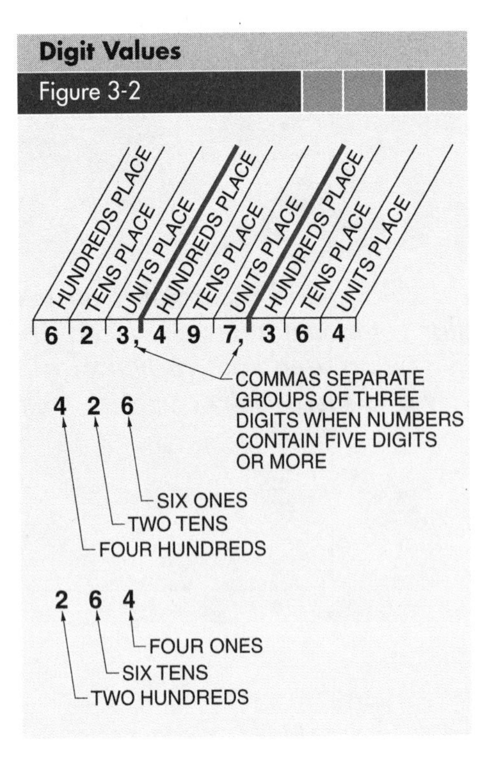

Figure 3-2. The value of a digit varies depending on its place in a whole number.

Adding Whole Numbers

Addition is the process of combining two or more whole numbers to obtain a total. A plus sign (+) indicates addition. A *sum* is the total that results from addition. For example, addition is used to determine the total length of a wall when the dimensions for several shorter lengths of wall segments are shown on a print.

Fundamentals. When whole numbers are added vertically, the numbers are placed above one another with the units places aligned. The units digit of each number is placed in the units place, the tens digit in the tens place, and so on. The columns are added from top to bottom, beginning with the units place. When the units sum is less than 10, the sum is recorded at the bottom of the column. When the units sum is greater than 10, the last digit of the sum is recorded and the remaining digit is carried to the next column to the left. This procedure is used in all columns to obtain a sum of the whole numbers. See Figure 3-3.

Applications. In printreading, whole numbers are added for a variety of applications. A common application of whole number addition is counting the number of items required for quantity takeoffs during estimating. For example, whole numbers are added to count the number of windows, doors, electrical switches, light fixtures, and other building components.

Dimensions on prints are commonly added to verify overall lengths. When adding feet and inches, inches are added first and then feet are added. Inches are added in base 12, with each 12″ equal to 1′-0″. Feet are added in base 10. For example, adding 6′-3″, 7′-6″, and 4′-6″ requires first adding the inches in base 12 (3″ + 6″ + 6″ = 15″ = 1′-3″), then adding the feet (6′ + 7′ + 4′ = 17′). The two subtotals are added to obtain the final sum (1′-3″ + 17′ = 18′-3″). See Figure 3-4.

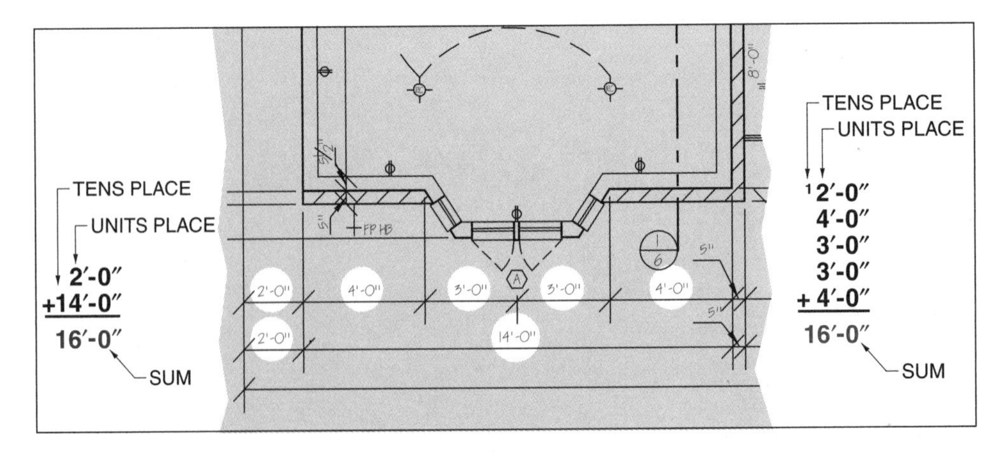

Figure 3-3. Addition of whole numbers begins with properly aligning the digits vertically.

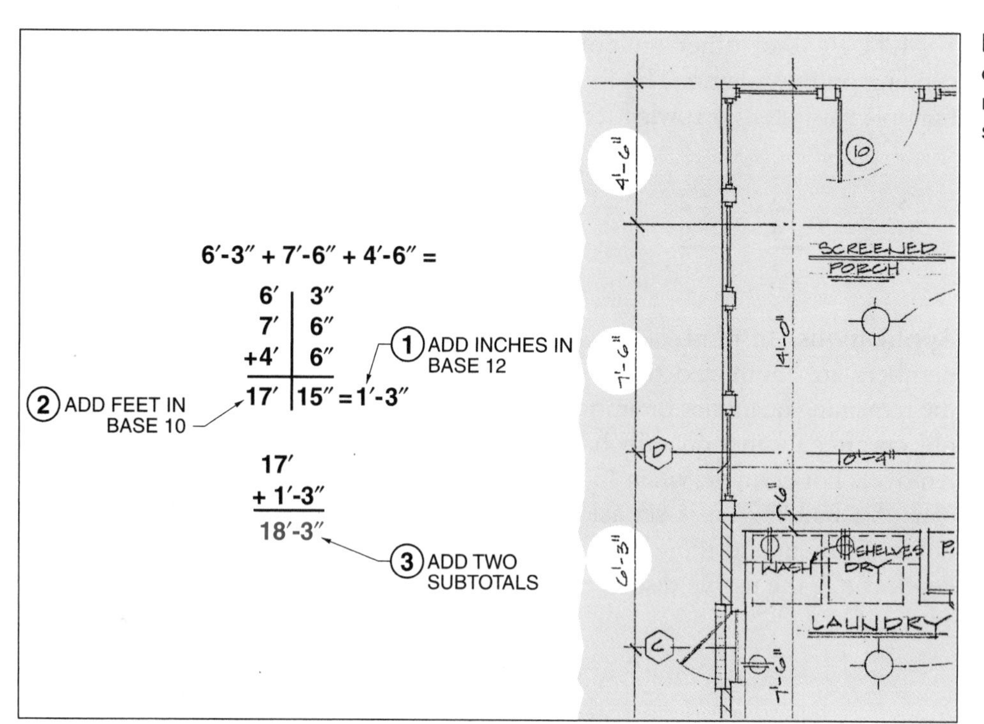

Figure 3-4. Addition of combined feet and inch measurements requires separate processes for each.

Subtracting Whole Numbers

Subtraction is the mathematical process of taking one number away from another number. Subtraction is the opposite, or inverse, mathematical operation of addition. The minus sign (–) indicates subtraction. The subtraction operation that begins with a quantity of three items and removes two to achieve a result of one is expressed as $3 - 2 = 1$.

Fundamentals. The *minuend* is the total number of units prior to subtraction. The *subtrahend* is the number of units to be removed from the minuend. The *result* is the difference between the minuend and subtrahend.

When whole numbers are subtracted vertically, the minuend is placed above the subtrahend with the units places aligned. Whole number subtraction starts with the units column on the right and moves from right to left through the columns. See Figure 3-5.

When the subtrahend digit in a column is larger than the unit in the minuend digit above, one unit is borrowed from the column immediately to the left. The borrowed unit may be

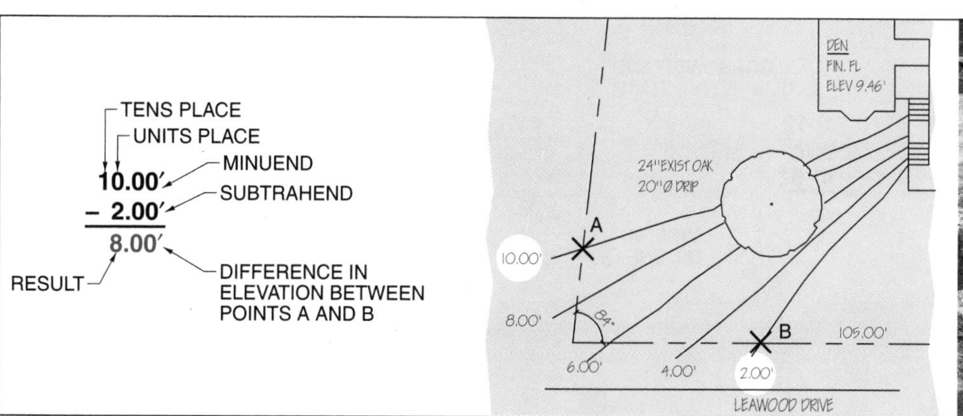

Figure 3-5. Subtraction is used on a construction site to determine differences in elevations.

David White Instruments

10, 100, 1000, or other multiples depending on the column. The following example illustrates borrowing ten units:

$$\begin{array}{r} 22 \\ -\ 9 \\ \hline \end{array} = \begin{array}{r} {}^{1}12 \\ -\ 9 \\ \hline 13 \end{array}$$

Arabic numbers are expressed by ten digits—0 through 9. Roman numerals are expressed by the letters I, V, X, L, C, D, and M.

Applications. In printreading, whole numbers are subtracted to determine the remaining quantities or amounts of objects after another quantity has been removed. For example, when 15′ of the rear of a building lot is set aside as a utility easement and the total property depth is 87′, the usable distance is 72′:

$$\begin{array}{r} 87' \\ -15' \\ \hline 72' \end{array}$$

Similar to the addition of feet and inches, in subtraction the inches are subtracted first in base 12 and then feet are subtracted in base 10. See Figure 3-6. For example, to determine the missing dimension on the print , 4′-8″ is subtracted from 7′-0″ (7′-0″ – 4′-8″). Inches are subtracted first. Since 8″ cannot be subtracted from 0″, 1′-0″ is converted to 12″ to allow for subtraction to proceed (12″ – 8″ = 4″). The remaining feet are then subtracted (6′ – 4′ = 2′). The two results are added (4″ + 2′ = 2′-4″).

Multiplying Whole Numbers

Multiplication is the mathematical process of adding a number to itself as many times as there are units indicated by the other number. The multiplication sign (×) is used to indicate multiplication. In effect, multiplication is an addition shortcut. For example, the notation of 10 × 8 is the same as adding ten to itself eight times (10 + 10 + 10 + 10 + 10 + 10 + 10 + 10 = 80).

Fundamentals. The *multiplicand* is the number being multiplied. The *multiplier* is the number that indicates the number of times the addition should occur. The *product* is the result of multiplication. From the example 10 × 8 = 80, 10 is the multiplicand, 8 is the multiplier, and 80 is the product.

A multiplication table is a chart that can be studied and memorized to quickly recall various products of multiplication. See Figure 3-7. Memorizing that 7 × 12 results in a product of 84 is quicker than performing the function longhand.

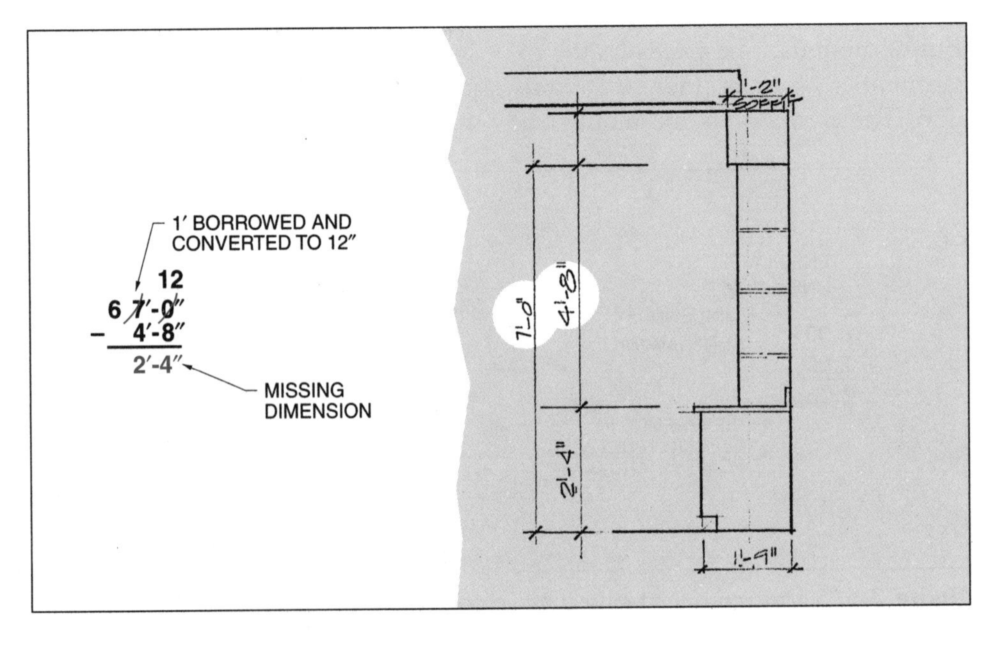

Figure 3-6. When borrowing in the subtraction of feet and inches, care is taken to borrow 12″ for each 1′.

MULTIPLICATION TABLE

1	2	3	4	5	6	7	8	9	10	11	12
2	4	6	8	10	12	14	16	18	20	22	24
3	6	9	12	15	18	21	24	27	30	33	36
4	8	12	16	20	24	28	32	36	40	44	48
5	10	15	20	25	30	35	40	45	50	55	60
6	12	18	24	30	36	42	48	54	60	66	72
7	14	21	28	35	42	49	56	63	70	77	84
8	16	24	32	40	48	56	64	72	80	88	96
9	18	27	36	45	54	63	72	81	90	99	108
10	20	30	40	50	60	70	80	90	100	110	120
11	22	33	44	55	66	77	88	99	110	121	132
12	24	36	48	60	72	84	96	108	120	132	144

7″ x 12″= 84″

Figure 3-7. Memorization of the multiplication table allows for quick and accurate calculations.

When multiple-digit whole numbers are multiplied vertically, the multiplicand is placed over the multiplier with the units places aligned. The first digit of the multiplier is multiplied across the entire multiplicand starting with the unit on the right, and the product is written in the first row below the horizontal line. The tens digit in the multiplier is then multiplied across the entire multiplicand, starting with the unit on the right, and the product is written in the second row below the horizontal line with the last digit of the product in the tens column. The two products are then added:

```
   365
  x 42
   730
 14 60
15,330
```

Any number multiplied by zero is zero. When there is a zero in the multiplier (for example, 407), only those numbers that have value are multiplied. The last digit of each separate product is placed under the digit used to calculate it. When all of the products have been calculated, the digits are added vertically. If the addition results in a sum of 10 or more in any given column, the ones digit is written in the proper place and the tens digit is carried to the next column:

```
    346
  x 407
  2 422
138 4
140,822
```

One method to ensure the product accuracy is to reverse the multiplicand and multiplier and perform the multiplication again. For example, when properly calculated 45 × 236 results in the same product as 236 × 45.

Another method to ensure multiplication accuracy is to divide the product by the multiplier. The quotient should equal the multiplicand value.

Applications. In printreading, whole numbers are multiplied to determine areas and volumes and to determine material quantities for multistory structures. For example, a 12-story apartment project with each floor containing 19 lavatories requires 228 lavatories:

```
  12
x 19
 108
 12
 228
```

Dividing Whole Numbers

Division is the mathematical process of determining the number of times one number is contained in another number. The division sign (÷) is used to indicate division. An example of a division operation is 6 ÷ 2 = 3. Three is contained in six two times. Division is the inverse mathematical operation of multiplication.

Fundamentals. The *dividend* is the number to be divided. The *divisor* is the number that the dividend is divided by. The *quotient* is the result of the division operation. When a dividend is not divided evenly by a divisor, a remainder, fraction, or decimal will occur.

When long division is performed, the dividend is placed inside the division bar and the divisor outside the bar.

The division operation is then carried out by dividing the divisor into the first number(s). See Figure 3-8. The number of times the divisor goes into the dividend is recorded on top of the division bar. The same operation is then carried out on the next number(s). If a remainder occurs, it is placed to the right of the quotient and noted as a remainder, placed over the divisor and expressed as a fraction, or expressed as a decimal.

An easy way to check the result of division is to multiply the quotient by the divisor. For example, 25 × 4 = 100 checks the division of the problem 100 ÷ 4 = 25.

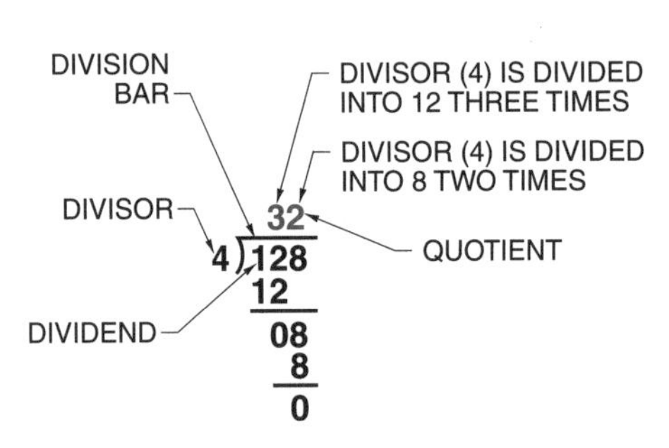

Figure 3-8. Proper alignment of units and accurate subtraction and multiplication are needed for accurate long division.

Applications. On prints, an architect might not provide specific dimensions to various features but will note them as being spaced equally or centered between two other features. Division is required to accurately locate the features. For example, two reinforced concrete pads and columns are equally spaced between two walls that are 36′ apart. Two pads equally spaced require that the distance between the walls be divided into three equal spaces. On-center (OC) spacing of the concrete pads is 12′-0″ (36′ ÷ 3 = 12′-0″ OC). See Figure 3-9.

Providing equal spacing between components with a given width requires a combination of multiplication, subtraction, and division. For example, four window units, each 5′ wide, are to be placed with exactly the same space between each of them and between adjoining walls that are 100′ apart. First, the overall width of the window units is determined by multiplying the width by the number of units (5′ × 4 = 20′).

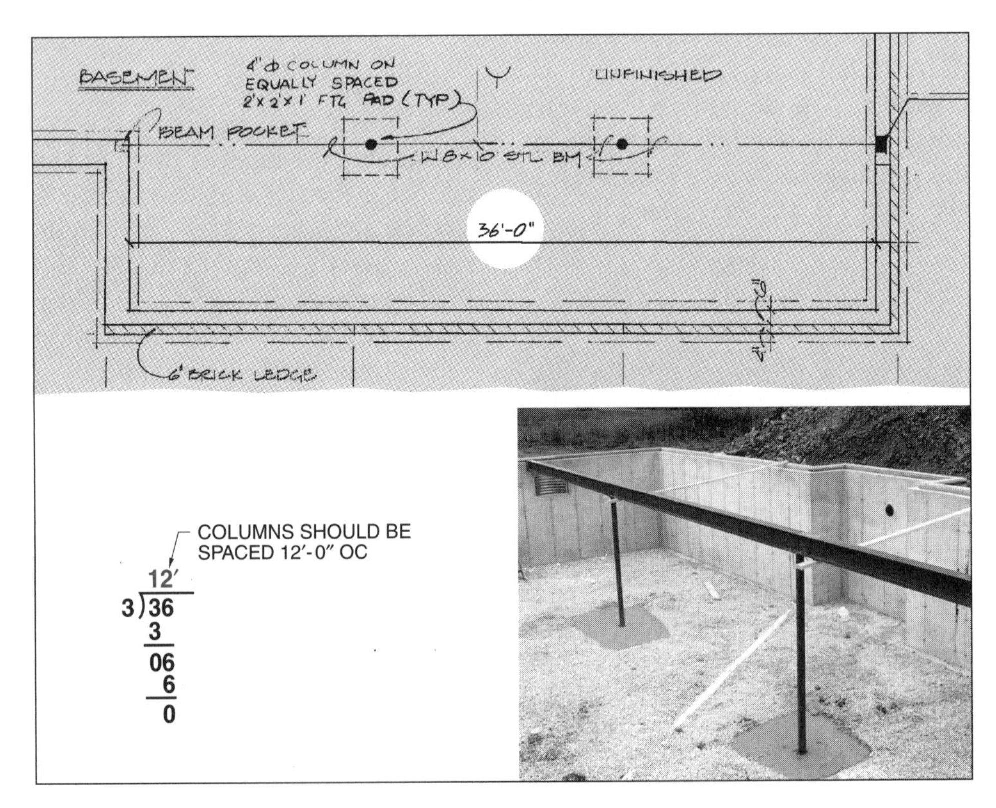

Figure 3-9. Division is used to locate various building components where equal spacing is required.

Next, the remaining space is determined by subtracting the overall width from the distance between the adjoining walls (100′ – 20′ = 80′). Finally, the exact spacing between the window units is determined by dividing the remaining space by the number of spaces between the four units (80′ ÷ 5 = 16′).

FRACTIONS AND DECIMALS

Fractions and decimals are a portion of a whole number. When expressing fractions, numbers are placed on each side of a horizontal or diagonal fraction bar, for example, ½. The number 2 indicates the total number of parts, and the number 1 indicates the number of parts included in the fraction. Decimals are represented by digits to the right of a decimal point that is placed to the right of a whole number. A decimal is expressed in base 10. The decimal of .5 indicates five-tenths of a whole number. See Figure 3-10.

Fractions are comprised of a denominator, numerator, and fraction bar. The *denominator* is the part of a fraction that indicates the total number of parts or divisions of a whole number. Denominators are placed below or to the right of the fraction bar. In ⅜, the denominator (8) indicates that a whole unit is divided into eight equal parts. The *numerator* is the part of a fraction that indicates the number of parts included in the fraction. Numerators are placed above or to the left of a fraction bar. In the ⅜ example, a numerator of 3 indicates that three of the eight equal parts are used to make up the total fraction. Fractions are common in linear measurement.

Decimal numbers are represented by digits to the right of a whole number, separated by a decimal point. A *decimal fraction* is a fraction with a denominator of 10, 100, 1000, 10,000, and so on. More digits in a decimal indicate greater accuracy. For example, 2.5 indicates two and five-tenths. A decimal of 2.568 indicates two and five hundred sixty-eight thousandths. Decimals are common in measuring elevations and heights.

> Elevations, such as those shown on plot plans, are typically expressed in decimal numbers.

> The addition of zeros to a number with a decimal point in front of it does not change its value.

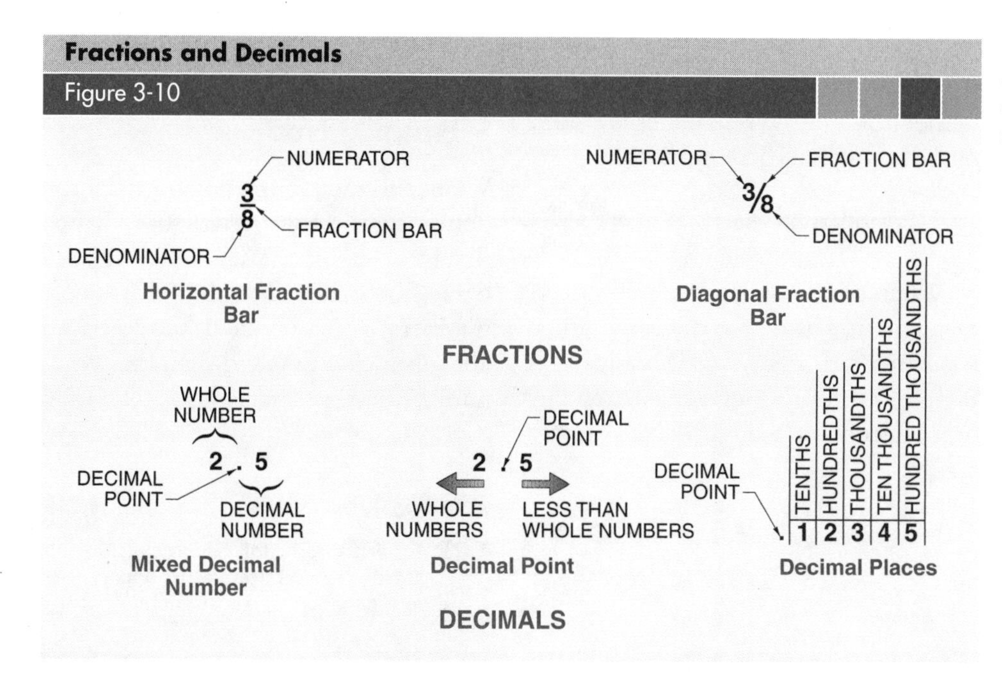

Figure 3-10. Smaller segments of whole numbers are described with fractions and decimals.

Adding Fractions

When adding fractions, the denominators of the fractions must be the same (common denominators). For example, 1/16 and 3/16 have common denominators and can be easily added, while 1/8 and 1/2 do not have common denominators. Fractions that do not have common denominators must be converted before adding.

Fundamentals. When adding a group of fractions with a common denominator, the fractions are arranged horizontally, the numerators are added, and the sum is placed over the denominator:

$$\frac{1}{8} + \frac{4}{8} + \frac{2}{8} = \frac{7}{8}$$

When adding groups of fractions, the sum of the numerators may be larger than the denominator:

$$\frac{1}{8} + \frac{3}{8} + \frac{5}{8} = \frac{9}{8}$$

A fraction should be reduced to its lowest terms after making a calculation. An improper fraction should be converted to a mixed number.

An *improper fraction* is a fraction with the numerator larger than the denominator. An improper fraction must be converted to a mixed number. A *mixed number* is a fraction containing a whole number and a fraction. When converting an improper fraction to a mixed number, the denominator is divided into the numerator and the remainder is placed over the original denominator:

$$\frac{9}{8} = 9 \div 8 = 1\frac{1}{8}$$

When adding fractions with different denominators, the fractions must be converted so that their equivalents contain common denominators. The numerators are then added to obtain a sum:

$$\begin{array}{rcl} \frac{1}{2} \times \frac{2}{2} & = & \frac{2}{4} \\ +\frac{1}{4} & = & \frac{1}{4} \\ \hline & & \frac{3}{4} \end{array}$$

In some trades, fractions with very dissimilar denominators may be encountered. The denominators must be multiplied together to determine a common denominator. The numerators are then multiplied by the denominator of the other fraction:

$$\begin{array}{rcl} \frac{3}{12} \times \frac{100}{100} & = & \frac{300}{1200} \\ +\frac{4}{100} \times \frac{12}{12} & = & \frac{48}{1200} \\ \hline & & \frac{348}{1200} \end{array}$$

Large fractions (such as 348/1200) are reduced by dividing both the numerator and denominator by a common number. For 348/1200, both 348 and 1200 can be divided by 12 to obtain 29/100.

Applications. On most construction prints, portions of an inch are expressed as sixteenths (1/16), eighths (1/8), quarters (1/4), and halves (1/2). Dimensions with different denominators are frequently added to obtain an overall dimension. See Figure 3-11. When necessary, fractions are converted to their equivalents containing common denominators and the numerators of the fractions are added. Inches are then added in base 12. Feet are added in base 10 and the three sums are added to determine overall length.

Subtracting Fractions

When subtracting fractions, a common denominator for the fractions must first be determined. Where mixed numbers (fractions combined with whole numbers) are subtracted, the fractions are subtracted first and then the whole numbers are subtracted.

To ensure subtraction accuracy, add the result and subtrahend. The sum should be equal to the original minuend.

Fundamentals. After a common denominator has been determined and the fractions have been converted to their equivalents, the numerators are subtracted and the difference is placed over the denominator. For example, $\frac{5}{16}$ is subtracted from $\frac{11}{16}$:

$$\frac{11}{16} - \frac{5}{16} = \frac{6}{16} \div \frac{2}{2} = \frac{3}{8}$$

In this case, the fraction is reduced by dividing both the numerator and denominator by 2 (6 ÷ 2 = 3 and 16 ÷ 2 = 8).

Applications. Fractions are frequently subtracted when determining required clearance or calculating omitted dimensions on a print. For example, it may be necessary to determine the distance between the base cabinet countertops and ceiling. See Figure 3-12. In some situations, it may be easier to perform calculations in inches (rather than feet and inches). In the example, the overall ceiling height is first converted to inches (8′ × 12″ = 96″ + ½″ = 96½″). Next, the floor-to-countertop height is subtracted from the overall ceiling height (96½″ – 35¾″). Then, the fraction(s) are converted to equivalent fraction(s) with a common denominator; in this case, ½″ = 2⁄4″. Since ¾″ cannot be subtracted from 2⁄4″, one full inch is borrowed, converted to a fraction (4⁄4), and added to the 2⁄4 to allow subtraction to proceed (2⁄4 + 4⁄4 = 6⁄4).

Wood Truss Council of America
Accurate measurement and math calculations are required in the construction of wood trusses.

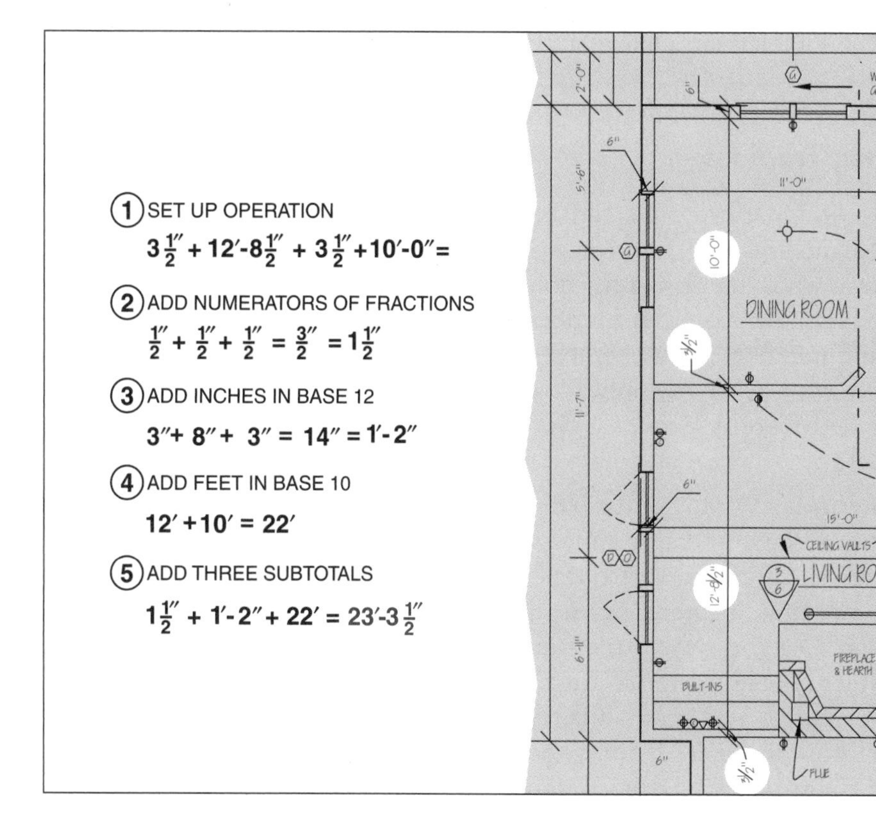

Figure 3-11. Accurate length calculations on prints require the addition of fractions of an inch, whole inches, and whole feet.

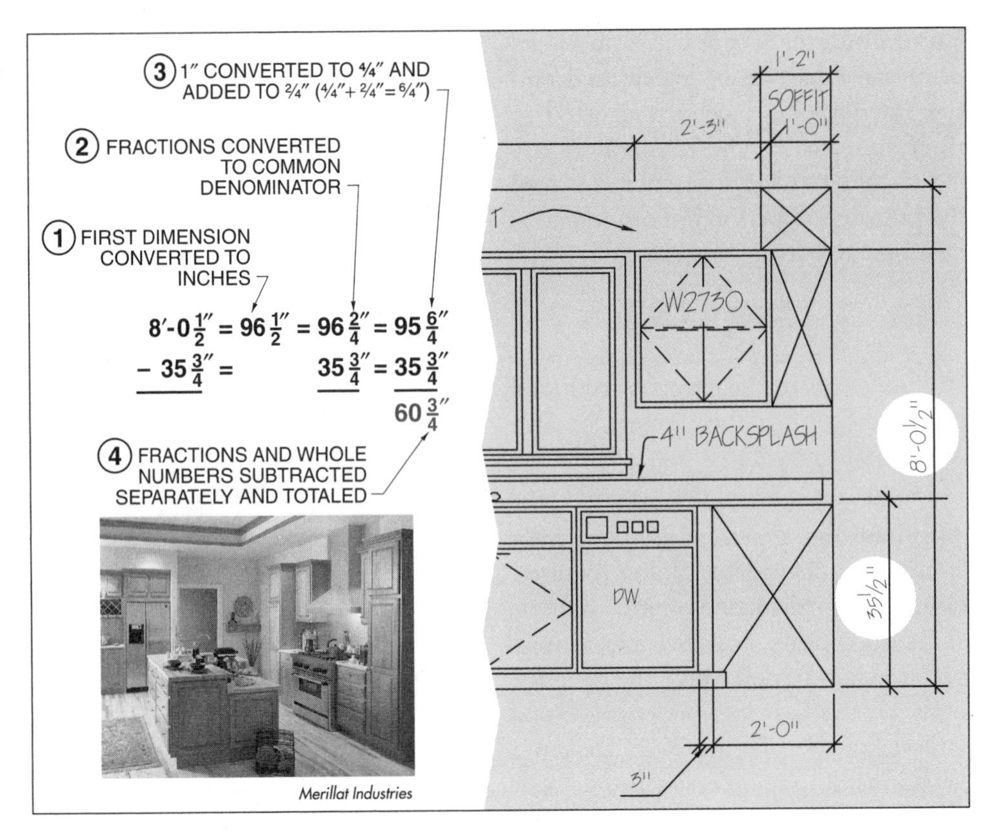

Figure 3-12. Subtraction of fractions, inches, and feet is utilized to calculate openings and remaining dimensions.

The fractions are subtracted (6⁄4 – 3⁄4 = 3⁄4) and then the remaining whole inches (95″ – 35″ = 60″) are subtracted. The subtotals are added (3⁄4″ + 60″ = 60 3⁄4″).

Multiplying Fractions

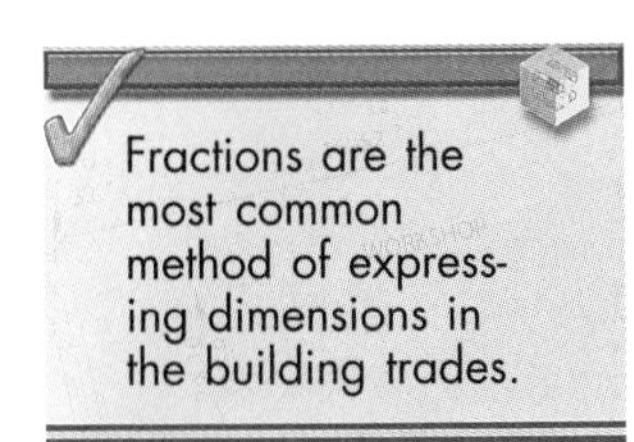

Fraction combinations that can be multiplied include two fractions, a fraction by a whole number, a fraction by a mixed number, a mixed number by a whole number, and two mixed numbers. Each type of multiplication requires a different process. See Figure 3-13.

Fundamentals. When multiplying two fractions, the numerator and denominator of one fraction are multiplied by the numerator and denominator of the other fraction. The product is reduced to its lowest terms as required. The product is reduced to lowest terms by dividing the numerator and denominator by the same number. For example, the product of 12⁄48 can be reduced to lowest terms (1⁄4) by dividing the numerator and denominator by 12 (12 ÷ 12 = 1; 48 ÷ 12 = 4).

When multiplying a fraction by a whole number, the fraction numerator is multiplied by the whole number and the denominator is left unchanged. The fraction is reduced to its lowest terms as required.

When multiplying a fraction by a mixed number, the mixed number is first converted to its improper fraction equivalent. The numerator and denominator of the two fractions are multiplied and reduced to lowest terms as required.

When multiplying a mixed number by a whole number, the mixed number is first converted to its improper fraction equivalent. The improper fraction numerator is then multiplied by the whole number and reduced to lowest terms as required.

Multiplying Fractions

Figure 3-13

① SET UP OPERATION

$\frac{1}{2} \times \frac{3}{4} =$

② MULTIPLY NUMERATORS AND DENOMINATORS. REDUCE AS REQUIRED

$\frac{1 \times 3}{2 \times 4} = \frac{3}{8}$

MULTIPLYING TWO FRACTIONS

① SET UP OPERATION

$\frac{3}{16} \times 2 =$

② MULTIPLY NUMERATOR BY WHOLE NUMBER. REDUCE AS REQUIRED

$\frac{3 \times 2}{16} = \frac{6}{16} = \frac{3}{8}$

MULTIPLYING FRACTIONS BY WHOLE NUMBERS

① SET UP OPERATION

$1\frac{3}{8} \times \frac{1}{2} =$

② CONVERT MIXED NUMBER TO IMPROPER FRACTION

$1\frac{3}{8} = \frac{11}{8}$

③ MULTIPLY NUMERATORS AND DENOMINATORS. REDUCE AS REQUIRED

$\frac{11 \times 1}{8 \times 2} = \frac{11}{16}$

MULTIPLYING MIXED NUMBERS BY FRACTIONS

① SET UP OPERATION

$2\frac{1}{2} \times 3 =$

② CONVERT MIXED NUMBER TO IMPROPER FRACTION

$2\frac{1}{2} = \frac{5}{2}$

③ MULTIPLY NUMERATOR BY WHOLE NUMBER. REDUCE AS REQUIRED

$\frac{5 \times 3}{2} = \frac{15}{2} = 7\frac{1}{2}$

MULTIPLYING MIXED NUMBERS BY WHOLE NUMBERS

① SET UP OPERATION

$2\frac{3}{4} \times 5\frac{1}{2} =$

② CONVERT MIXED NUMBERS TO IMPROPER FRACTIONS

$2\frac{3}{4} = \frac{8}{4} + \frac{3}{4} = \frac{11}{4}$

$5\frac{1}{2} = \frac{10}{2} + \frac{1}{2} = \frac{11}{2}$

③ MULTIPLY NUMERATORS AND DENOMINATORS. REDUCE AS REQUIRED

$\frac{11 \times 11}{4 \times 2} = \frac{121}{8} = 15\frac{1}{8}$

MULTIPLYING MIXED NUMBERS

Figure 3-13. Different mathematical operations are required for multiplication of various combinations of fractions, whole numbers, and mixed numbers.

When multiplying two mixed numbers, the mixed numbers are first converted to their fraction equivalents. The numerators and denominators are multiplied and the fraction is reduced as required.

Applications. A common application of multiplying fractions and mixed numbers is the calculation of the total rise of stairs. See Figure 3-14. The architect may note that a stairway has 14 risers, each with a unit rise of 7⅜″. The total rise of the stairs is 8′-7¼″.

Dividing Fractions

Dividing fractions is the opposite, or inverse, of multiplying fractions. When dividing fractions, the divisor is inverted (denominator placed over the numerator) and the fractions are then multiplied as previously described.

Calculators are often used on a job site to perform mathematical functions.

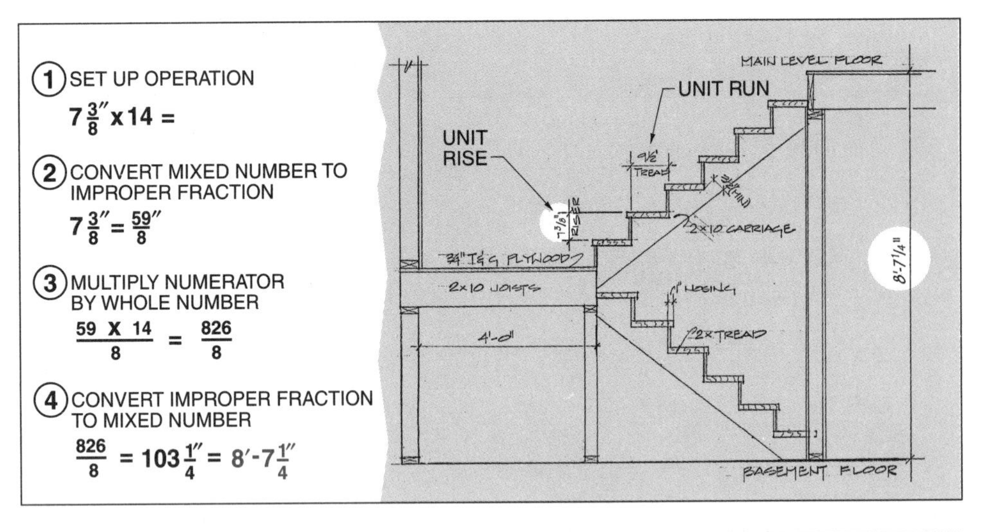

Figure 3-14. Multiplication of fractions is used to determine a stair landing height where a unit rise is indicated.

The numerator and denominator of a fraction can be multiplied or divided by the same number (except zero) without changing the value of the fraction.

Fraction combinations that can be divided include two fractions, a fraction by a whole number, a whole number by a fraction, a mixed number by a whole number, and two mixed numbers. Each type of fractional division requires a different process. See Figure 3-15.

Fundamentals. The dividend is the number to be divided. The divisor is the number that the dividend is divided by. The quotient is the result of the division operation. To divide two fractions, the numerator and denominator in the divisor are first inverted. For example, when dividing ⅞ by ¼, the divisor (¼) is inverted to 4/1. The numerator and denominator of the dividend and the inverted divisor are then multiplied.

When dividing a fraction by a whole number, the whole number is converted to a fraction by placing it over 1. The divisor is inverted and the numerators and denominators are multiplied. The fraction is reduced as required. A similar process is used to divide a whole number by a fraction.

When dividing a mixed number by a whole number, the mixed and whole numbers must be converted to improper fractions. The divisor is then inverted and the fractions are multiplied and reduced to lowest terms as required. The same process is used to divide two mixed numbers.

A fraction can easily be divided by 2 by doubling the denominator.

Applications. The most common application of division in printreading is the centering or equal spacing of features, fixtures, or other items. See Figure 3-16. For example, centering a door opening between two walls that are 3′-6½″ apart requires dividing 42 ½″ (3′-6½″) by 2. The door is centered on a mark 21¼″ from either wall.

Another common application of division of fractions and mixed numbers is the determination of riser heights when constructing stairways. For example, a stairway with a total rise of 36⅞″ requires five equal-height risers of 7⅜″:

$$36\frac{7}{8} = \frac{295}{8}$$

$$\frac{295}{8} \div 5 = \frac{295}{8} \div \frac{5}{1}$$

$$\frac{295}{8} \times \frac{1}{5} =$$

$$\frac{295 \times 1}{8 \times 5} =$$

$$\frac{295}{40} = 7\frac{15}{40} = 7\frac{3}{8}''$$

Adding and Subtracting Decimals

The two types of decimal numbers are proper decimal numbers and mixed decimal numbers. A *proper decimal number* is a decimal number that does not

Dividing Fractions
Figure 3-15

DIVIDING TWO FRACTIONS

1. SET UP OPERATION
$\frac{7}{8} \div \frac{1}{4} =$
2. INVERT DIVISOR
$\frac{7}{8} \times \frac{4}{1} =$
3. MULTIPLY NUMERATORS AND DENOMINATORS
$\frac{7 \times 4}{8 \times 1} = \frac{28}{8}$
4. REDUCE AS REQUIRED
$\frac{28}{8} = 3\frac{4}{8} = 3\frac{1}{2}$

DIVIDING FRACTIONS BY WHOLE NUMBERS

1. SET UP OPERATION
$\frac{3}{4} \div 3 =$
2. CONVERT WHOLE NUMBER TO FRACTION
$3 = \frac{3}{1}$
3. INVERT DIVISOR
$\frac{3}{4} \times \frac{1}{3} =$
4. MULTIPLY NUMERATORS AND DENOMINATORS
$\frac{3 \times 1}{4 \times 3} = \frac{3}{12}$
5. REDUCE AS REQUIRED
$\frac{3}{12} = \frac{1}{4}$

DIVIDING WHOLE NUMBERS BY FRACTIONS

1. SET UP OPERATION
$6 \div \frac{3}{4} =$
2. CONVERT WHOLE NUMBER TO FRACTION
$6 = \frac{6}{1}$
3. INVERT DIVISOR
$\frac{6}{1} \times \frac{4}{3} =$
4. MULTIPLY NUMERATORS AND DENOMINATORS
$\frac{6 \times 4}{1 \times 3} = \frac{24}{3}$
5. REDUCE AS REQUIRED
$\frac{24}{3} = 8$

DIVIDING MIXED NUMBERS BY WHOLE NUMBERS

1. SET UP OPERATION
$8\frac{3}{4} \div 4 =$
2. CONVERT MIXED NUMBER TO IMPROPER FRACTION
$\frac{35}{4} \div 4 =$
3. CONVERT WHOLE NUMBER TO FRACTION AND INVERT
$\frac{35}{4} \times \frac{1}{4} =$
4. MULTIPLY NUMERATORS AND DENOMINATORS
$\frac{35 \times 1}{4 \times 4} = \frac{35}{16}$
5. REDUCE AS REQUIRED
$\frac{35}{16} = 2\frac{3}{16}$

DIVIDING MIXED NUMBERS

1. SET UP OPERATION
$1\frac{1}{4} \div 2\frac{1}{2} =$
2. CONVERT MIXED NUMBERS TO IMPROPER FRACTIONS
$\frac{5}{4} \div \frac{5}{2} =$
3. INVERT DIVISOR
$\frac{5}{4} \times \frac{2}{5} =$
4. MULTIPLY NUMERATORS AND DENOMINATORS
$\frac{5 \times 2}{4 \times 5} = \frac{10}{20}$
5. REDUCE AS REQUIRED
$\frac{10}{20} = \frac{1}{2}$

Figure 3-15. Different mathematical operations are required for division of various combinations of fractions, whole numbers, and mixed numbers.

have a whole number, such as .9. A *mixed decimal number* is a decimal number consisting of a whole number and a decimal number separated by a decimal point. For example, 4.75 is a mixed decimal number.

Fundamentals. When decimal numbers are added or subtracted vertically, the decimal points of the numbers must be aligned. Aligning the decimal points will align whole numbers over whole numbers, tenths over tenths, hundredths over hundredths, and so on. After the numbers are aligned, addition or subtraction proceeds in the same manner as addition or subtraction of whole numbers.

Applications. Dimensions such as elevations and distances around a building lot are commonly expressed in feet and hundredths of a foot on elevation drawings. Determining distances between various building points and differences in elevations requires the addition and subtraction of decimals. See Figure 3-17. For example, the difference in elevation of two floor levels of a house is calculated by subtracting the elevation at the lowest grade point from the elevation at the highest grade point. If the highest reading is 13.46′ and the lowest point is 9.46′, the difference in elevation is 4.00′ (13.46′ – 9.46′ = 4.00′).

Multiplying Decimals

When multiplying decimals, the number of decimal digits to the right of the decimal point in the multiplicand and multiplier must carefully be determined. When multiplication is complete, the same number of decimal digits must be counted off in the product. Otherwise, multiplication of decimals is the same as multiplication of whole numbers.

Fundamentals. When decimals are multiplied vertically, the multiplicand is placed above the multiplier. See Figure 3-18. The first digit of the multiplier is multiplied across the entire multiplicand and the product is written in the first row below the horizontal line. The tens digit in the multiplier is then multiplied across the entire multiplicand and the product is written in the second row below the horizontal line with the last digit of the product in the tens column. The last digit of each separate product is placed under the digit used to calculate it. When all of the subtotals have been calculated, the digits are added vertically. If the addition results in a sum of 10 or more in any given column, the ones digit is written in the proper position and the tens digit is carried to the next column. The number of digits following the decimal point in the multiplicand and multiplier

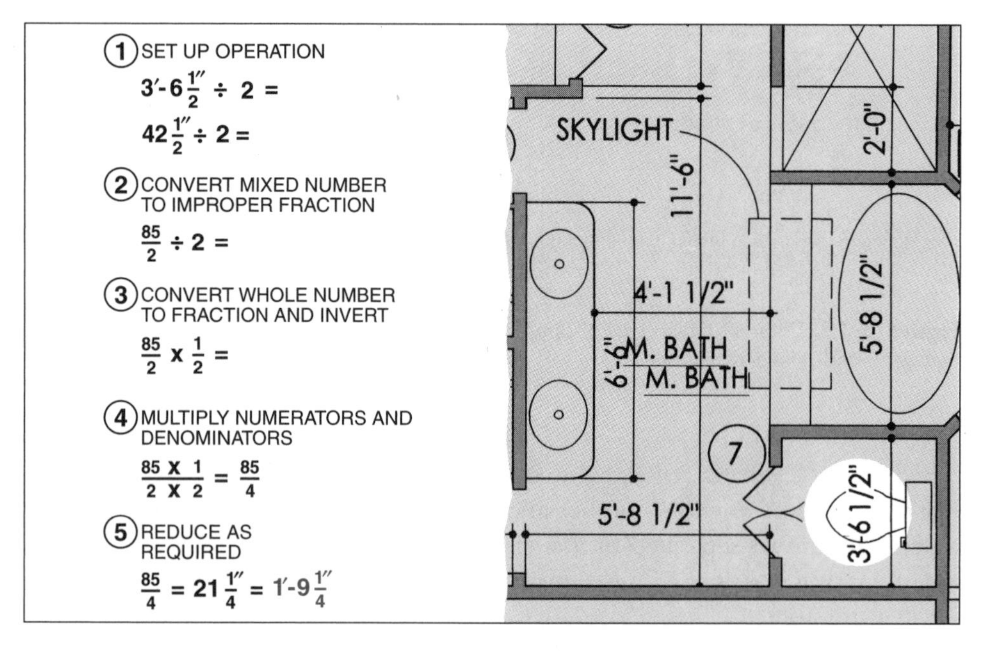

Figure 3-16. Centering a fixture may require the division of a mixed number of feet, inches, and fractions of an inch by 2.

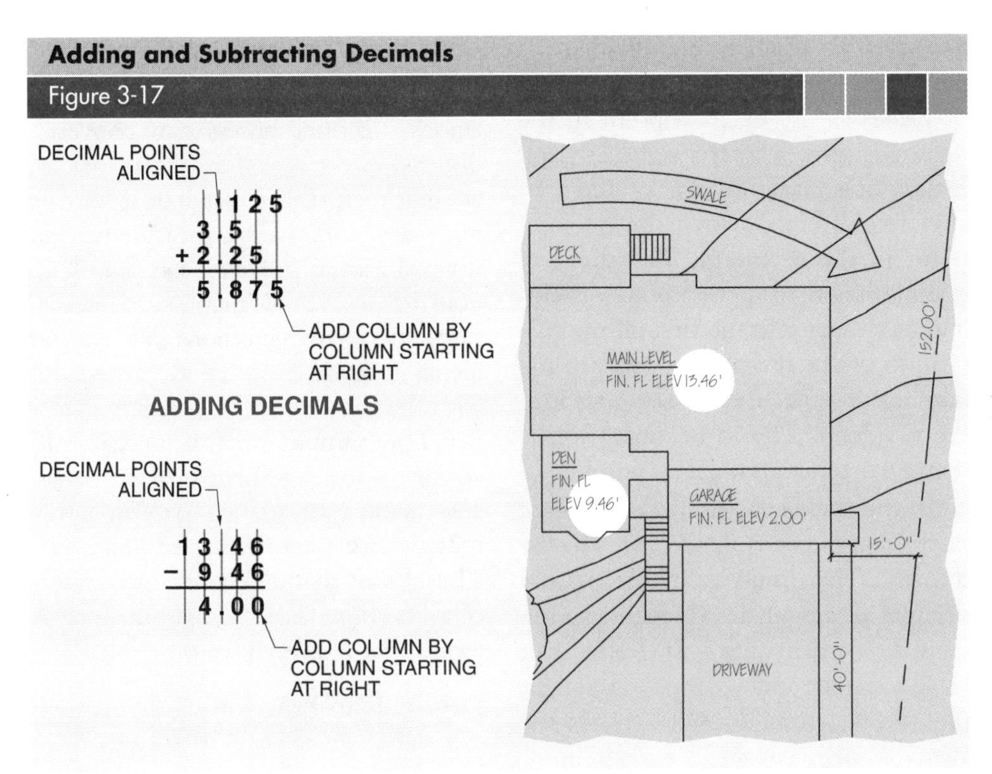

Figure 3-17. Elevation differences expressed as mixed decimal numbers are calculated using decimal addition and subtraction.

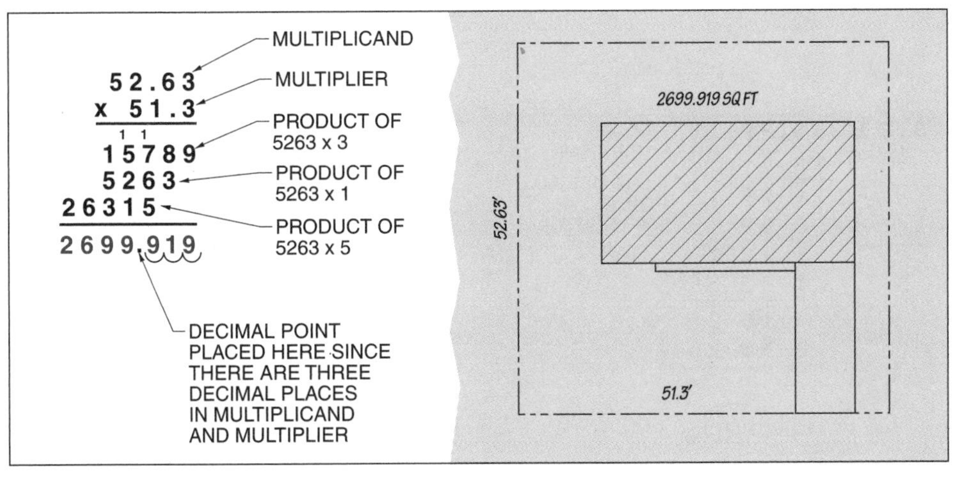

Figure 3-18. The decimal point must be properly located in the final product when multiplying decimal numbers.

are counted. In the answer, the same number of digits starting at the digit on the right is counted off and the decimal point is placed before the last counted digit.

Applications. Calculations such as area and volume are performed using decimal numbers. For example, the area of a building lot measuring 100.25′ × 53.5′ can be calculated by multiplying the decimal numbers (100.25′ × 53.5′ = 5363.375 sq ft).

Dividing Decimals

Dividing decimals is the same as dividing whole numbers. The decimal point must be properly placed for an accurate calculation. When performing long division, the decimal point of the quotient is positioned directly above the decimal point in the divisor.

Fundamentals. When performing long division of decimals, the dividend is placed inside the division bar and the divisor outside the bar. See Figure 3-19.

The decimal point in the dividend is moved to the right the same number of places as the decimal point in the divisor is moved to the right. A decimal point is placed on top of the division bar directly above the decimal point in the dividend. The division operation is then carried out by dividing the divisor into the first number(s). On top of the division bar, record the number of times the divisor goes into the dividend. The same operation is then carried out on the next number(s).

If the dividend has fewer decimal places that the divisor, add zeros to the dividend. There must be at least as many decimal places in the dividend as the divisor. As many zeros as desired can be added since adding zeros to the right of a decimal point does not change the value of the decimal (3.5 = 3.5000).

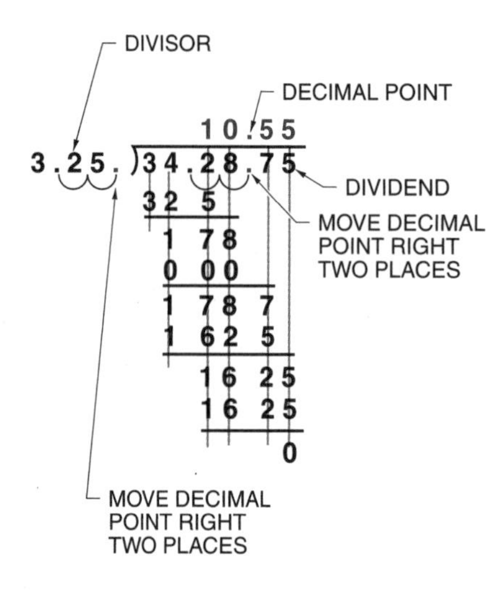

Figure 3-19. The decimal point must be properly located in the divisor, dividend, and quotient when performing long division.

Applications. Decimal division is commonly used in operations such as calculating elevations or distances found on a plot plan. For example, determining one-half the height of an elevation of 32.87′ requires dividing 32.87 by 2 (32.87′ ÷ 2 = 16.435′).

Percentages are converted to decimals by moving the decimal point left two places.

Converting Fractions and Decimals

When performing mathematical calculations, it is often necessary to convert a dimension expressed in one format to another format. For example, it may be necessary to convert a measurement in decimal format to feet, inches, and fractions of an inch. It may also be necessary to perform calculations with feet and inches that are more easily carried out with the dimensions in a different format. For example, it may be necessary to convert a measurement in feet, inches, and fractions of an inch to decimal format before performing calculations. The ability to quickly and accurately convert dimensions from one format to another is invaluable on a job site.

Decimals to Fractions. Decimals are calculated on a base 10 system. In a base 10 system, decimals are converted to fractions by placing the numbers to the right of the decimal point over 10, 100, 1000, or another base 10 unit. For example, .875 is converted to $\frac{875}{1000}$, which can be reduced in lowest terms to $\frac{7}{8}$. When decimal numbers include a whole number, the digits to the right of the decimal point are converted as previously shown and then the whole number is added. For example, 4.125 is converted to $4\frac{1}{8}$ ($.125 = \frac{125}{1000} = \frac{1}{8}$; $\frac{1}{8} + 4 = 4\frac{1}{8}$). See Figure 3-20.

Fractions to Decimals. When performing mathematical operations with fractions, it may be preferable to convert the fractions to decimals for easier calculation. When converting a fraction to its decimal equivalent, the numerator is divided by the denominator. For example, $\frac{5}{8}$ is converted to .625 (5 ÷ 8 = .625). When adding a series of fractions with different denominators, such as $\frac{5}{16}$, $\frac{1}{2}$, and $\frac{7}{8}$, each is converted to a decimal and the addition operation is performed ($\frac{5}{16} = .3125$; $\frac{1}{2} = .5$; $\frac{7}{8} = .875$; .3125 + .5 + .875 = 1.6875). The decimal can then be converted back to its fraction equivalent.

Converting Fractions and Decimals

Figure 3-20

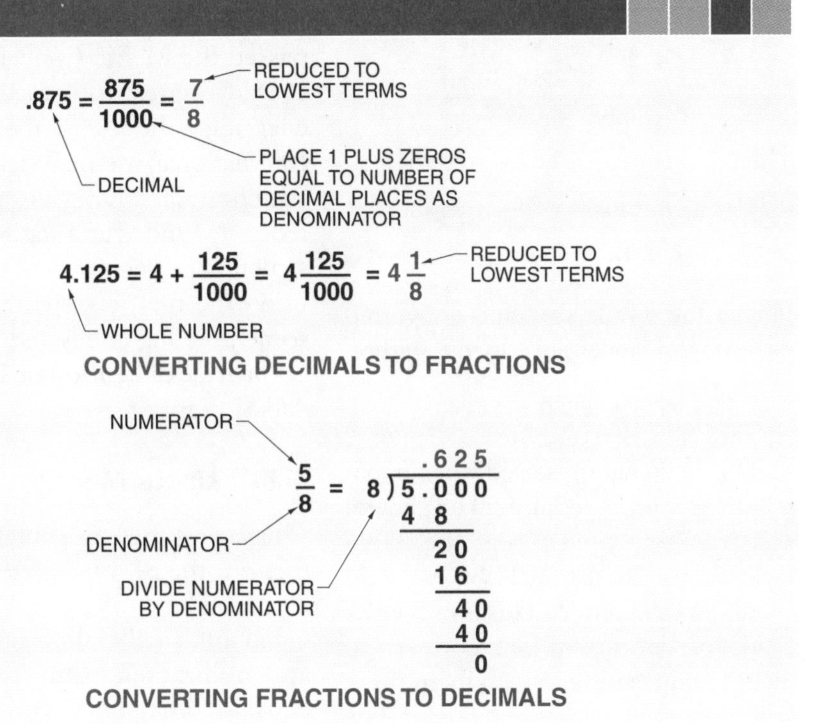

Figure 3-20. Some job site calculations are easier to perform when fractions and decimals can be readily converted.

Feet and Inches to Fractions and Decimals. Unlike decimals, which are in the base 10 system, inches are in the base 12 system. Inches are converted to their decimal foot equivalent by dividing the inch measurement by 12. For example, 4″ = .333′ (4 ÷ 12 = .333). When fractions of an inch are involved, the fractional inch is converted to its decimal inch equivalent, the converted value is added to the whole inch(es) and divided by 12. For example, 4⅜″ = .36′:

$$4\frac{3}{8} = 4 + \frac{3}{8}$$

$$\frac{3}{8} = .375$$

$$4 + .375 = 4.375$$

$$4.375 \div 12 = .36$$

$$4\frac{3}{8}'' = .36'$$

When feet, inches, and fractions of an inch are converted to decimal foot equivalents, three separate calculations are required. First, the fraction is converted to decimal feet, then the inch value is converted to decimal feet, and finally the foot value is added to the first two values. Each ⅛″ equals .0096′. For example, 5′-8¾″, when converted to its decimal foot equivalent, equals 5.7242′.

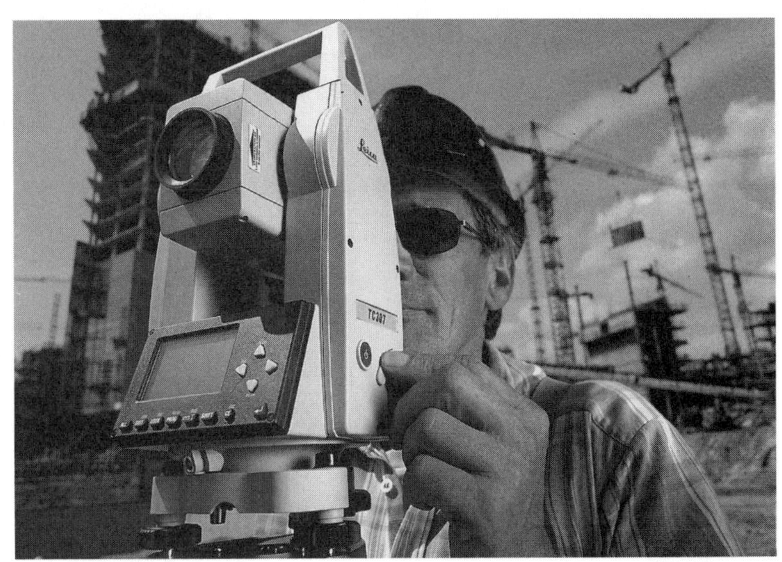

Leica Geosystems

Total station instruments used for site layout provide very accurate dimensional data in decimal format.

First the fraction is converted to decimal feet:

$$\frac{3''}{4} = \frac{6''}{8}$$

$$6 \times .0096 = .0576'$$

Then the inches are converted to decimal feet:

$$8'' \div 12 = .6666'$$

Finally the whole feet and converted decimal feet values are added together:

$$5' + .0576' + .6666' = 5.7242'$$

A construction calculator is an efficient means of converting feet and inches to fractions and decimals.

The .0096 factor is used when maximum accuracy is required. For general construction applications, the .0096 factor is rounded to .01′ (1⁄8″ = .01′) for easier calculation. This is known as field conversion. When converting inches and fractions of an inch, three separate calculations are required. For example, when converting 6 3⁄8″ to its decimal foot equivalent, the fraction is converted to decimal feet by multiplying by .01 (3 × .01 = .03′). The inch value is then converted to decimal feet (6 ÷ 12 = .5′). Finally, the converted fraction value is added to the converted decimal foot value (.03′ + 5′ = .53′).

Southern Forest Products Association

Metric lumber and panel sizes are based on actual standard lumber sizes and are expressed in millimeters (mm).

METRIC MEASUREMENT

In 1975, the United States Congress passed the Metric Conversion Act. In 1998, Congress took further action to encourage the use of the SI (Système International) metric system in the construction field. Drawings prepared for federally funded and state-funded construction projects may be dimensioned using the metric system. In addition, some building codes and manufacturers provide dimensions in both English and SI metric formats.

Metric Linear Measurement

The *meter* is the basic unit of measurement in the SI. One meter is equal to 39.37″ (1 m = 39.37″). A meter is divided into smaller units of centimeters and millimeters. One centimeter is equal to 1⁄100 meter (1 cm = .01 m) and one millimeter is equal to 1⁄1000 meter (1 mm = .001 m). See Figure 3-21. Centimeters and millimeters are used to express small length dimensions, such as the diameter of a bolt. Meters are used to express longer dimensions, such as the length of a property line.

The *kilogram* is the basic unit of weight in the SI. A kilogram is divided into smaller units including grams, centigrams, and milligrams. One gram is equal to .035 oz. One gram is equal to 1⁄1000 kilogram (1 g = .001 kg). One centigram is equal to 1⁄100 gram (1 cg = .01 g) and one milligram is equal to 1⁄1000 gram (1 mg = .001 g).

Metric Linear Measurement

Figure 3-21

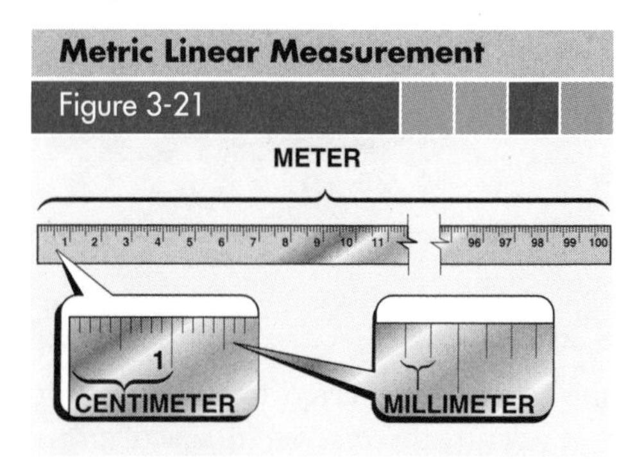

Figure 3-21. The metric system uses base 10 for all calculations.

Metric Conversion. One of the main advantages of the SI is that it uses a base 10 system. One metric unit is converted to another metric unit by multiplying or dividing by multiples of 10, which moves the decimal point to the right or left. When a metric unit is converted to the next smaller unit, the value to be converted is multiplied by 10. For example, to determine the millimeter equivalent of 13.5 cm, the centimeter value is multiplied by 10 (13.5 cm × 10 = 135 mm), which moves the decimal point one place to the right. When a metric unit is converted to the next larger unit, the value to be converted is divided by 10. For example, to determine the meter equivalent of 320 cm, the centimeter value is divided by 100 (320 cm ÷ 100 = 3.2 cm).

Conversion tables are used to convert common SI units to English units and English units to SI units. See Figure 3-22. Tables for other SI and English system conversions exist as well.

AREA AND VOLUME

Area is a two-dimensional surface measurement. Area is expressed in square units, such as square inches, square feet, and square yards. Area is commonly used to express measurements such as room size (400 sq ft [square feet]) or window area (432 sq in. [square inches]). *Volume* is the three-dimensional capacity of a space. Volume is expressed in cubic units, such as cubic feet or cubic yards. Volume is commonly used to express measurements such as the amount of concrete (3 cu yd [cubic yards]) or amount of ventilation (1800 cfm [cubic feet per minute]).

Calculating Area

The formula used to calculate the area of a specific space depends on the shape of the space. One formula is used to calculate the areas of squares and rectangles, while other formulas are used to calculate the areas of triangles and circles. See Figure 3-23.

When calculating the area of a square or rectangular shape, the following formula is applied:

$A = w \times h$ (or l)

where

A = area

w = width

h (or l) = height (or length)

For example, a floor measuring 12′ wide by 18′ long has an area of 216 sq ft:

$A = w \times l$

$A = 12 \times 18$

$A = 216$ sq ft

When calculating the area of a triangular shape, ½ is multiplied by the base length by the height:

$A = \frac{1}{2}bh$

where

A = area

½ = constant

b = base

h = height

The SI metric system was the first standardized system of measurement. Gabriel Mouton is credited with originating it in 1670.

ENGLISH/METRIC CONVERSION			
To Convert	**To**	**Multiply By**	**Example**
inches (in.)	millimeters (mm)	25.4	7″ x 25.4 = 177.8 mm
inches (in.)	centimeters (cm)	2.54	2″ x 25.4 = 5.08 cm
feet (ft)	centimeters (cm)	30.48	10″ x 30.48 = 304.8 cm
feet (ft)	meters (m)	.3048	6′ x .3048 = 1.8288 m
yards (yd)	centimeters (cm)	91.44	3 yd x 91.44 = 274.32 cm
yards (yd)	meters (m)	.9144	6 yd x .9144 = 5.4864 m
millimeters (mm)	inches (in.)	.03937	320 mm x .03937 = 12.5984″
centimeters (cm)	inches (in.)	.3937	22 cm x .3937 = 8.6614″
meters (m)	feet (ft)	3.281	6.5 m x 3.281 = 21.3265′
meters (m)	yards (yd)	1.0937	11 m x 1.0937 = 12.0307 yd

Figure 3-22. Conversion of English and metric values is accomplished using conversion tables.

For example, a triangular area with a base of 4′ and a height of 8′ has an area of 16 sq ft:

$A = \frac{1}{2}bh$

$A = \frac{1}{2} \times 4 \times 8$

$A = 16$ sq ft

Two formulas may be used to calculate the area of a circular shape, one using the radius measurement and the other using the diameter measurement. When calculating the area using the radius, pi (π) is multiplied by the radius squared:

$A = \pi r^2$

where

A = area

π = 3.1416 (constant)

r = radius

An ellipse has two diameters that define its shape—a major and a minor diameter. Area of an ellipse is calculated by multiplying .7854 by the major diameter by the minor diameter.

The symbol π is a conversion factor equal to 3.1416. For example, a circular area with a radius of 14′ has an area of 615.7536 sq ft:

$A = \pi r^2$

$A = 3.1416 \times 14^2$

$A = 3.1416 \times 14 \times 14$

$A = 3.1416 \times 196$

$A = 615.7526$ sq ft

When calculating the area using the diameter, .7854 is multiplied by the diameter squared:

$A = .7854d^2$

where

A = area

.7854 = constant

d = diameter

For example, a circular area with a diameter of 22″ has an area of 380.1336 sq in.:

$A = .7854d^2$

$A = .7854 \times 22^2$

$A = .7854 \times 22 \times 22$

$A = 380.1336$ sq in.

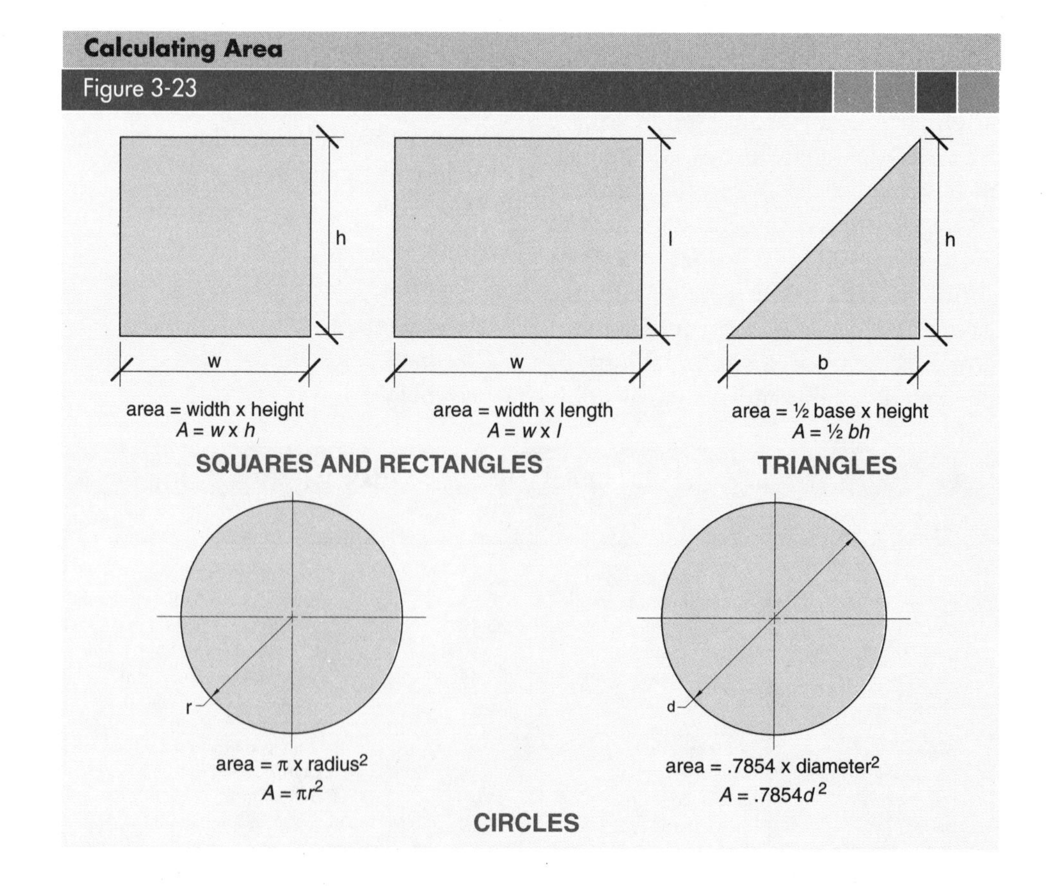

Figure 3-23. Area formulas depend on the shape of the two-dimensional object.

Applications. Area calculations are commonly used to determine material quantities for finish materials such as ceiling tile, paint, carpet, and roofing. Standard ceiling tiles are 2′ × 4′, each covering 8 sq ft (2′ × 4′ = 8 sq ft). After measuring the ceiling and determining the total ceiling area, the ceiling area (in sq ft) is divided by 8 to roughly calculate the number of ceiling tiles required.

The coverage area is commonly provided for paint or other finishes. The coverage area per gallon of paint is compared to the area to be painted in order to calculate the number of gallons required. For example, a gallon of paint covers approximately 400 sq ft. A wall measuring 10′ × 20′ requires approximately ½ gal. of paint (10′ × 20′ = 200 sq ft; 200 sq ft ÷ 400 sq ft = .5 = ½ gal.).

Carpet coverage is commonly calculated in square yards (3′ × 3′ = 9 sq ft = 1 sq yd). After determining the total floor area, the area (in sq ft) is divided by 9 to determine the amount of carpet required to cover the floor (in sq yd). For example, if the room measures 12′ × 15′, 20 sq yd of carpet is required to cover the floor (12′ × 15′ = 180 sq ft; 180 sq ft ÷ 9 = 20 sq yd).

Area calculations are also used to determine the amount of roofing materials required. Roofing materials, such as asphalt shingles, are commonly specified as squares. A roofing square is equal to 100 sq ft. After determining the total roof area, the area (in sq ft) is divided by 100 to determine the number of roofing squares required for the job. For example, if one side of a gable roof measures 20′ × 45′, 18 roofing squares are required to cover the roof (20′ × 45′ × 2 = 1800 sq ft; 1800 sq ft ÷ 100 = 18 squares).

Calculating Volume

Volume is the three-dimensional capacity of a space. Therefore, three dimensions are used to define the space—width, height, and depth. When determining volume, the three dimensions—width, height, and depth—are multiplied together. See Figure 3-24.

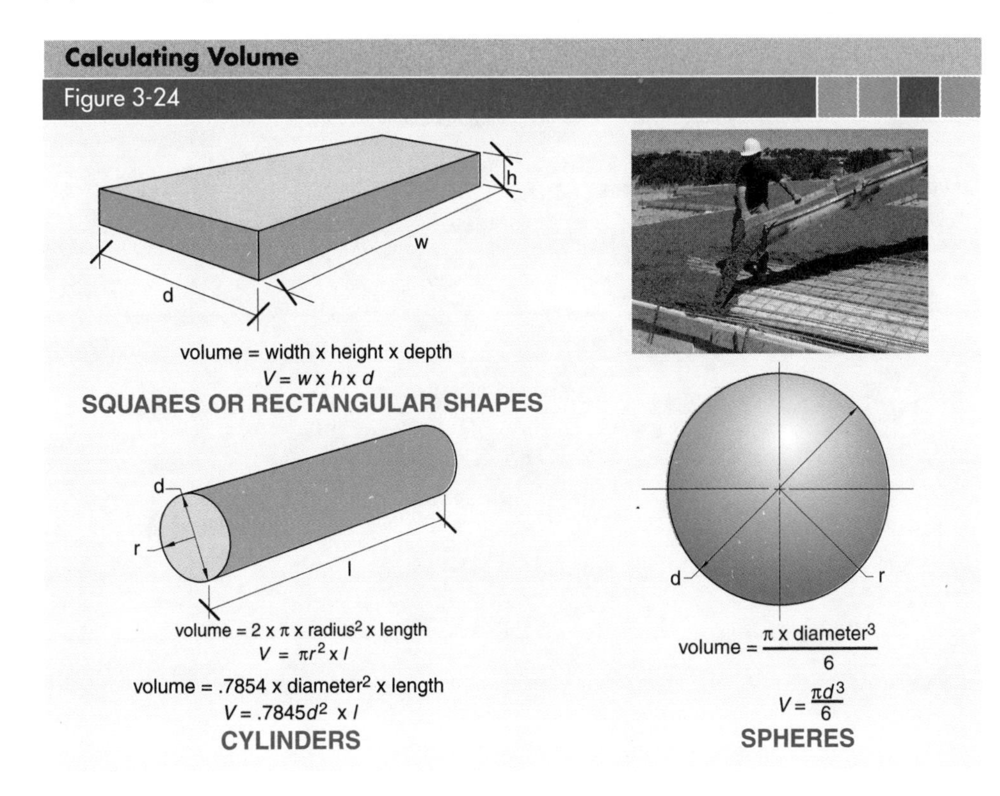

Figure 3-24. Volume calculations provide cubic measurements. Concrete is typically ordered in cubic yards (cu yd).

When calculating the volume of a square or rectangular shape, the following formula is applied:

$V = w \times h \times d$

where

V = volume

w = width

h = height

d = depth

For example, a footing measuring 1′ × 2′ × 10′ has a volume of 20 cu ft (1′ × 2′ × 10′ = 20 cu ft).

$V = w \times h \times d$

$V = 1 \times 2 \times 10$

V = 20 cu ft

One cubic yard is equal to 27 cubic feet (1 cu yd = 27 cu ft).

When determining the volume of a cylinder, the area of one end of the cylinder is calculated by multiplying pi by the radius squared ($A = \pi r^2$) or by multiplying .7854 by the diameter squared ($A = .7854d^2$). The area is multiplied by the length of the cylinder to calculate the volume:

$V = \pi r^2 \times l$

where

V = volume

π = 3.1416

r = radius

l = length

or

$V = .7854d^2 \times l$

where

V = volume

.7854 = constant

d = diameter

l = length

For example, a cylinder with a radius of 14′ and a length of 26′ has a volume of 16,009.59 cu ft:

$V = \pi r^2 \times l$

$V = 3.1416 \times 14^2 \times 26$

$V = 3.1416 \times 14 \times 14 \times 26$

V = 16,009.59 cu ft

or

$V = .7854d^2 \times l$

$V = .7854 \times 28^2 \times 26$

$V = .7854 \times 28 \times 28 \times 26$

V = 16,009.59 cu ft

Concrete is specified for a job in cubic yards (cu yd).

When calculating the volume of a sphere, π is multiplied by the diameter cubed and the product is divided by 6:

$$V = \frac{\pi d^3}{6}$$

where
V = volume
π = 3.1416
d = distance
6 = constant

For example, a sphere with a diameter of 26′ has a volume of 9202.79 cu ft:

$$V = \frac{\pi d^3}{6}$$

$$V = \frac{3.1416 \times 26^3}{6}$$

$$V = \frac{3.1416 \times 26 \times 26 \times 26}{6}$$

$$V = \frac{55{,}215.761}{6}$$

V = 9202.79 cu ft

The symbol π denotes the ratio of the circumference of a circle to its diameter. The circumference is determined by multiplying π by the diameter ($C = \pi d$).

Applications. Volume calculations are used when determining the amount of soil removed from excavations, specifying the amount of concrete for a slab, and determining the air-handling requirements for a heating, ventilating, and air conditioning (HVAC) system for a building. Volumes of soil and concrete are expressed in cubic yards.

A cubic yard, which is equal to 27 cu ft, is a volume measuring 3′ × 3′ × 3′. (3′ × 3′ × 3′ = 27 cu ft = 1 cu yd). To determine the number of cubic yards of soil or concrete, the cubic foot volume is divided by 27. For example, 6.389 cu yd of concrete is required for a 6″ thick concrete slab measuring 15′ × 23′:

6″ × 15′ × 23′ =
.5′ × 15′ × 23′ = 172.5 cu ft
172.5 cu ft ÷ 27 = 6.389 cu yd

In HVAC applications, the amount of air to be moved is calculated and expressed in cubic feet per minute (cfm). Fans, air handlers, and other HVAC equipment are commonly rated in cfm capacity. When operating at 100% efficiency, a fan rated at 1700 cfm moves 1700 cu ft of air per minute.

Annual production of concrete in the United States is approximately 5,000,000,000 cu yd.

CertainTeed Corporation
Air flow, measured in cubic feet per minute (cfm), is calibrated with a flow hood.

Trade Math 3

Review Questions

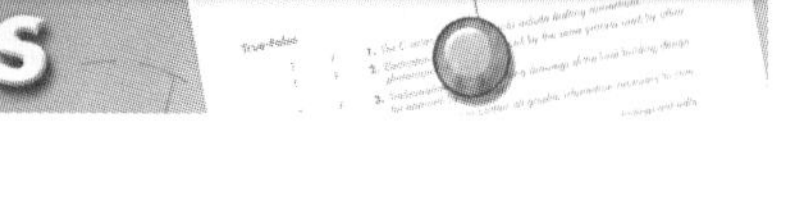

Name ______________________ Date ______________

Computation 3-1

1. 18 + 81

2. 125 + 47

3. 5744 + 588

4. 10,588 − 95

5. 56,565 + 875

6. 42 − 17

7. 652 − 89

8. 7505 − 99

9. 4231 − 877

10. 35,611 − 577

11. 12 × 12

12. 14 × 85

13. 16 × 36

14. 456 × 19

15. 1567 × 4329

16. $3\overline{)18}$

17. $6\overline{)45}$

18. $13\overline{)150}$

19. $18\overline{)5750}$

20. $73\overline{)6548}$

Multiple Choice

_______________ **1.** A ___ is a number that does not have a fractional or decimal part.

A. multiplicand
B. proper decimal number
C. whole number
D. minuend

_______________ **2.** ___ is the mathematical process of taking one number away from a larger number.

A. Addition
B. Subtraction
C. Multiplication
D. Division

_______________ **3.** The ___ is the result of multiplication.

A. multiplicand
B. quotient
C. multiplier
D. product

_______________ **4.** The ___ is the part of a fraction placed below or to the right of the fraction bar.

A. divisor
B. dividend
C. numerator
D. denominator

_______________ **5.** A decimal fraction has a denominator of ___.

A. 10
B. 100
C. 1000
D. all of the above

_______________ **6.** The sum of 42″ and 16″ is ___″.

A. 2.625
B. 26
C. 58
D. 672

_______________ **7.** The SI metric system is a base ___ system.

A. 6
B. 8
C. 10
D. 12

_______________ **8.** The ___ is the basic unit of linear measurement in the SI metric system.

A. kilogram
B. meter
C. gram
D. kilometer

______ **9.** Area is a ___-dimensional surface measurement.

A. one
B. two
C. three
D. four

______ **10.** A ___ is the position that a digit occupies; it represents the value of the digit.

A. column
B. location
C. place
D. none of the above

______ **11.** An 18″ square by 14′ concrete column contains ___ cu ft and requires ___ cu yd of concrete.

A. 31.5; 1.17
B. 35.84; 1.33
C. 252; 9.33
D. 4536; 168

______ **12.** A built-up flat roof measuring 68′ × 248′ is ___ sq ft.

A. 316
B. 368
C. 14,880
D. 16,864

Computation 3-2

Reduce to lowest terms when necessary.

1. $\frac{1}{8} + \frac{3}{8} =$

2. $\frac{5}{16} + \frac{13}{16} =$

3. $\frac{1}{2} + \frac{3}{8} =$

4. $\frac{5}{8} + \frac{15}{16} =$

5. $\frac{13}{32} + 5\frac{3}{4} =$

6. $\frac{5}{8} - \frac{1}{8} =$

7. $\frac{7}{8} - \frac{1}{2} =$

8. $5\frac{1}{4} - \frac{3}{8} =$

9. $16\frac{5}{8} - 8\frac{7}{8} =$

10. $756\frac{5}{32} - 89\frac{1}{2} =$

11. $\frac{1}{8} \times 6 =$

12. $\frac{5}{16} \times 12 =$

13. $7\frac{7}{8} \times 14\frac{1}{2} =$

14. $18\frac{3}{4} \times 22\frac{5}{8} =$

15. $975\frac{5}{16} \times 5\frac{1}{2} =$

16. $625 \div 5\frac{1}{2} =$

17. $18\frac{3}{4} \div 2 =$

18. $56\frac{7}{8} \div 7 =$

19. $465\frac{1}{2} \div 4 =$

20. $15 \div 7\frac{1}{2} =$

True-False

T F **1.** A fraction with the numerator larger than the denominator is an improper fraction.

T F **2.** An odd number can be divided by two without a remainder or decimal occurring.

T F **3.** Multiplication is the mathematical process of determining the number of times one number is contained in another number.

T F **4.** In the fraction $\frac{3}{8}$, 3 is the numerator.

T F **5.** A mixed number consists of a whole number and a fraction.

T F **6.** A quotient is the result of a division operation.

T F **7.** The dividend is the number to be divided.

T F **8.** A proper decimal number includes a whole number.

T F **9.** One millimeter is equal to .01 meter.

T F **10.** Volume is a three-dimensional capacity of space.

Computation 3-3

1. $\begin{array}{r} 12.5 \\ +\ 6 \\ \hline \end{array}$

2. $\begin{array}{r} 13.85 \\ +\ 5.7 \\ \hline \end{array}$

3. $\begin{array}{r} 125.65 \\ +\ 7524.8 \\ \hline \end{array}$

4. $\begin{array}{r} 432.66 \\ +\ .85 \\ \hline \end{array}$

5. $\begin{array}{r} 1085.655 \\ +\ 145.39 \\ \hline \end{array}$

6. $\begin{array}{r} 55.6 \\ -\ .9 \\ \hline \end{array}$

7. $\begin{array}{r} 652.5 \\ -\ 18.75 \\ \hline \end{array}$

8. $\begin{array}{r} 12.855 \\ -\ 6.9 \\ \hline \end{array}$

9. $\begin{array}{r} 8321.15 \\ -\ 57.875 \\ \hline \end{array}$

10. $\begin{array}{r} 9211.333 \\ -\ 55.67 \\ \hline \end{array}$

11. $\begin{array}{r} 45.5 \\ \times\ 8.5 \\ \hline \end{array}$

12. $\begin{array}{r} 16.875 \\ \times\ 12 \\ \hline \end{array}$

13. $\begin{array}{r} 854.675 \\ \times\ 13.5 \\ \hline \end{array}$

14. $\begin{array}{r} 75.67 \\ \times\ 87.333 \\ \hline \end{array}$

15. $\begin{array}{r} 6584.75 \\ \times\ 32.125 \\ \hline \end{array}$

16. $5.5\overline{)14.75}$

17. $12.875\overline{)75.67}$

18. $6\overline{)125.333}$

19. $186.3\overline{)7489.675}$

20. $7.5\overline{)53,244.8}$

Completion

_______________ **1.** One meter is equal to ___″.

_______________ **2.** A room measuring 15′ × 25′ contains ___ sq ft.

_______________ **3.** The ___ is the total that results from addition.

_______________ **4.** The ___ is the total number of units prior to subtraction.

_______________ **5.** Inches are added in base ___.

_______________ **6.** The sign used to indicate multiplication is ___.

_______________ **7.** A(n) ___ decimal number includes a whole number and decimal number separated by a decimal point.

_______________ **8.** One gram is equal to ___ kilogram.

_______________ **9.** Pi is abbreviated with the ___ symbol.

_______________ **10.** When determining volume, ___ dimensions are multiplied together.

_______________ **11.** An excavation measuring 15′-9″ × 73′ × 159′ contains ___ cu yd of soil.

Computation 3-4

Convert the following decimals to fractions.

_______________ **1.** .50

_______________ **2.** .125

_______________ **3.** .625

_______________ **4.** .875

_______________ **5.** .25

Convert the following fractions to decimals.

_______________ **6.** $\frac{3}{16}$

_______________ **7.** $\frac{3}{8}$

_______________ **8.** $\frac{5}{16}$

_______________ **9.** $\frac{13}{32}$

_______________ **10.** $\frac{3}{4}$

Trade Math 3

Trade Competency Test

Name ______________________________ Date ______________

Identification 3-1

______________ **1.** 63

______________ **2.** 132

______________ **3.** 21

______________ **4.** 36

______________ **5.** 96

______________ **6.** 48

______________ **7.** 44

______________ **8.** 110

______________ **9.** 66

______________ **10.** 45

______________ **11.** 81

______________ **12.** 100

______________ **13.** 121

MULTIPLICATION TABLE

1	2	3	4	5	6	7	8	9	10	11	12
2	4	6	8	10	12	14	16	18	20	22	24
3	6	9	12	15	18	A	24	27	30	33	36
4	8	12	16	20	24	28	32	36	40	B	48
5	10	15	20	25	30	35	40	C	50	55	60
6	12	18	24	30	D	42	48	54	60	E	72
7	14	21	28	35	42	49	56	F	70	77	84
8	16	24	32	40	G	56	64	72	80	88	H
9	18	27	36	45	54	63	72	I	90	99	108
10	20	30	40	50	60	70	80	90	J	110	120
11	22	33	44	55	66	77	88	99	K	L	132
12	24	36	48	60	72	84	96	108	120	M	144

Computation

1. $\begin{array}{r} 255 \\ +\ 88 \\ \hline \end{array}$

2. $\begin{array}{r} 7899 \\ +\ 722 \\ \hline \end{array}$

3. $\begin{array}{r} 64 \\ -\ 45 \\ \hline \end{array}$

4. $\begin{array}{r} 865 \\ -\ 76 \\ \hline \end{array}$

5. $\begin{array}{r} 18 \\ \times 90 \\ \hline \end{array}$

6. $\begin{array}{r} 563 \\ \times\ 75 \\ \hline \end{array}$

7. $6 \overline{)52}$

8. $18 \overline{)235}$

9. $\frac{7}{16} + \frac{11}{16} =$

10. $\frac{1}{2} + \frac{7}{8} =$

11. $\frac{3}{8} - \frac{1}{16} =$

12. $12\frac{3}{4} - \frac{7}{32} =$

13. $\frac{3}{8} \times 12 =$

14. $17\frac{3}{8} - 11\frac{1}{4} =$

15. $21\frac{1}{4} \div 18 =$

16. $35\frac{7}{16} \div 4\frac{3}{8} =$

17. $\begin{array}{r} 12.355 \\ +\ 11.69 \\ \hline \end{array}$

18. $\begin{array}{r} 514.85 \\ +\ 1543.4 \\ \hline \end{array}$

19. $\begin{array}{r} 5.62 \\ -\ .39 \\ \hline \end{array}$

20. $\begin{array}{r} 3662.432 \\ -\ \ 12.87 \\ \hline \end{array}$

21. $\begin{array}{r} 5.65 \\ \times\ 7.3 \\ \hline \end{array}$

22. $\begin{array}{r} 459.667 \\ \times\ \ 38.25 \\ \hline \end{array}$

23. $3.33 \overline{)55.125}$

24. $47 \overline{)587.575}$

25. $3.3 \overline{)696}$

Identification 3-2

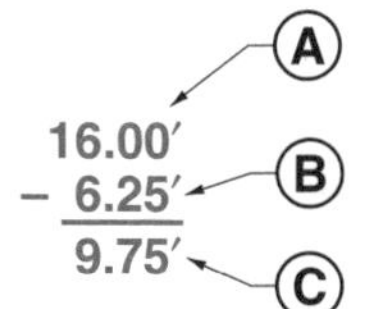

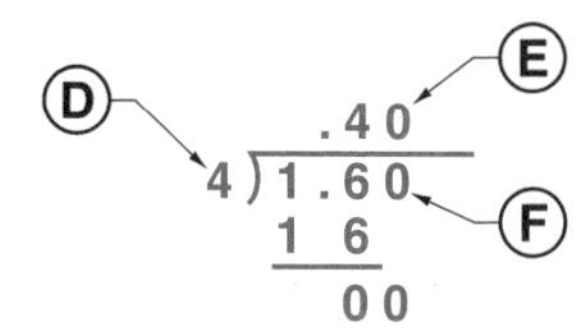

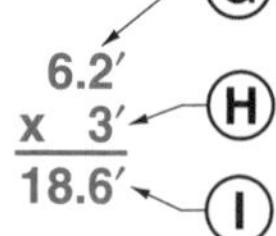

______________________ **1.** Divisor

______________________ **2.** Multiplier

______________________ **3.** Subtrahend

______________________ **4.** Dividend

______________________ **5.** Quotient

______________________ **6.** Minuend

______________________ **7.** Multiplicand

______________________ **8.** Result

______________________ **9.** Product

Completion

______________________ **1.** A room measuring 24′-0″ × 22′-6″ requires ___ sq ft of ceiling tile and ___ sq yd of carpet.

______________________ **2.** A rectangular building lot measures 75′ × 167′ and contains ___ sq ft.

______________________ **3.** A porch measuring 8′-3″ × 18′-6″ has ___ sq ft of floor space.

______________________ **4.** A 36″ square column footing 18″ deep requires ___ cu yd of concrete.

______________________ **5.** A basement excavation measuring 68′-8″ × 59′-6″ × 11′-4″ contains ___ cu yd of soil.

PRINTREADING

Refer to Stearns Residence—Main Level Floor Plan on page 87.

Completion

______ **1.** The Master Bedroom has an area of ___ sq ft.

______ **2.** The rear wall of the Living Room projects ___ beyond Bedroom #2.

______ **3.** The window in the Office is ___ away from the Garage.

______ **4.** The C window is centered along the 24′-7″ dimension of the Garage wall, placing it ___ from outside of the wall.

______ **5.** Bedroom #3 has an area of ___ sq ft and requires ___ sq yd of carpet.

Multiple Choice

______ **1.** The #6 door is centered along the 5′-0″ wall next to the overhead door. The center of the #6 door is ___ from the edge of the house.

A. 1′-6″
B. 2′-0″
C. 2′-6″
D. 3′-6″

______ **2.** The Kitchen island is ___ sq ft.

A. 1.75
B. 8.75
C. 9.25
D. 9.36

______ **3.** One section of the Kitchen cabinets measures 2′-0″ x 10′-0″ x 32″. The volume of the cabinets is ___ cu ft.

A. 20.6
B. 53.3
C. 65.3
D. 640

______ **4.** The A window in Bedroom #3 is centered along the wall. The total dimension of the wall is ___.

A. 10′-8″
B. 11′-7″
C. 11′-11″
D. 12′-3″

______ **5.** The total distance from the edge of the Porch to the Garage wall is ___.

A. 26′-0½″
B. 27′-3½″
C. 27′-6½″
D. 28′-0½″

Refer to printreading questions on page 86.

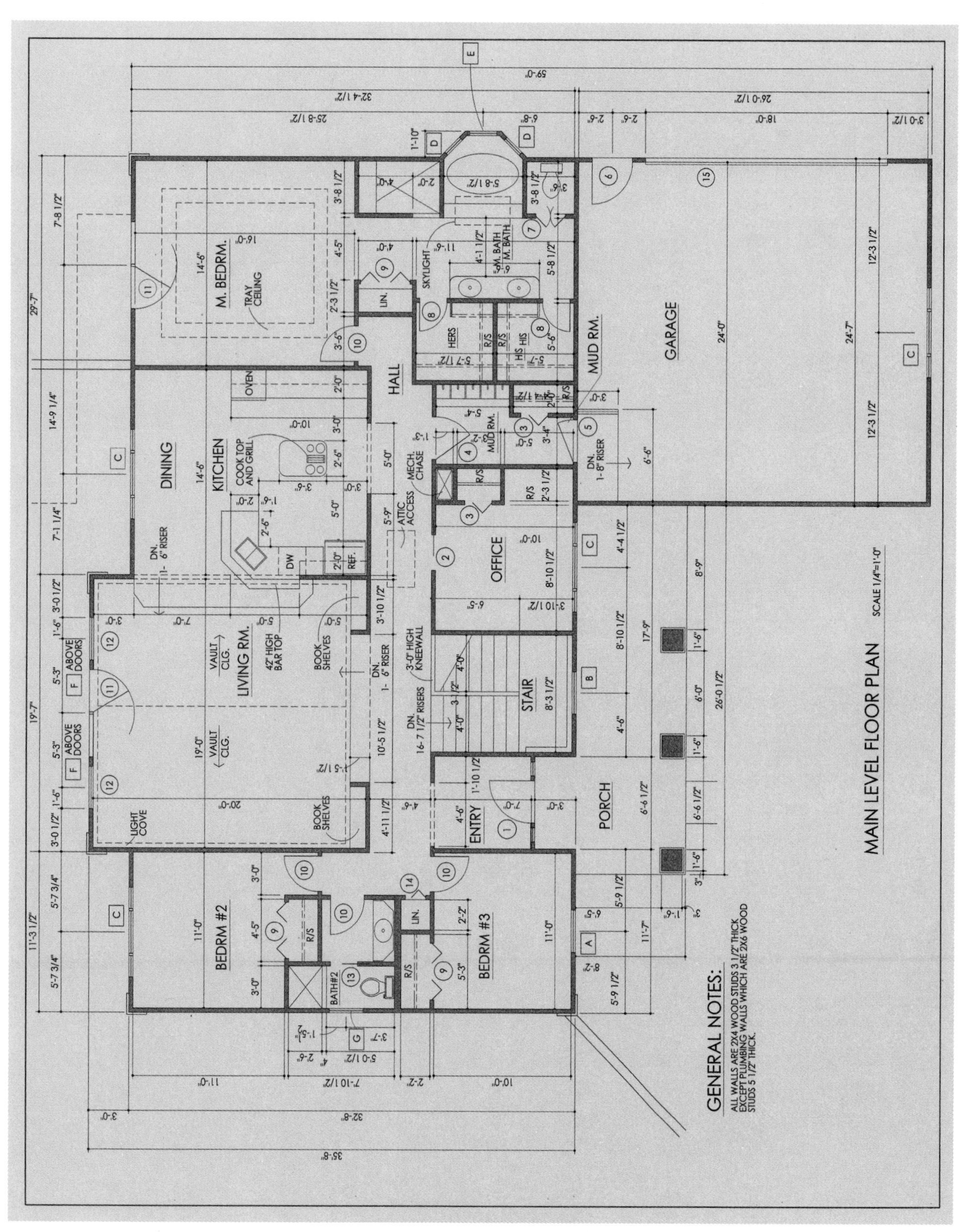

STEARNS RESIDENCE

Symbols and Abbreviations

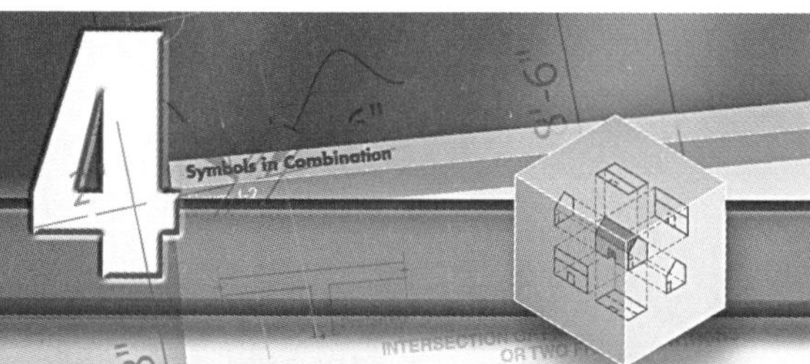

Symbols and abbreviations save time when producing working drawings for plans and also conserve space on plans. Symbols and abbreviations are standardized to aid in the accurate interpretation of plans. In addition to standard architectural material and graphic symbols, specialized symbols are also available for electrical, plumbing, and heating, ventilating, and air conditioning systems.

SYMBOLS

A *symbol* is a pictorial representation of a structural or material component required to complete a construction project. Prints use symbols to show materials, equipment, and building parts. See Figure 4-1. Symbols conserve space on prints, are easy to draw, and are easily recognized by experienced printreaders. The American National Standards Institute (ANSI), in conjunction with recognized trade associations, has standardized symbols in the various trade areas. CAD systems use symbol libraries from which the desired symbol is selected. Templates are commercially available for symbols drawn by the conventional method.

Symbols for the same material may be drawn differently from plan to elevation to section views. For example, the symbol for concrete block is drawn differently for each view. Other materials may have no standardized symbol for a particular view. For example, there is no standardized symbol for earth on a plan view.

Various symbols are used in combination to show the relationship of building materials. For example, a plan view of a fireplace in a brick veneer wall requires that the plan symbols for wood framing and brick be used when

> ANSI Y32.9, *Graphic Symbols for Electrical Wiring and Layout Diagrams used in Architecture and Building Construction*, provides standardized symbols commonly used in the electrical trade. Specialized symbols are noted on prints with notations.

APA—The Engineered Wood Association

Building materials are represented on prints with symbols and abbreviations.

showing the wall. In addition, plan symbols showing the face brick, common brick, firebrick, and tile for the hearth are required. See Figure 4-2.

Openings for exterior walls and interior partitions are shown on elevation and plan views. The symbols used to show the doors and windows are drawn differently on each of these views. For example, the symbol for a double-hung window on an elevation view is drawn as if looking directly at the surface of the window. The symbol for the same window on a plan view is drawn as if looking directly down at the top of the window. See Figure 4-3.

Electrical symbols are composed of graphic elements to which letters or numbers can be added. For example, a lighting outlet is shown as a circle. If the outlet is designed for a specific task, the appropriate letter is added to designate that task. For example, a lighting outlet with a lampholder is shown with an L. Electrical switches are shown with letters and numbers. The letter S denotes the switch and the numbers indicate single-pole, double-pole, and so on. Various low-voltage systems and their components, such as alarms and voice/data/video (VDV) systems, are shown on floor plans with specialized symbols or on separate shop drawings. In many instances, a generic symbol is shown and notes identify the device or component.

Standardized symbols for plumbing include graphic elements with letters, as required, to show fixtures. Symbols for piping and valves include stylized and simplified line drawings.

Standard plumbing, piping, and valve symbols are designated per ASME Y32.3.3, *Graphical Symbols for Pipe Fittings, Valves, and Piping.*

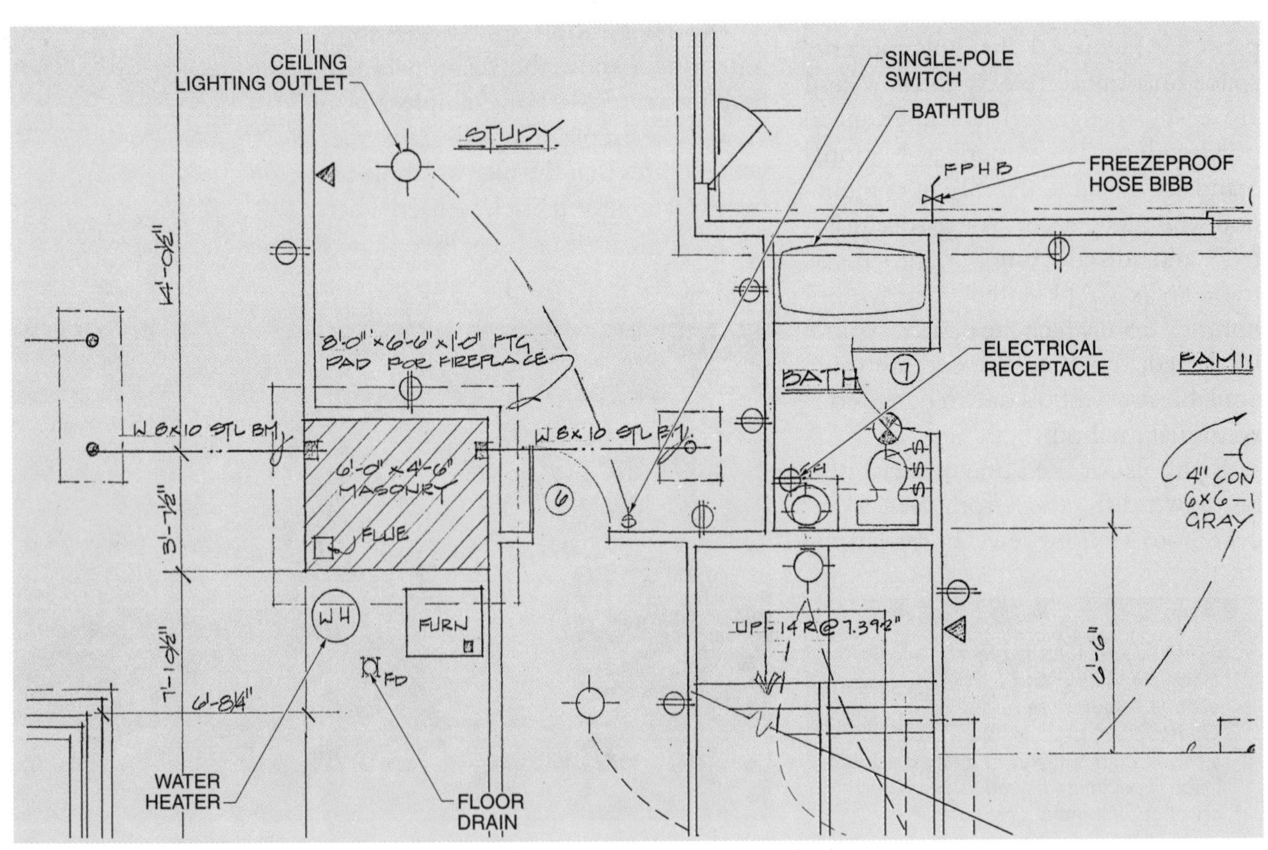

Figure 4-1. Prints use standardized symbols to show building materials and fixtures.

Symbols in Combination
Figure 4-2

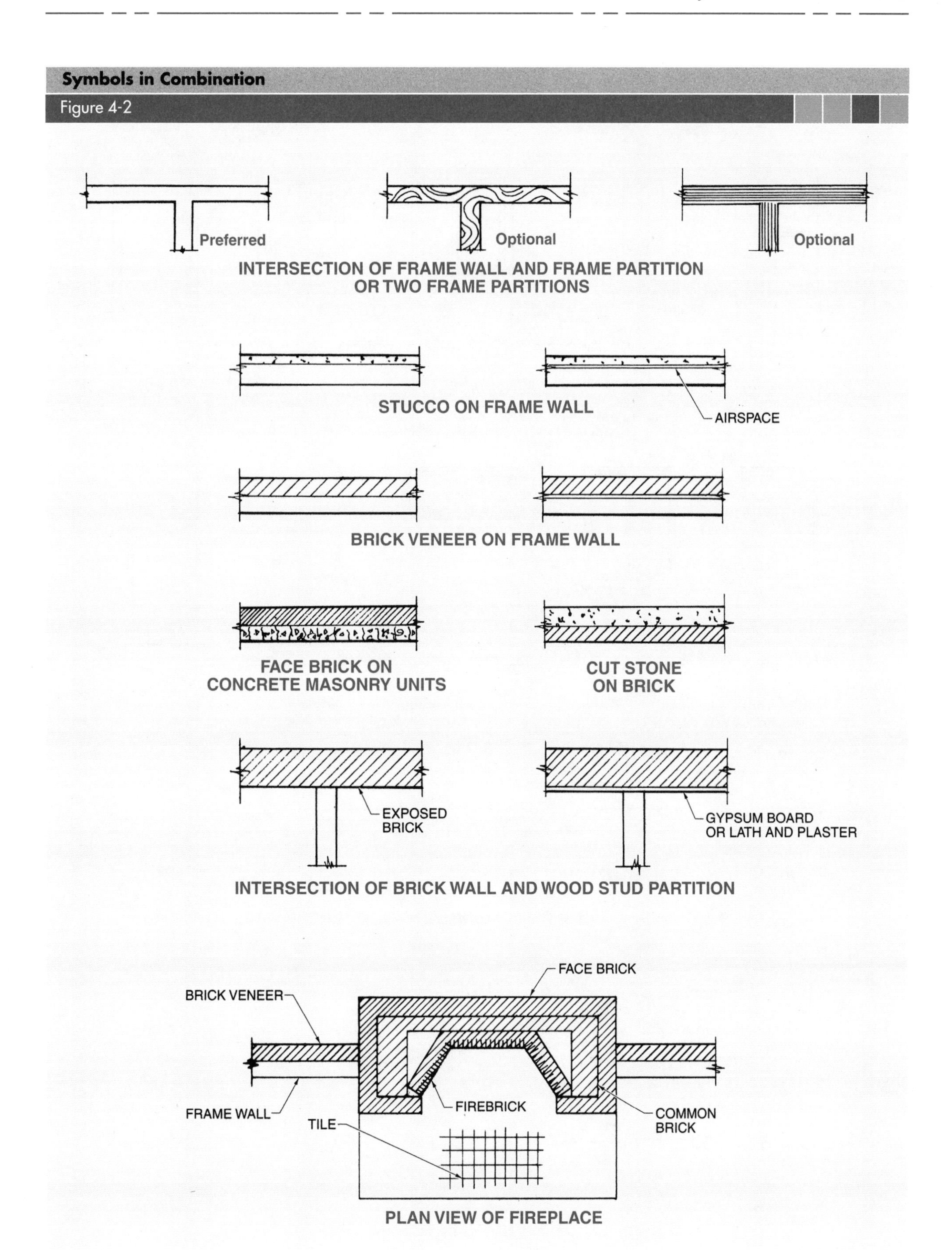

Figure 4-2. A symbol may be used in combination with other symbols.

Exterior Walls and Interior Partitions

Figure 4-3

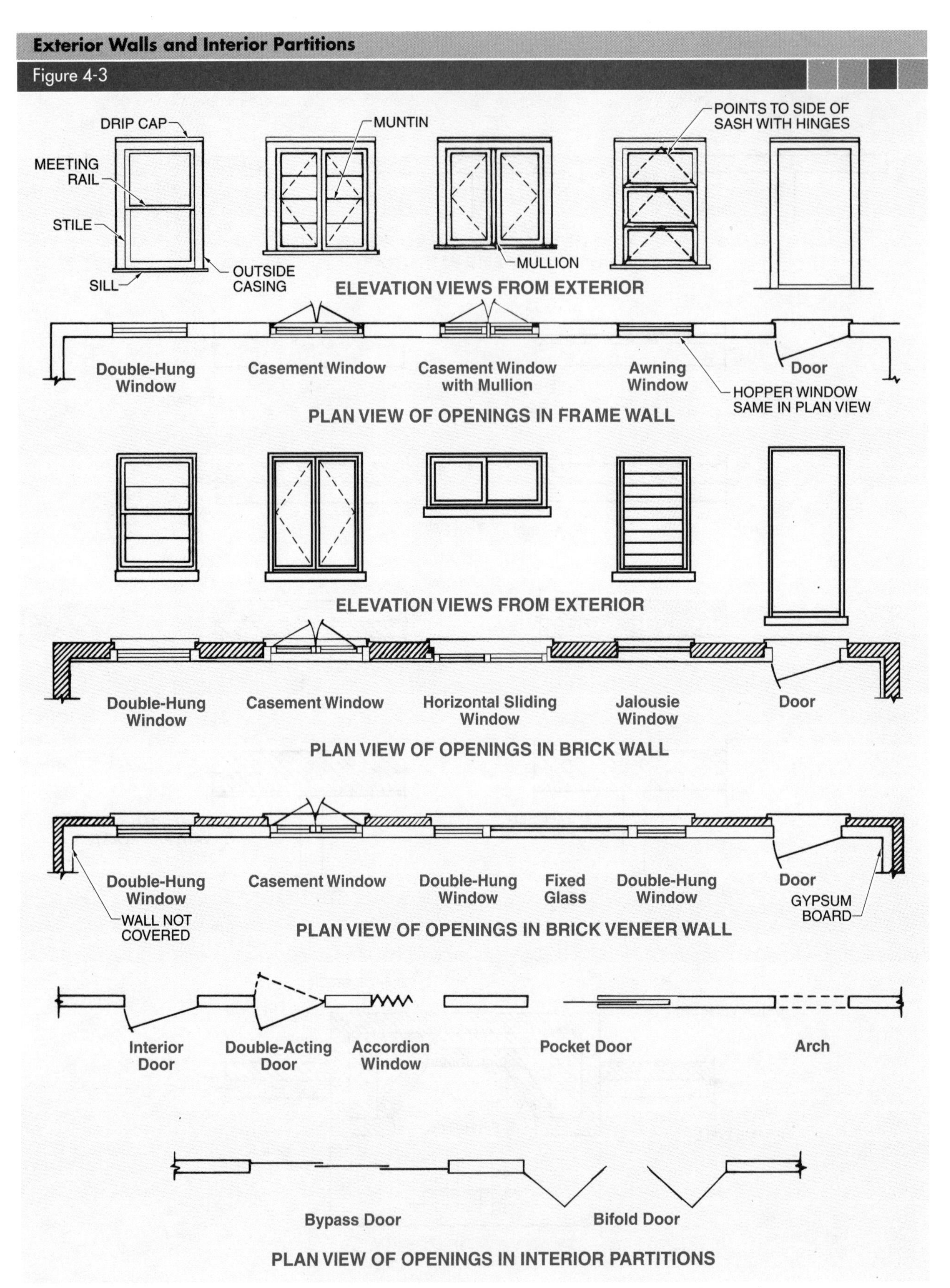

Figure 4-3. Different symbols are used on elevations and floor plans to show the same window or door.

ABBREVIATIONS

An *abbreviation* is a letter or series of letters or words denoting a complete word. Like symbols, abbreviations save time and conserve space on plans. Standardized abbreviations are developed by standards organizations such as ANSI. Abbreviations are used to denote materials, fixtures, and areas and to provide simplified instructions to tradesworkers. See Appendix.

Certain materials, fixtures, and other features may be referred to by their acronym. An *acronym* is an abbreviated word formed from the first letter of each word that describes the article. For example, the acronym GFCI refers to a ground-fault circuit interrupter, a type of electrical receptacle.

The same abbreviation may be used to denote different items. For example, R is the abbreviation for range, riser, and room. Generally, the location of the abbreviation will indicate its intent. Abbreviations that form a word are followed by a period to avoid confusion with the word. For example, SEW. is the abbreviation for sewer and KIT. is an abbreviation for kitchen.

Some words have more than one abbreviation. Both FIN. and FNSH are abbreviations for finish. When two abbreviations are given for the same word, the first abbreviation is preferred. In addition, some architects may use abbreviations that are not standardized but are fairly obvious based on their location and use.

Symbols and abbreviations are used together to provide comprehensive information in a relatively small space. For example, the shape of a closet is shown on a plan view by parallel lines, which symbolize the framed partitions. The door symbol shows the swing of the door or a sliding door, and an abbreviation, either C, CL, or CLOS, designates the closet. See Figure 4-4.

Prints may include a legend of symbols used on the drawings.

Door swing, or door hand, is indicated on prints with the door symbol.

Symbols and Abbreviations

Figure 4-4

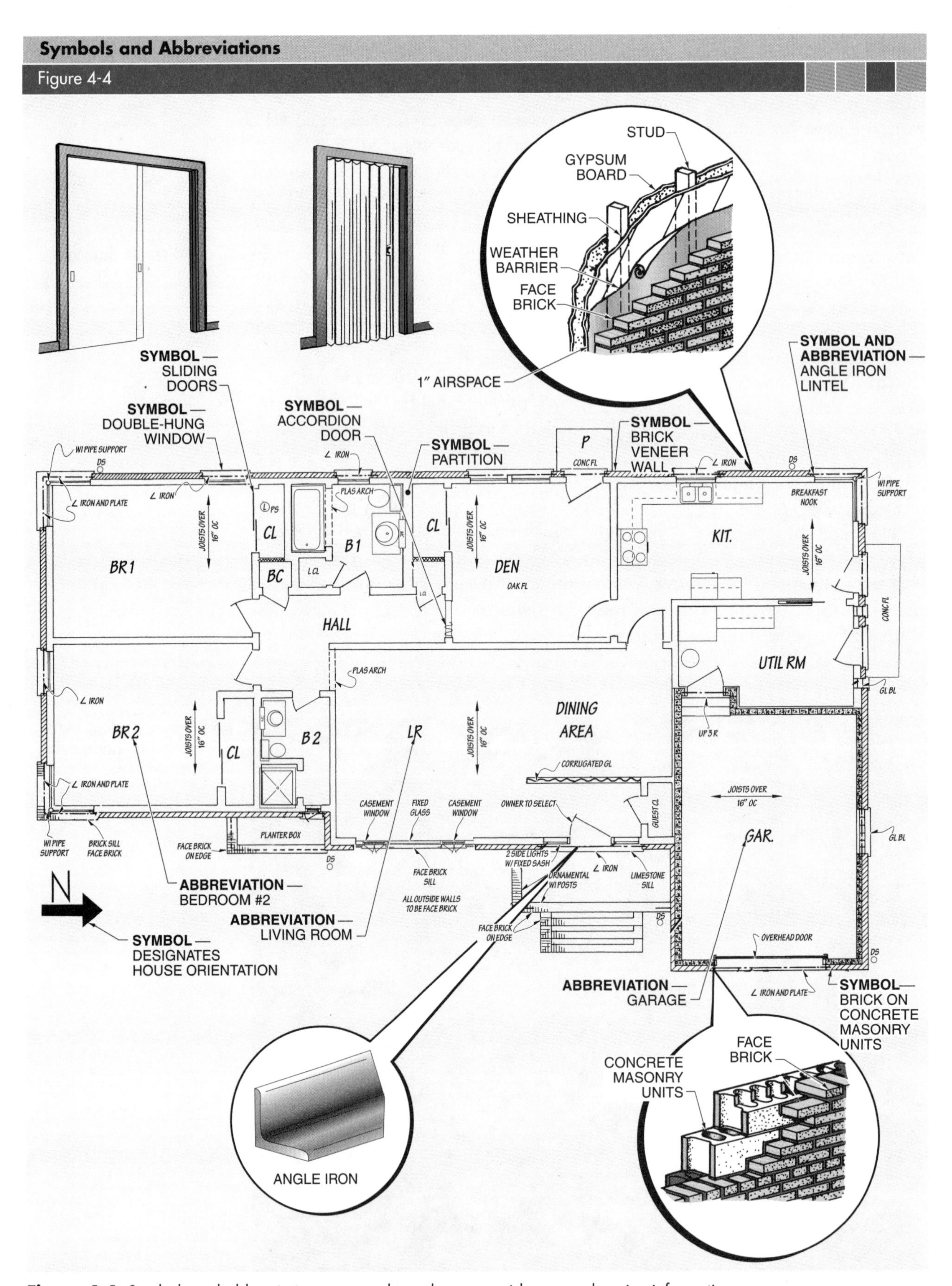

Figure 4-4. Symbols and abbreviations are used together to provide comprehensive information.

Symbols and Abbreviations 4

Sketching

Name ______________________ Date ______________

Sketching 4-1

Complete the chart by sketching the missing symbols. Refer to Appendix.

	ELEVATION	PLAN	SECTION
CONCRETE	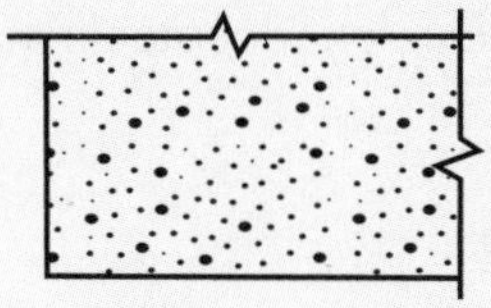	LIGHTWEIGHT ①	SAME AS PLAN VIEW
CONCRETE MASONRY UNITS	②	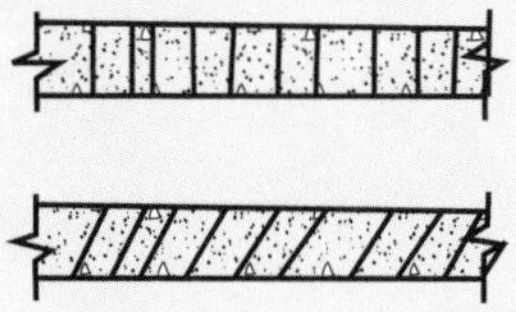	③
WOOD	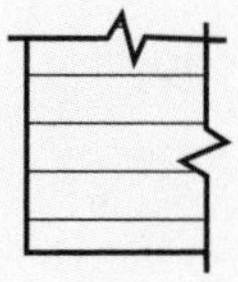 SIDING PANEL ④	STUD ⑤	OR ROUGH MEMBER ⑥ TRIM MEMBER ⑦
STRUCTURAL CLAY TILE	⑧	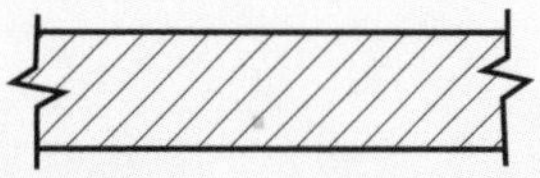	SAME AS PLAN VIEW
GLASS	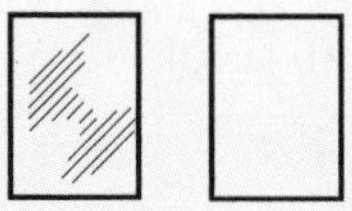	⑨	 SMALL SCALE LARGE SCALE ⑩

Sketching 4-2

Sketch the symbol indicated. Refer to Appendix.

For example: Three-way switch S_3

1. Finish grade
2. Bush
3. Fence
4. Natural grade
5. Point of beginning
6. Property line
7. Water heater
8. Hose bibb
9. Ceiling lighting outlet
10. Junction box
11. Range outlet
12. Single-pole switch
13. Duplex receptacle outlet
14. Lighting outlet (wall)
15. Motor
16. Exposed radiator
17. Supply duct
18. Return duct
19. Thermostat
20. Water closet

Symbols and Abbreviations 4

Review Questions

Name ______________________________ Date ______________

Identification 4-1

Refer to Appendix.

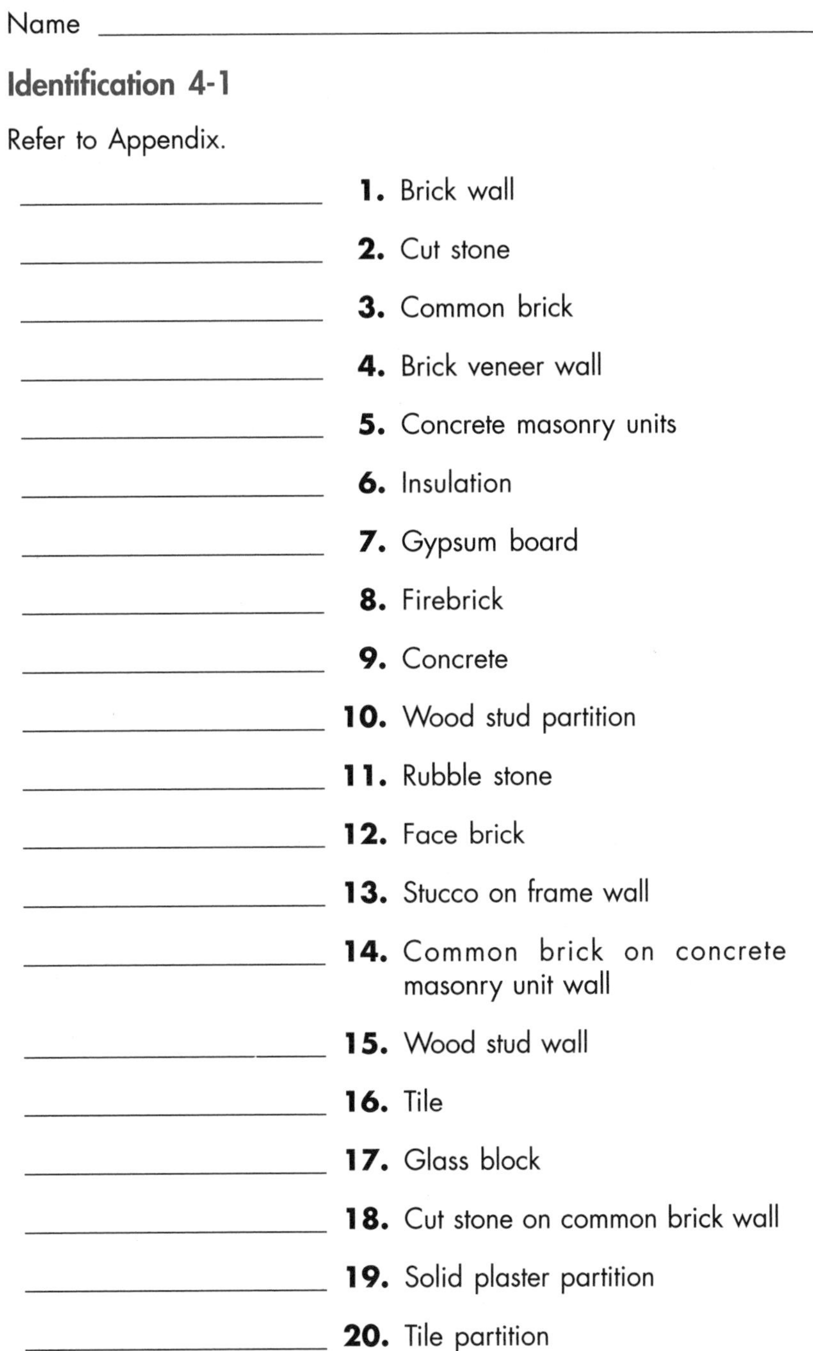

_______________ **1.** Brick wall

_______________ **2.** Cut stone

_______________ **3.** Common brick

_______________ **4.** Brick veneer wall

_______________ **5.** Concrete masonry units

_______________ **6.** Insulation

_______________ **7.** Gypsum board

_______________ **8.** Firebrick

_______________ **9.** Concrete

_______________ **10.** Wood stud partition

_______________ **11.** Rubble stone

_______________ **12.** Face brick

_______________ **13.** Stucco on frame wall

_______________ **14.** Common brick on concrete masonry unit wall

_______________ **15.** Wood stud wall

_______________ **16.** Tile

_______________ **17.** Glass block

_______________ **18.** Cut stone on common brick wall

_______________ **19.** Solid plaster partition

_______________ **20.** Tile partition

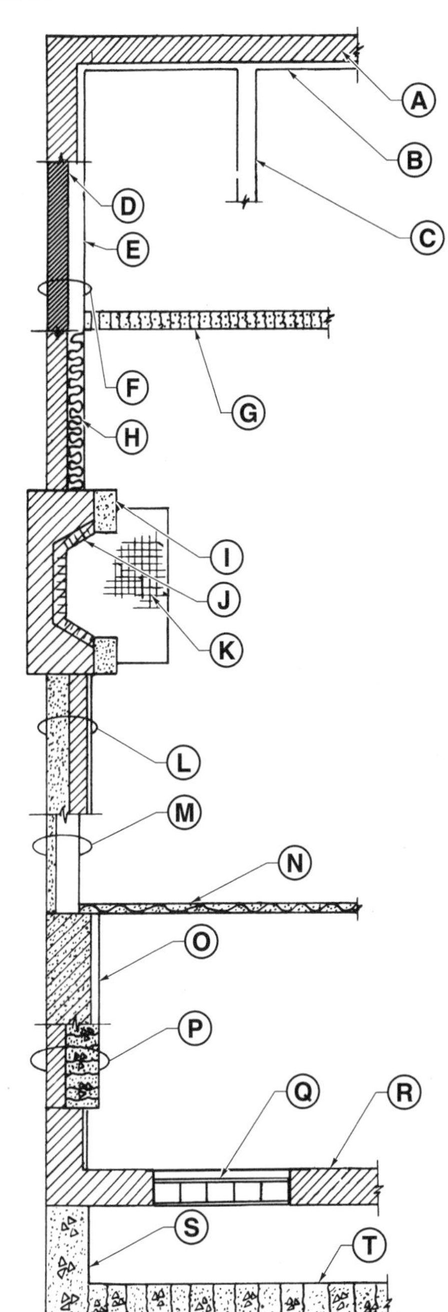

Abbreviations 4-1

Write the word(s) for the abbreviation(s) in each notation.

_______________ **1.** AL SASH

_______________ **2.** 16″ C TO C

_______________ **3.** OAK FIN. FLR

_______________ **4.** WALL-HUNG WC

_______________ **5.** 4″ CI

_______________ **6.** 2400 SQ FT

_______________ **7.** BR 1

_______________ **8.** ½″ GYP BD

_______________ **9.** COMMON BRK

_______________ **10.** 14 R UP

_______________ **11.** ⅝″ SC

_______________ **12.** REFR CAB.

_______________ **13.** PRCST CONC

_______________ **14.** T&G SIDING

_______________ **15.** 4 × 10 BM

Identification 4-2

Refer to Appendix.

_______________ **1.** Duplex receptacle outlet

_______________ **2.** Range outlet

_______________ **3.** Lighting panel

_______________ **4.** Motor

_______________ **5.** Home run

_______________ **6.** Natural grade

_______________ **7.** Earth

_______________ **8.** Point of beginning

_______________ **9.** Three-way switch

_______________ **10.** Pushbutton

R

Ⓐ Ⓑ Ⓒ

Ⓓ Ⓔ Ⓕ

S3

Ⓖ Ⓗ Ⓘ

M

Ⓙ

PRINTREADING

Refer to Smith Residence—Floor Plan on page 103.

Completion 4-1

Complete the chart to list the number of outlets, fixtures, and switches shown in each room.

	○	$○_{FL}$	$Ⓛ_{PS}$	⊖	⊖ (half-filled)	$⊖_{R}$	$⊖_{GFCI}$	$⊖_{WP\ GFCI}$	S	S_3	S_4	Ⓕ	Ⓒ	Ⓣ
EXAMPLE: LIVING ROOM-DINING AREA	2			6	1				2	2				1
1. BATHROOM 2														
2. BEDROOM 1														
3. BEDROOM 1 CLOSET														
4. BEDROOM 2														
5. BEDROOM 2 CLOSET														
6. HALL														
7. BATHROOM 1														
8. DEN														
9. DEN CLOSET														
10. KITCHEN														
11. UTILITY ROOM														
12. GARAGE														
13. EXTERIOR														
14. TOTAL														

NOTE: INCLUDE EXAMPLE IN TOTAL.

True-False

T F **1.** The hot water tank is located in the Kitchen.

T F **2.** Soil stacks are located in partitions behind toilets.

T F **3.** Four hose bibbs are shown.

T F **4.** LT indicates a light table in the Utility Room.

T F **5.** Medicine cabinets are located over lavatories in the bathrooms.

T F **6.** The house is heated by a forced warm air system.

T F **7.** Nine warm air registers are shown on the floor plan.

T F **8.** Nine cold air registers are shown on the floor plan.

T F **9.** The Kitchen range is gas operated.

T F **10.** A cabinet is shown over the refrigerator.

T F **11.** Seven stainless steel thresholds are shown on the floor plan.

T F **12.** Vinyl flooring is used in the Kitchen and Utility Room.

T F **13.** Ceramic tile flooring is used in the bathrooms.

T F **14.** Ceiling joists are spaced 16″ OC.

T F **15.** A planter box is shown on the east wall.

Multiple Choice

______ **1.** The ___, ___, and ___ have windows facing east.
- A. UTIL RM; BR 1; LR
- B. LR; KIT.; UTIL RM
- C. BR 1; BR 2; DEN
- D. LR; BR 2; B 2

______ **2.** Not counting the UTIL RM, HALL, B 1, B 2, and GAR., the house has ___ rooms.
- A. three
- B. four
- C. five
- D. six

______ **3.** The shortest route from BR 1 to the front door is ___.
- A. HALL, DEN, LR
- B. HALL, LR
- C. HALL, BR 2, LR
- D. none of the above

______ **4.** Exterior walls, excluding the garage, are ___.
- A. concrete block
- B. brick on concrete block
- C. brick veneer
- D. siding

______ **5.** ___ exterior walls are shown for the garage.
- A. Framed
- B. Brick veneer
- C. Concrete block
- D. Brick on concrete block

______ **6.** Windows in the LR are ___ windows.
- A. double-hung
- B. awning
- C. fixed
- D. jalousie

______ **7.** Lintels over windows are ___.
- A. angle iron, and angle iron and plate
- B. angle iron only
- C. plate only
- D. none of the above

_______________ **8.** Ceiling joists over the LR run ___.

A. north and south
B. east and west
C. north and south and east and west
D. randomly

_______________ **9.** Ceiling joists over the GAR. ___.

A. run east and west
B. are spaced 16″ OC
C. are not indicated on the floor plan
D. none of the above

_______________ **10.** Not including the GAR. door, ___ exterior doors are shown on the floor plan.

A. two
B. three
C. four
D. five

_______________ **11.** Glass block is located in the GAR. and the ___.

A. DEN
B. KIT.
C. UTIL RM
D. DINING AREA

_______________ **12.** Windows with fixed sashes are located on ___.

A. both sides of the DEN door
B. both sides of the front door
C. one side of the front door
D. one side of the rear door

_______________ **13.** B 1 has a(n) ___ window near the CL.

A. hopper
B. casement
C. awning
D. double-hung

_______________ **14.** An accordion door is shown ___.

A. between the HALL and BR 1
B. between the HALL and B 2
C. for each L CL in the HALL
D. between the DEN and KIT.

_______________ **15.** Hose bibbs are shown on the ___ Elevation(s).

A. East
B. East and West
C. North
D. North and South

_______________ **16.** Regarding the GAR., ___.

A. a casement window is shown on the north wall
B. three risers lead to the UTIL RM door
C. the 6″ concrete floor is brush finished
D. none of the above

_______________ **17.** Regarding the windows, ___.

A. corner windows have angle iron and plate lintels
B. three corner windows are shown
C. glass block sills are to be face brick
D. all of the above

_______________ **18.** Regarding the plumbing, ___.

A. each bathroom contains a tub-shower combination
B. B 2 is adjacent to the DEN
C. access panels are located in closets
D. all of the above

_______________ **19.** Regarding the scale, ___.

A. the Smith Residence is drawn to the scale of ⅛″ = 1′-0″
B. the Smith Residence is drawn to the scale of ¼″ = 1′-0″
C. no scale is shown
D. none of the above

_______________ **20.** Regarding the masonry, ___.

A. a face brick sill is shown beneath the LR windows
B. all exterior walls are face brick
C. face brick on edge finishes the planter box
D. all of the above

Completion 4-2

_______________ **1.** The water heater is located in the ___ room.

_______________ **2.** The front of the Smith Residence faces ___.

_______________ **3.** Tub and shower walls are finished with ceramic tile to a height of ___.

_______________ **4.** A(n) ___ sill is shown at the front door.

_______________ **5.** The L CL near the DEN has two shelves and one ___.

_______________ **6.** ___ floors are shown in the living room.

_______________ **7.** Glass block sills are ___ brick.

_______________ **8.** Three ornamental ___ posts are noted on the front porch.

_______________ **9.** A(n) ___ door separates the KIT. from the UTIL RM.

_______________ **10.** The Smith Residence is a(n) ___-story dwelling.

Refer to printreading questions on page 99.

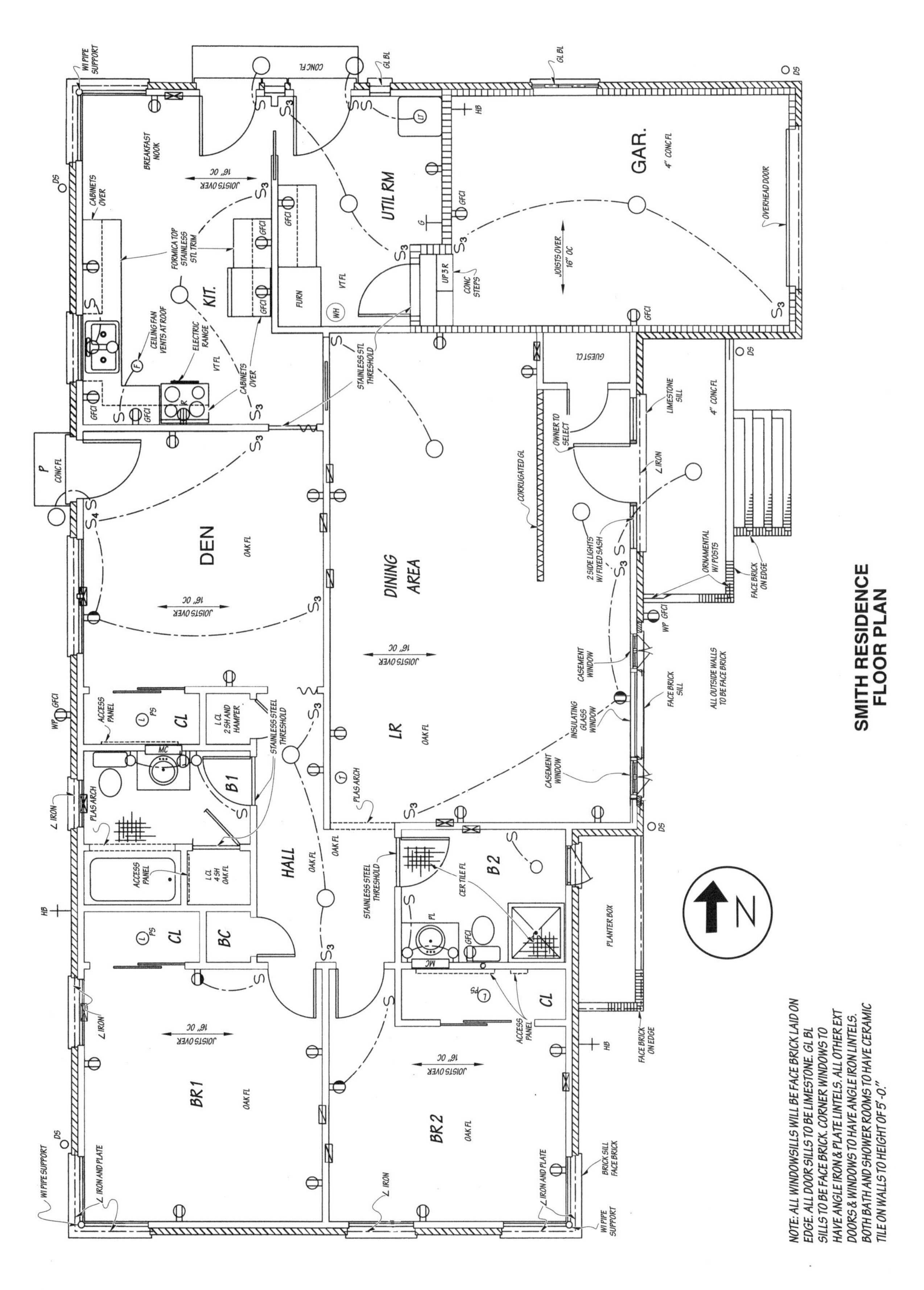

Symbols and Abbreviations 4

Trade Competency Test

Name ______________________________ Date ______________

PRINTREADING

Refer to Hughes Residence—First Floor Plan on page 108.

True-False

T F **1.** Rafters in the GAR. are spaced 24″ OC.

T F **2.** Three WP receptacles are shown on the floor plan.

T F **3.** Six duplex convenience outlets, of which five are split-wired, are shown in the LR.

T F **4.** The door leading from the house into the GAR. is metal covered.

T F **5.** All dimensions are to centers of studs.

T F **6.** Flush doors are shown in the LAV.

T F **7.** The ceiling outlet in the GAR. is controlled by a three-way switch.

T F **8.** Three hose bibbs are shown.

T F **9.** The scale of the floor plan is ¼″ = 1′-0″.

T F **10.** The Kitchen door leading to the outside is designated as an E door.

Multiple Choice

______________ **1.** The first floor plan shows a(n) ___.

A. complete Bathroom
B. enclosed Entry passage
C. open Living Room-Dining Room combination
D. separate Garage

______________ **2.** Access to the second floor is gained by a stairway ___.

A. partially open to the Entry
B. open to the Living Room
C. with 13 risers
D. that makes a 180° turn

_______________ **3.** Access to the basement is gained by ___.

A. winding stairs from the Entry
B. stairs with 14 risers
C. an exterior stairway
D. a stairway open to the Entry

_______________ **4.** Regarding floor levels, the ___.

A. Living and Garage areas are the same level
B. Garage floor is lower than the Living Area
C. Garage and Entry passage floor levels are the same
D. Garage floor is level

_______________ **5.** The windows in the Living Room-Dining Room area are all ___.

A. casement windows
B. casement windows with mullions
C. insulating glass
D. casement or fixed-sash windows

_______________ **6.** The Kitchen and Lavatory windows are ___.

A. double hung
B. casement
C. casement with mullion or double hung
D. casement without mullion or double hung

_______________ **7.** Doors designated B on the floor plan are ___.

A. swinging or sliding (pocket) doors
B. panel doors
C. of different sizes
D. metal clad

_______________ **8.** Doors designated C and D are ___.

A. louvered doors
B. identical
C. all 1¾″ thick
D. flush doors

_______________ **9.** Metal thresholds are provided ___.

A. only at outside doors
B. only at doors where flooring changes from one material to another
C. at the Lavatory doorway
D. at Garage doors

_______________ **10.** Regarding flooring, the ___.

A. house has either oak or ceramic tile
B. Garage floor and the Entry passage floor have the same material and base
C. fireplace hearth is tile
D. Entry has a slate floor

_______________ **11.** Regarding the Living Room-Dining Room area, ___.

A. a wood base and plaster cornice are shown
B. there are no ceiling outlets
C. all convenience outlets are controlled by wall switches
D. a pass-through is provided to the Kitchen

_____ **12.** Closets have ___.
- A. sliding doors
- B. shelves and rods
- C. shelves
- D. lights with pull switches

_____ **13.** The Kitchen has ___.
- A. a countertop range
- B. a ceramic tile floor and base
- C. an electric oven
- D. wall cabinets on three walls

_____ **14.** The Lavatory has ___.
- A. one ceiling light
- B. a ceramic tile floor and base only
- C. a soil stack located in the partition behind the toilet
- D. one light at the medicine cabinet

_____ **15.** Regarding Kitchen cabinets, ___.
- A. ceramic tile countertops are shown
- B. a dishwasher is shown to the right of the sink
- C. a freestanding range is supplemented by a built-in wall oven
- D. none of the above

Completion

_____ **1.** The finished opening for the overhead door is ___ wide.

_____ **2.** The fireplace has a(n) ___ hearth.

_____ **3.** The KIT. measures ___ × ___.

_____ **4.** A kitchen exhaust fan is located on the ___ wall of the KIT.

_____ **5.** ___ and ___ closets are located in the LAV.

_____ **6.** The fireplace opening is ___ wide.

_____ **7.** Wood columns supporting the roof over the ENTRY PASSAGE are ___″ square.

_____ **8.** A(n) ___ door is shown between the DR and KIT.

_____ **9.** The front door to the house is designated ___.

_____ **10.** ___ B doors are required.

_____ **11.** The garage floor is pitched ___″ to the garage door.

_____ **12.** Trusses in the GAR. are 24″ ___.

_____ **13.** The oven and range are ___ operated.

_____ **14.** ___ glass is shown on the fixed sash window in the DR.

_____ **15.** All exterior receptacles are WP and ___ protected.

Refer to printreading questions on page 105.

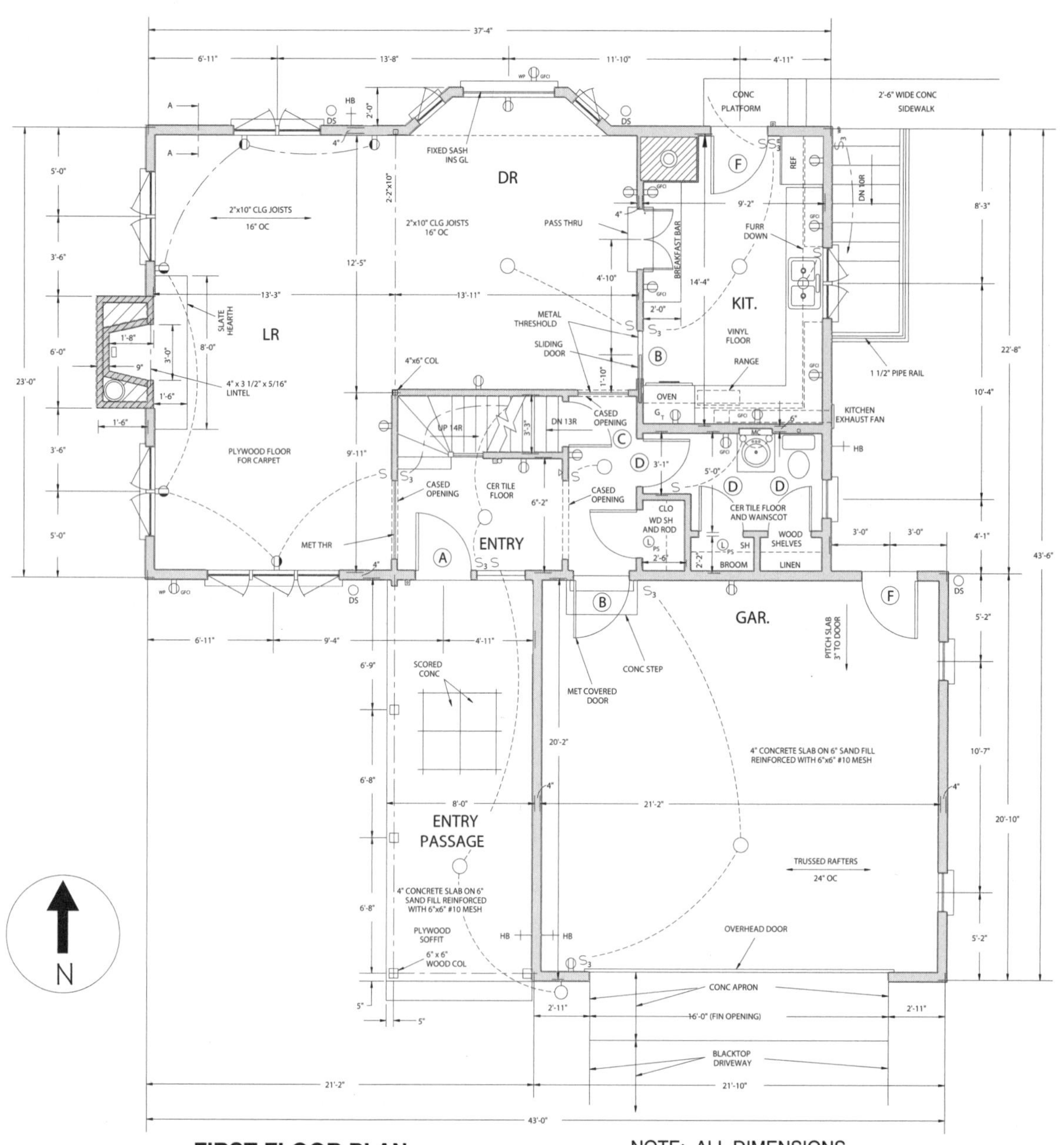

FIRST FLOOR PLAN

SCALE: 1/4" = 1'-0"

NOTE: ALL DIMENSIONS ARE TO FACE OF STUDS

DOOR SCHEDULE

MARK	SIZE	AM'T REQ'D	REMARKS	MARK	SIZE	AM'T REQ'D	REMARKS
A	3'-0" x 6'-8" x 1 3/4"	1	EXTERIOR FLUSH DOOR	D	2'-4" x 6'-8" x 1 3/8"	4	FLUSH DOORS
B	2'-8" x 6'-8" x 1 3/4"	2	FLUSH DOORS 1-SLIDING 1-METAL COVERED	D_1	2'-4" x 6'-8" x 1 3/8"	1	LOUVERED
				E	1'-3" x 6'-8" x 1 3/8"	1	BIFOLD LOUVERED
C	2'-6" x 6'-8" x 1 3/8"	4	FLUSH DOORS	F	2'-10" x 6'-8" x 1 3/4"	2	EXTERIOR 2-LIGHTS
C_1	2'-6" x 6'-8" x 1 3/8"	2	LOUVERED	G	2'-8" x 6'-8" x 1 3/4"	1	EXTERIOR 2-LIGHTS

HUGHES RESIDENCE

Plot Plans 5

Local municipalities develop and provide information required for survey plats. Public streets and utilities are shown on survey plats. Plot plans show the building and improvement locations on a property. The authority having jurisdiction issues permits and conducts inspections to ensure that minimum standards consistent with local ordinances are met.

BUILDING CODES

Building codes are used to enforce minimum building and safety standards and to ensure the use of proper materials and sound construction methods. The authority having jurisdiction in the area enforces local building codes, which are commonly based on model codes. A *model code* is a national building code developed through conferences between building officials and industry representatives around the country. Model codes can be adopted in part or in entirety by states or local communities. For example, the *International Residential Code* is a model code developed by the Building Officials and Code Administration, Inc. (BOCA), the International Congress of Building Officials (ICBO), and the Southern Building Code Congress International (SBCCI).

Building permits must be obtained after the plans are completed by the architect and before construction begins. Plans are submitted to local building officials for review and approval prior to issuance of a building permit. The building permit must be displayed on the job site throughout construction. Inspectors enforce the local building codes by checking materials and construction methods during construction. The final inspection is made after the house is completed. The house may be occupied only after it passes the final inspection and an occupancy permit is issued. See Figure 5-1.

A residence undergoes many inspections throughout the course of construction.

Plot plans provide information required for properly locating a building on a lot.

Figure 5-1. Local building codes are based on model codes. Building permits are issued before construction begins.

BUILDING CODE

CITY OF JACKSON

BUILDING PERMIT

Address

Lot #

General Contractor

INSPECTIONS

Sanitary Lateral Date

Foundation Date

Plumbing Rough Date

Electrical Rough Date

Building Framing Date

Mechanical Systems Date

Plumbing Final Date

Electrical Final Date

Building Final Date

Permit Issued

Permit Expires

Building Official

BUILDING PERMIT

SURVEY PLATS

Basic information for drawing a plot plan is found on a survey plat drawn by a licensed surveyor. A *survey plat* is a map showing a division of land, such as a portion of a quarter section of a township that is subdivided into streets and lots. A *township* is a square area that is six miles long on each side, or 36 square miles. Townships are subdivided into sections and quarter sections.

Townships are grouped based on a gridwork of North-South and East-West lines crisscrossing the United States. Groups of townships form squares that are 36 miles on each side. A *meridian* is the North-South line that defines a group of townships on the grid. A *baseline* is the East-West line that defines a group of townships on the grid. Individual townships are identified by their location in relation to meridians and baselines. For example, a township three rows North of a baseline and five rows East of a meridian is designated T3N, R5E.

Townships are divided into sections. A *section* is a division of a township and is one mile long on each side. Township sections are numbered 1 through 36 beginning at the northeast corner of the township. Sections are further subdivided into quarter sections. These are designated NE, SE, SW, and NW. See Figure 5-2.

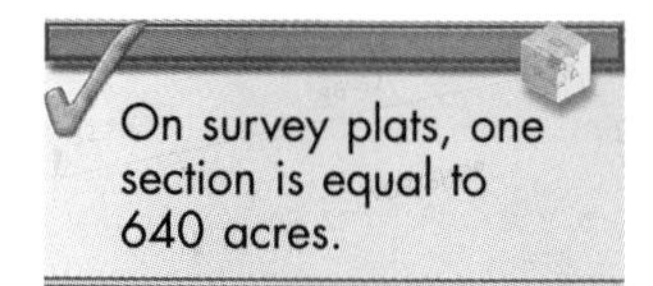

On survey plats, one section is equal to 640 acres.

Model building codes are developed through conferences between building officials and industry representatives. The *Uniform Building Code, Southern Standard Building Code, Basic National Building Code, National Building Code,* and *International Building Code* are the primary model codes used in the United States and Canada.

PLOT PLANS

A survey plat contains legal descriptions of land. For example, the survey plat shows the lot and the location of easements and public utilities. An *easement* is a strip of privately owned land set aside for placement of public utilities. Information from survey plats of sections and quarter sections is used when drawing plot plans.

A *plot plan* is a scaled view that shows the shape and size of the building lot; location, shape, and overall size of a house on the lot; and the finish floor elevation. In addition, locations of streets and utilities are shown. The plot plan is oriented from a point of beginning. A *point of beginning (POB)* is a location point from which horizontal dimensions and vertical elevations are made. This point is used to accurately locate all lot corners and to locate the house on the lot.

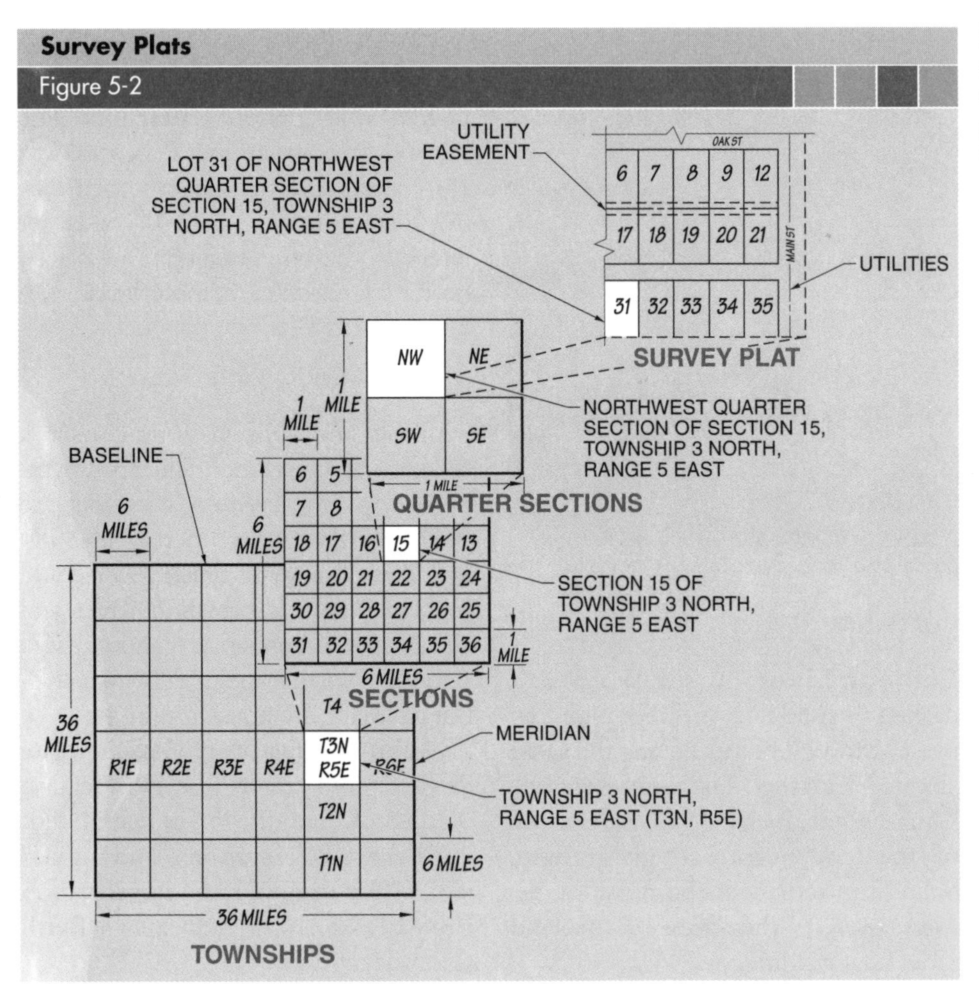

Figure 5-2. Townships are subdivided into sections and quarter sections. The survey plat shows streets, utilities, and easements.

Scale

A civil engineer's scale is used to draw plot plans. Plot plans are commonly drawn to the scale of 1″ = 10′-0″ or 1′ = 20′-0″ depending on the lot size and sheet size of the plan. Decimal dimensions are used on plot plans. For example, a property line that is 120′-6″ long is designated 120.50′, and a finish floor level of 5′-0″ is designated 5.00′. When locating a building on a lot, tradesworkers use tape measures showing feet, inches, and fractional inch increments. Therefore, decimal dimensions must be converted into fractions. See Appendix.

Measurements must be properly read and used when laying out the position of a building on a lot.

Elevations

An *elevation* is a vertical measurement above or below the point of beginning. (Note that the use of the word "elevation" here is different from an elevation view of a house.) The point of beginning is shown on the plot plan. A benchmark on the actual lot represents the point of beginning. A *benchmark* is a stake driven into the ground or a point in the street or along the curb that is used as the point of beginning. For example, a benchmark with an elevation of 112′ indicates that the point is 112′ above the reference point used by the community. The reference point for a community may be in relation to mean sea level or to a specific point in the community.

Total station instruments may be used to lay out a building site. Total stations combine digital data processing and survey technology.

A job site must be properly prepared prior to construction.

Contour Lines. A *contour line* is a dashed or solid line on a plot plan that passes through points having the same elevation. Dashed lines represent the natural grade. *Natural grade* is the slope of the land before rough grading. Solid lines represent the finish grade. *Finish grade* is the slope of the land after final grading. See Figure 5-3.

Closely spaced contour lines denote steeply sloped lots. Contour lines spaced farther apart denote flatter lots. The spacing of contour lines is generally in 1′-0″ intervals. For steeply sloped lots, the spacing may be in 2′-0″ or larger intervals. Elevation notations are placed on the high side of contour lines.

Symbols and Abbreviations

Symbols and abbreviations conserve space on plot plans. See Figure 5-4. Symbols show roads, streets, sidewalks, the point of beginning, property lines, utilities, natural and/or finish grades, and other pertinent information. Trees and shrubbery to remain are shown with symbols. North is indicated with a symbol to orient the house on the lot.

Abbreviations on plot plans relate to roads and streets, the lot, location of the house on the lot, utilities, building materials, elevations, and similar items. For example, the abbreviation FIN. FL EL 10.25′ indicates a finish floor elevation of 10′-3″.

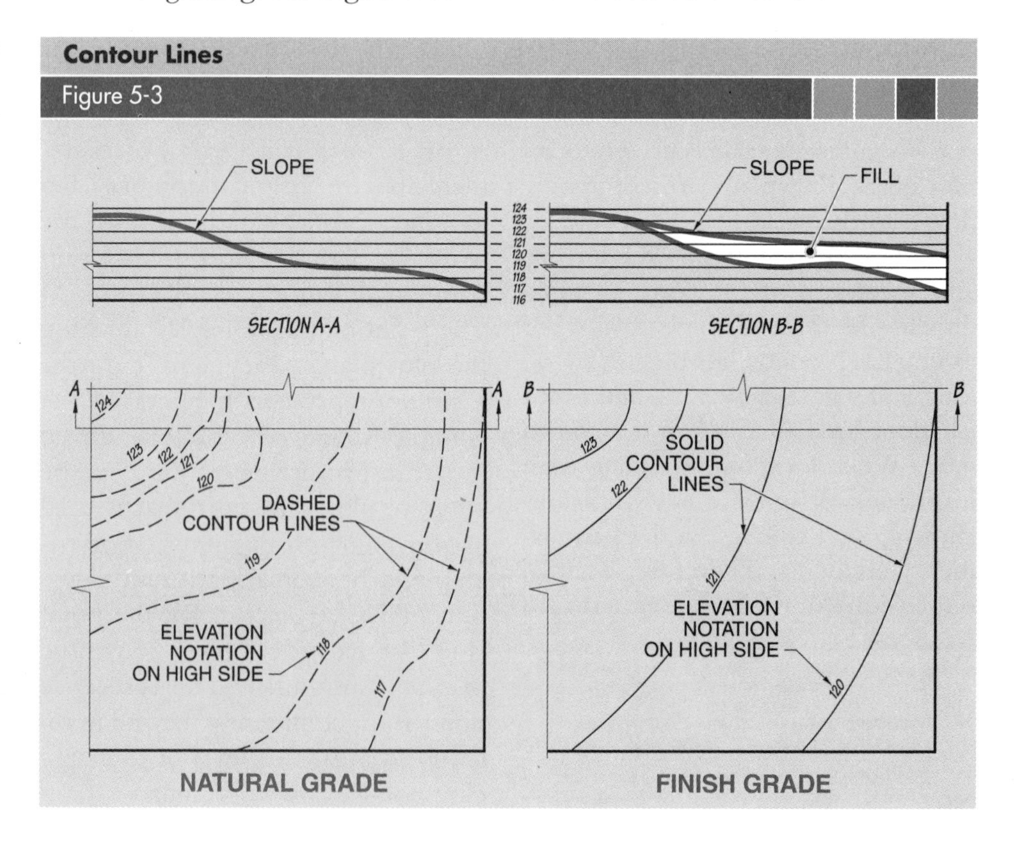

Figure 5-3. Contour lines show the slope of a lot.

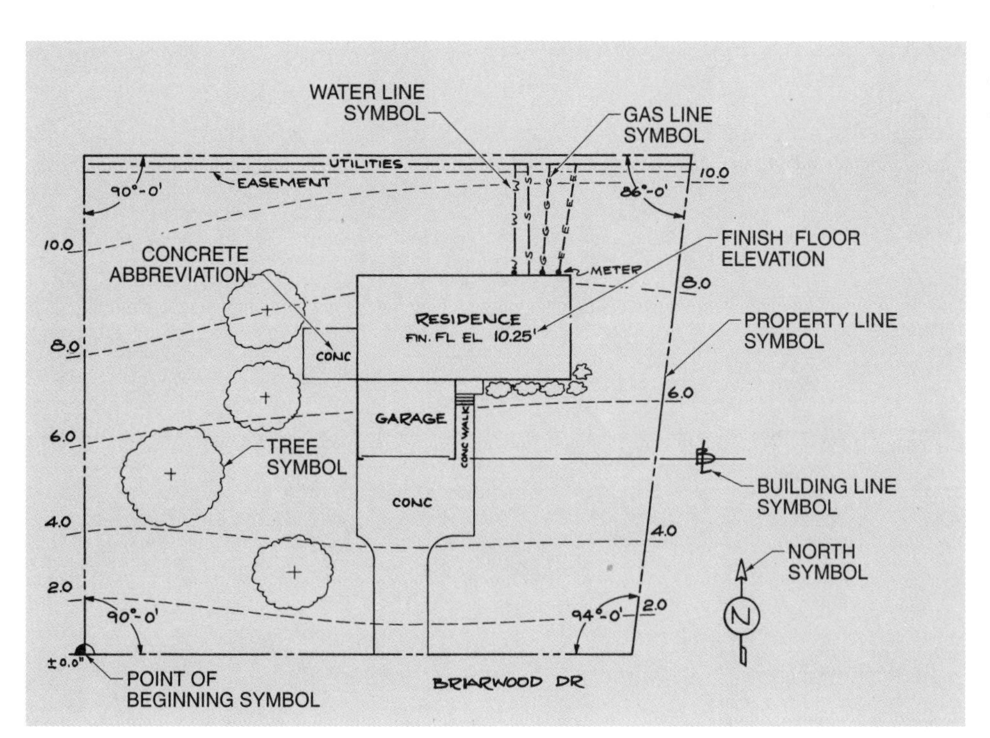

Figure 5-4. Symbols and abbreviations conserve space on plot plans.

READING PLOT PLANS

The plot plan for the Wayne Residence is drawn to the scale of 1″ = 20′-0″. See Figure 5-5. The lot, which is located on Country Club Fairways in Columbia, Missouri, is pentagonal. Dimensions for property lines and notations indicating the corner angles are given. For example, the angle of the southeast corner of the lot is 104°-23′-25″.

A 9′-0″ utility easement is shown between the curb and the property line along the street. An 8′ utility easement is shown along the northwest property line. The point of beginning is the southwest corner of the lot. The corner of the Garage is 25′-0″ north and 11′-0″ east of the property line. The finish floor elevation is 725′-0″. A concrete drive and 4′-0″ concrete walk are shown at the front of the house.

Contour lines show the finish grade of the lot. Numbers on the contour lines indicate elevations. The lot slopes down from west to east. The elevation near the southwest corner of the house is almost 726′ while the elevation at the southeast corner of the house is 718′. Retaining walls are shown on the west and north sides of the house. The closely spaced contour lines from the retaining wall on the north side of the house represent a steep slope in the finish grade.

Symbols are used to show four trees that are to remain on the lot. In addition, symbols show foundation shrubbery along the front of the house between the front entrance and the Garage.

Load-bearing capacity, which is the ability of soil to support weight, affects the amount of settlement that occurs.

Trus Joist, A Weyerhaeuser Business

Exterior walls are erected after concrete for the foundation has set and hardened sufficiently.

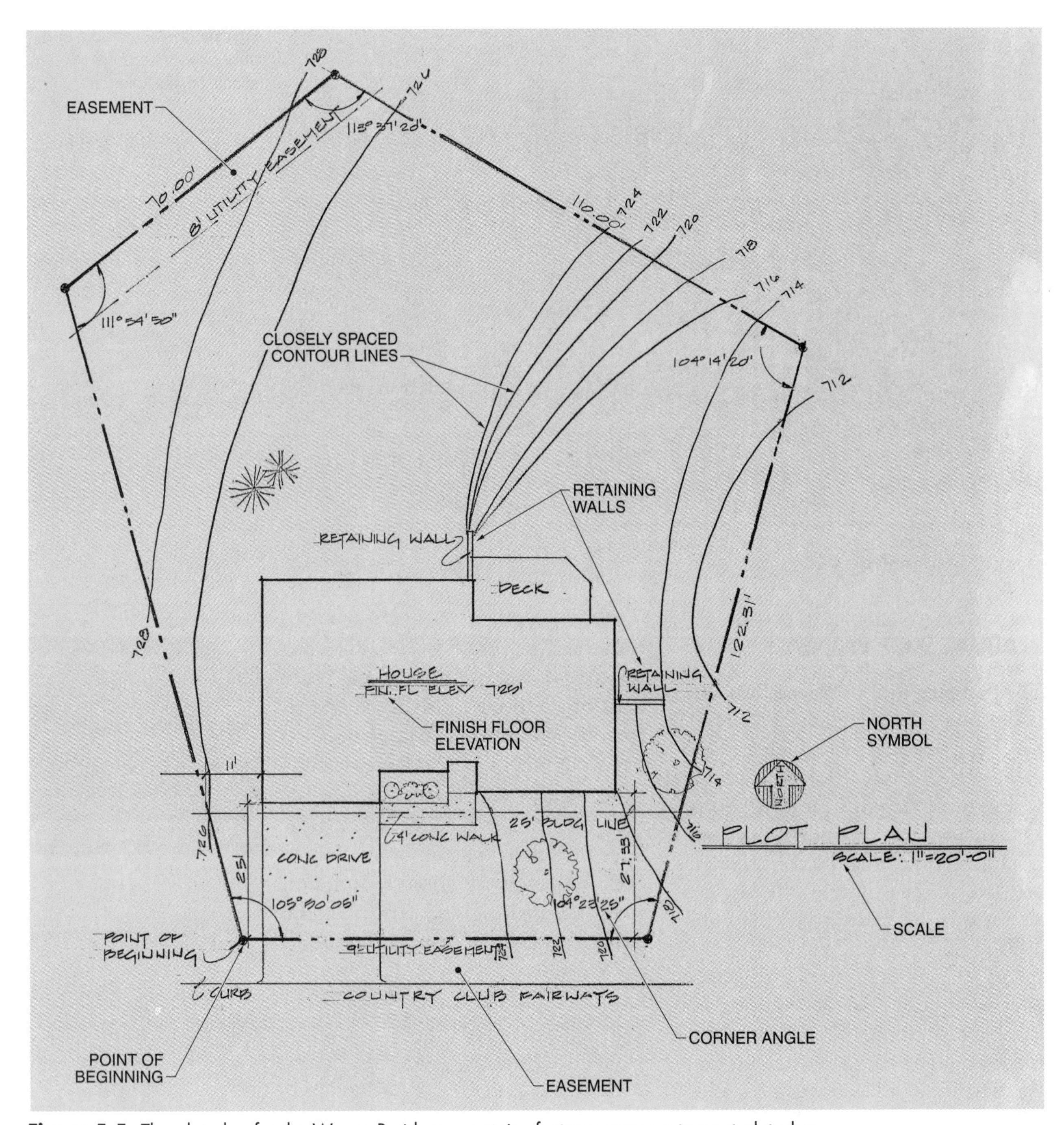

Figure 5-5. The plot plan for the Wayne Residence contains features common to most plot plans.

Plot Plans 5

Sketching

Name ______________________________ Date ______________

Sketching 5-1

Sketch plot plan symbols as indicated. Refer to Appendix.

① BUILDING LINE

② POINT OF BEGINNING

③ PROPERTY LINE

④ TREE

⑤ UTILITY METER

⑥ IMPROVED ROAD

⑦ UNIMPROVED ROAD

⑧ FIRE HYDRANT

⑨ NORTH

⑩ POWER POLE AND GUY

⑪ STREET SIGN

⑫ FENCE

Sketching 5-2

Sketch and/or letter the following to complete the plot plan.

1. 512′ (elevation notation)
2. Existing trees (symbol)
3. 150.5′ (dimension)
4. STORM DRAIN (symbol and notation)
5. WATER MAIN (symbol and notation)
6. GAS MAIN (symbol and notation)
7. POINT OF BEGINNING (notation)
8. 12.5′ (dimension)
9. FIRST FL ELEV 513.5′ (notation)

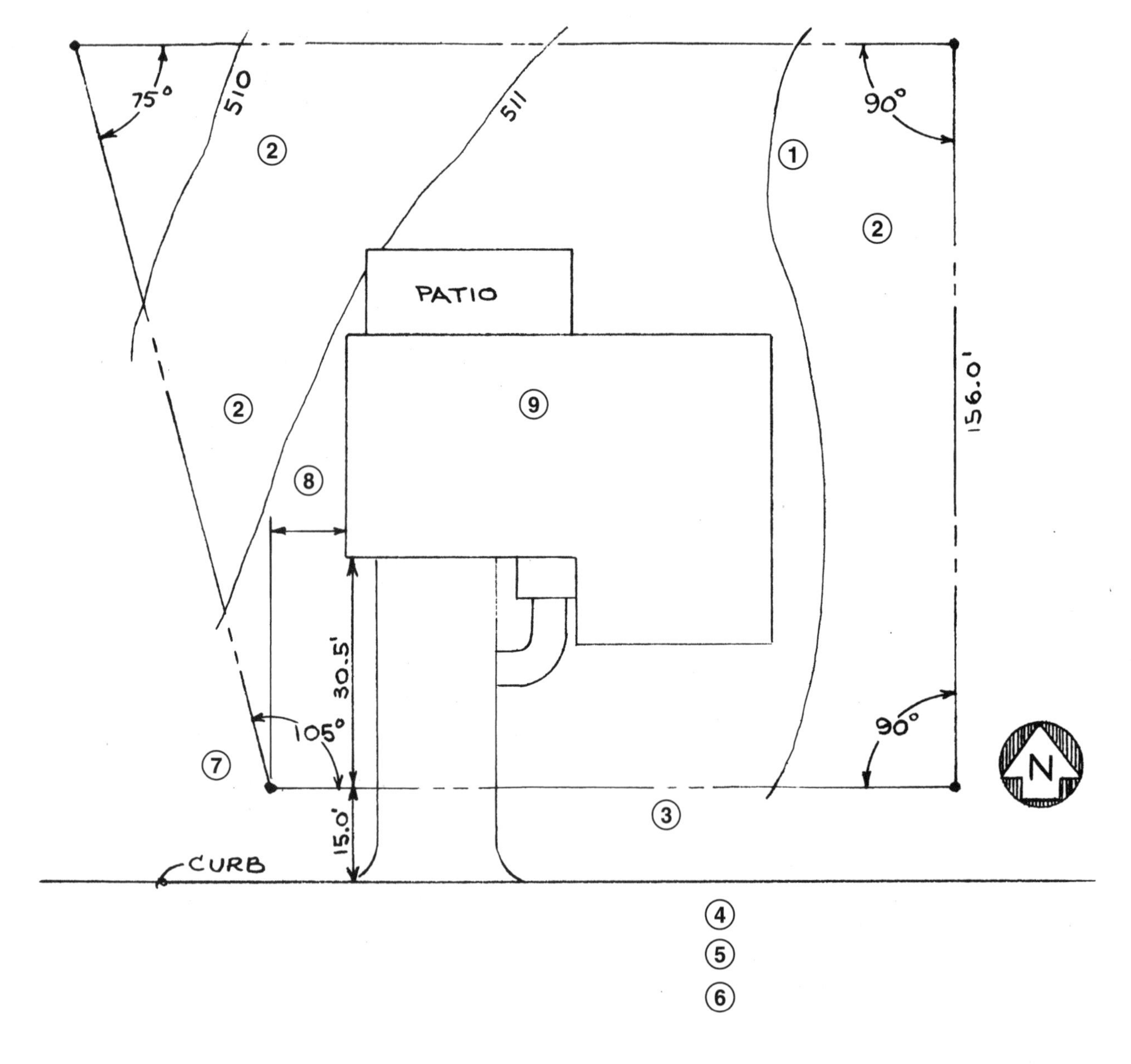

PLOT PLAN

Plot Plans 5

Review Questions

Name ______________________ Date ______________

True-False

T F **1.** Public streets and utilities are shown on plot plans.

T F **2.** The authority having jurisdiction in an area inspects the house during construction.

T F **3.** Building permits must be obtained before the plans are completed.

T F **4.** Townships are square areas one mile long on each side.

T F **5.** Baselines on maps showing townships run north and south.

T F **6.** A section contains one square mile.

T F **7.** Easements are private property containing public utilities.

T F **8.** Property dimensions on plot plans are generally given in feet and decimal parts of a foot.

T F **9.** The decimal .675 equals ⅝.

T F **10.** Solid contour lines show the natural grade of a lot.

T F **11.** Closely spaced contour lines on a plot plan represent a steep slope.

T F **12.** Trees and shrubbery should not be shown on plot plans.

T F **13.** The North symbol should always be included on a plot plan.

T F **14.** Utilities on plot plans may be shown with symbols and notations.

T F **15.** Spacing of contour lines on plot plans is generally in 1′-0″ intervals.

Multiple Choice

______________ **1.** A(n) ___ scale is used to draw plot plans.

A. architect's
B. civil engineer's
C. mechanical engineer's
D. tradesworker's

______________ **2.** A township contains ___ square miles.

A. 1
B. 6
C. 16
D. 36

____________________ **3.** A township four rows North of the baseline and two rows East of the meridian is designated ___.

A. T2N, R2E
B. R3E, T3N
C. T2N, R3N
D. none of the above

____________________ **4.** Townships are divided into ___.

A. sections
B. communities
C. towns and farms
D. none of the above

____________________ **5.** A plot plan shows the ___.

A. size of the building lot
B. public streets adjacent to the building lot
C. finish floor elevation
D. all of the above

____________________ **6.** A four-sided building lot with two 90° corners and one 83° corner also contains one ___° corner.

A. 89
B. 92
C. 97
D. 107

____________________ **7.** A plot plan showing a finish floor elevation of 132′-10″ is ___ above the point of beginning at an elevation of 128.5′.

A. 3′-8″
B. 4′-4″
C. 4′-5″
D. 4′-8″

____________________ **8.** A rectangular building lot measuring 126′-0″ × 212′-0″ contains ___ sq ft.

A. 252
B. 338
C. 22,424
D. 26,712

____________________ **9.** A(n) ___ is a vertical measurement above or below the point of beginning.

A. height
B. baseline
C. elevation
D. meridian

____________________ **10.** A survey plat is a map showing a ___ of land.

A. division
B. building lot
C. slope
D. all of the above

Plot Plans 5

Trade Competency Test

Name ______________________ Date ______________

Refer to Wayne Residence—Plot Plan, Sheet 6.

Completion

______________ **1.** The concrete walk is ___′ wide.

______________ **2.** The finish floor elevation is ___′.

______________ **3.** Solid ___ lines show the finish grade of the lot.

______________ **4.** The southwest property line of the lot meets the south property line at a(n) ___ angle.

______________ **5.** The utility easement along the street is ___′ wide.

______________ **6.** The plot plan is drawn to the scale of ___.

______________ **7.** A deck is shown near the ___ corner of the house.

______________ **8.** The southwest corner of the house is ___ from the west property line.

______________ **9.** The concrete walk leads from the front entry to the ___.

______________ **10.** A(n) ___′ utility easement runs along the northwest property line.

Multiple Choice

______________ **1.** Retaining walls are visible on the ___ Elevations of the house.

A. North and East
B. North and West
C. South and East
D. South and West

______________ **2.** Regarding the northwest property line, ___.

A. it is 70.0′ long
B. a 104°-14′-20″ angle is shown
C. it is the longest property line
D. none of the above

_______________ **3.** Regarding the point of beginning, ___.

A. it is 12.0′ from the curb
B. it is 716.0′ above sea level
C. a concrete drive is shown on that corner of the lot
D. all of the above

_______________ **4.** Regarding the utility easement, ___.

A. two utility easements are shown
B. utilities are located on the north side of the property only
C. utilities are located on the south side of the property only
D. utilities are located on the west side of the property only

_______________ **5.** Regarding the driveway, ___.

A. the Garage is located on the east side of the house
B. a brick paver driveway is shown
C. a concrete driveway is shown
D. an asphalt driveway is shown

_______________ **6.** Regarding the lot, ___.

A. the front yard is the largest
B. a steep slope runs north from the rear retaining wall
C. trees are shown on four sides of the house
D. none of the above

_______________ **7.** Regarding the building line, ___.

A. it is set back 25′-0″ from the curb
B. the roof overhang cannot exceed 2′-0″ to comply
C. it is located on the north side of the house
D. all of the above

_______________ **8.** Regarding the lot, ___.

A. the highest elevation shown is 111′-0″
B. elevation does not vary over 2′-0″
C. the natural grade is to remain
D. the lowest elevation is on the east side of the house

_______________ **9.** Regarding the lot, ___.

A. the house is on a corner lot
B. the house faces Country Club Fairways
C. the basic shape is hexagonal
D. none of the above

_______________ **10.** Regarding the title block for Sheet 6, the plans were ___.

A. revised
B. drawn by FGH
C. developed by Hulen & Hulen Designs
D. all of the above

Floor Plans

6

Plan Views

Floor plans are the most commonly used prints in a set of plans. The overall shape and size of the house are shown on the floor plans. The sizes and relationships of rooms are given. All other prints relate to the floor plan to show construction details, locations, size, fixtures, and equipment. Dimensions, symbols, notations, and abbreviations are used to provide trade information. Floor plans are generally drawn to the scale of ¼″ = 1′-0″ to allow adequate detail to be shown.

FLOOR PLANS

A *floor plan* is a scaled view of the various floors in a house looking directly down from a horizontal cutting plane taken 5′-0″ above each finished floor. Floor plans show the broad aspects of shapes, sizes, and relationships of rooms and the layout of auxiliary space such as hallways, stairs, and closets. The layout of attached garages, carports, patios, and decks is also shown on a floor plan. Floor plans are generally the first set of plans to be drawn and the most used set on the construction site.

Dimensions on floor plans give the overall size of the house and location of all wall offsets, partitions, doors, and windows. Wall and partition thicknesses are indicated and room sizes are shown. The location and sizes of closets, fireplaces, stairs, and other features are also dimensioned.

Symbols on floor plans represent materials, equipment, and fixtures. Floor tile, glass block, and common or face brick are examples of materials that are shown with symbols. Electrical equipment, such as receptacles and outlets, and plumbing equipment, such as hose bibbs and floor drains, are shown with symbols. Fixtures such as kitchen sinks, bathroom lavatories, bathtubs, showers, and water closets are also shown with symbols.

Abbreviations on floor plans conserve space and help prevent cluttering. Common abbreviations used to designate rooms are LR (living room), BR (bedroom), DR (dining room), and B (bath). In addition, abbreviations are used to designate materials, equipment, and fixtures.

Notations (notes) on floor plans may be general, typical, or specific. A *general note* is a notation that refers to an entire set of prints. One example of a general note is ALL DIMENSIONS TO FACE OF STUDS, which refers to all dimensioned studs in the building. A *typical note* is a notation that refers to all similar items on the prints. For example, the notation PROVIDE FURRED SOFFIT OVER ALL KITCHEN WALL CABINETS refers to all kitchen wall cabinets shown on the floor plan. A *specific note* is a notation that refers to a certain item only. For example, the notation CER TILE FLR (ceramic tile

While standards for symbols and abbreviations have been established to represent and indicate structural and material components, symbols and abbreviations may vary between sets of prints.

floor) refers to that floor area only. Note that abbreviations may be used in notations to conserve space. See Figure 6-1.

Scale

Floor plans are drawn to scale. All rooms, hallways, stairs, and other areas are drawn to the same scale so that they are in a consistent size relationship to each other. The scale for the floor plan is generally given as a general note near the name of the plan. For example, the notation SCALE: ¼″ = 1′-0″ may be placed below FLOOR PLAN. The scale may also be given in the title block if the floor plan is the only drawing on the sheet or if all drawings on the sheet were drawn at the same scale.

Plan Views

Figure 6-1

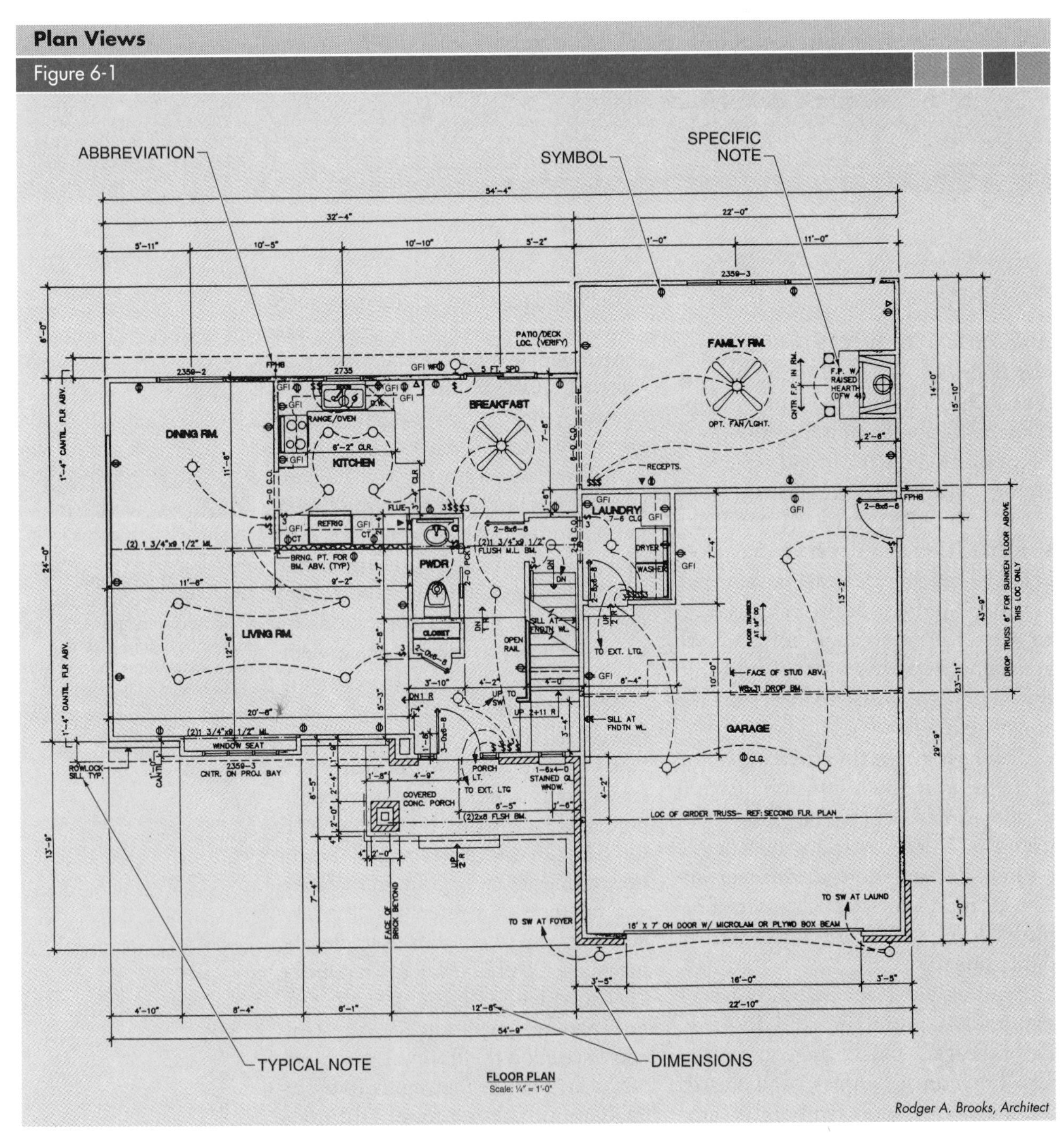

Rodger A. Brooks, Architect

Figure 6-1. Floor plans are scaled views of the various floors of a structure.

The most common scale for residential floor plans is ¼″ = 1′-0″ (1⁄48 scale). The ¼″ = 1′-0″ scale yields a floor plan large enough to convey sufficient information on standard size drawing sheets. Dimensions on scaled drawings give the actual size of the rooms, hallways, and other features. For larger houses and commercial buildings, a ⅛″ = 1′-0″ (1⁄96 scale) or 3⁄16″ = 1′-0″ (1⁄64 scale) scale may be used.

Cutting Planes

Floor plans are horizontal sections made by an imaginary cutting plane taken through the house 5′-0″ above each finished floor. The cutting plane passes through walls, partitions, upper window sashes, kitchen wall cabinets, medicine cabinets, and other built-in-place or installed features. Solid lines show features below the cutting plane. Hidden (dashed) lines show features above the cutting plane. For example, kitchen base cabinets are shown with solid lines while kitchen wall cabinets are shown with hidden lines. See Figure 6-2.

Floor plans may also include information about structural members that support the roof or floor above. For example, the note 2 × 10 JOISTS OVER on a first floor plan indicates that the joists are overhead and support the second floor.

Orientation

Floor plans are generally drawn, based on orthographic projection concepts, with the front of the house toward the bottom of the sheet. An orthographic projection contains at least two, or more commonly, three views.

However, floor plans may be oriented to fit the drawing sheet. For example, when a house is long and narrow, the front is usually placed facing the right edge of the drawing sheet. When floor plans for other floors are developed, they should be oriented in the same direction on the drawing sheet.

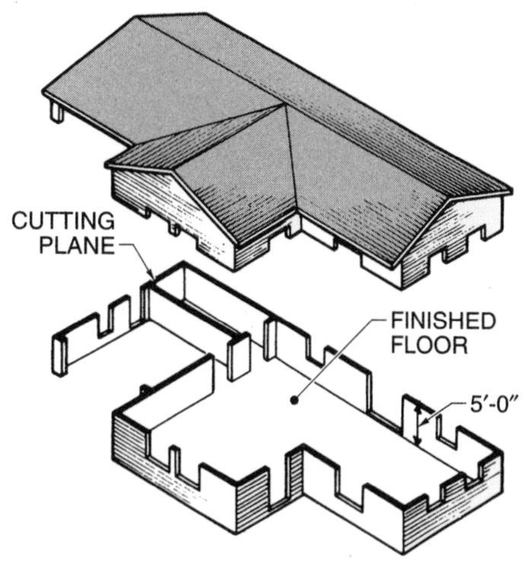

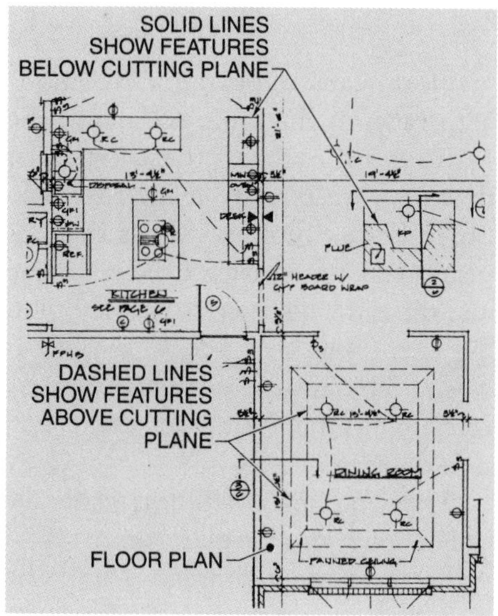

Figure 6-2. The cutting plane is positioned 5′-0″ above the finished floor. Solid lines represent features below the cutting plane. Dashed lines show features above the cutting plane.

Relationship

Floor plans for various floors are related to each other. Structural provisions are made so the load of floors and partitions is transferred to supporting members or partitions immediately below. Stairs are designed so they start on one floor and end in the proper place on the floor above or below. Heating ducts are designed to start at the furnace, pass through first floor partitions, and end at registers in the desired locations in second floor partitions or floors.

Roof rafters or trusses are positioned directly over exterior wall studs where possible so the roof weight is properly transferred to the foundation.

James Hardie Building Products

Exterior elevations are used to visualize the location of building features such as doors, windows, and dormers.

Floor plans and exterior elevations are drawn to the same scale, and the features on each of these types of drawings are directly related to each other. For example, windows that appear on the floor plan are the same size and the same distance from the building corners as they are on the elevations. The house is visualized by referring to the floor plans and elevation drawings.

The same information is generally only shown in one place on a set of prints to avoid redundancy and the potential for errors on the drawings. For example, a stairway that rises from the first to the second floor is not shown completely on the floor plans for either story. The first floor plan shows the exact location of the bottom riser and a few treads. A *riser* is the vertical portion of a stair step. A *tread* is the horizontal portion of a stair step. The stairway is terminated on a break line. The second floor plan shows the top riser location and a few of the descending stair treads that also terminate on a break line. A notation, such as 16 R UP or 16 R DN indicates the total number of risers for the stairway. A window that appears on a stairway landing can be shown on the first and second floor plan but is commonly only shown on one of these floor plans.

After the first floor plan has been drawn, it is used to locate walls, windows, stairways, and other features when drawing the second floor plan. Layering on CAD drawings allows an architect to quickly and easily maintain the proper relationship of floor plans for a multistory house. Layering on CAD drawings also allows plans to be developed from basic floor plans for specific trade areas. See Figure 6-3. For example, a basic floor plan is drawn first, and notes and symbols are added. Plumbing information may then be added to create a floor plan for the plumbing subcontractor. Electrical information may be added to the basic floor plan for the electrical subcontractor.

> Plumbing plans are typically shown in an isometric view.

SIMPLIFIED FLOOR PLANS

Simplified sketches of floor plans do not contain dimensions or detailed information. Symbols for materials of construction, windows, doors, and other basic building components are modified to permit visualization of room shapes and relationships without distracting detail. A set of prints should be studied in much the same order as a person might inspect a house. For houses with more than one floor plan, the first (main) floor plan is studied first. In a one-story house with a full basement, the floor plan (main floor plan) is studied first and the basement (foundation) plan is studied next.

Pictorial sketches are useful in determining room shapes and relationships when studying floor plans. Such sketches show basic shapes and relationships but provide little detail. Overall construction concepts can be developed from the sketches.

Layering of CAD Drawings
Figure 6-3

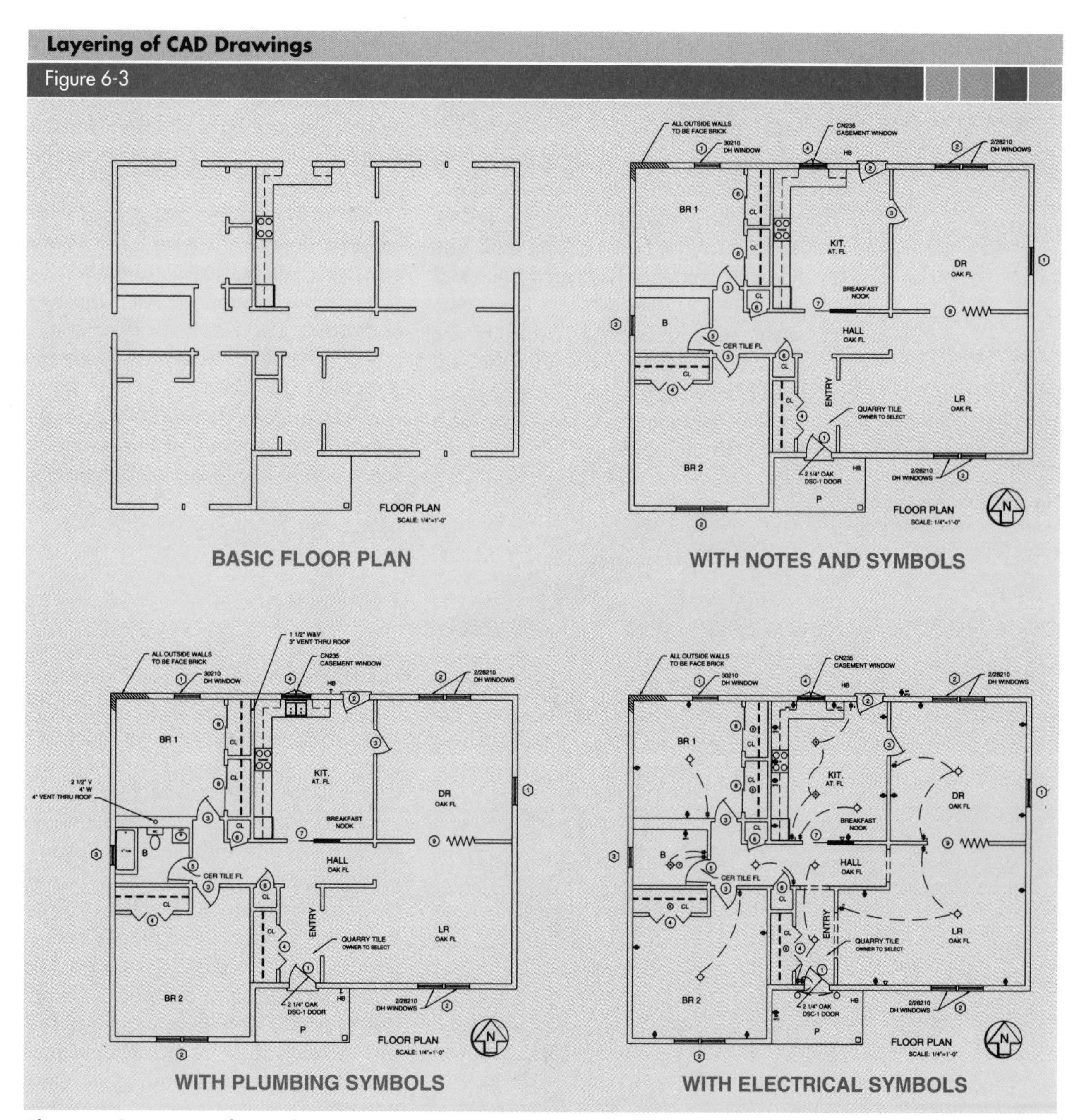

Figure 6-3. Layering of CAD drawings allows plans with specific types of information to be developed.

One-Story House

A one-story house has one floor plan. Two floor plans are required if the house has a basement—one floor plan for the first floor and one plan for the basement. A floor plan is drawn as if a cutting plane had been passed through the house 5′-0″ above the finished floor and the upper part had been removed so that the printreader can look directly down on the floor. See Figure 6-4.

For the one-story house shown in Figure 6-4, entry to the house is through either the Living Room door or Kitchen door. The main entry is the Living Room door. A coat closet (CL) to the right of the Living Room

The attic of a one-story house is usually too small for living purposes.

door serves as a windbreak and divides the entry area from the Dining Area. Picture windows are placed along the front of the house.

The Dining Area is between the Living Room and the Kitchen. It is not enclosed by partitions. One window is shown in the Dining Area wall. The Kitchen contains L-shaped base and wall cabinets. Base cabinets are shown with solid lines. Wall cabinets are shown with dashed lines since they are above the cutting plane. A window on the wall opposite the cabinets provides light and ventilation.

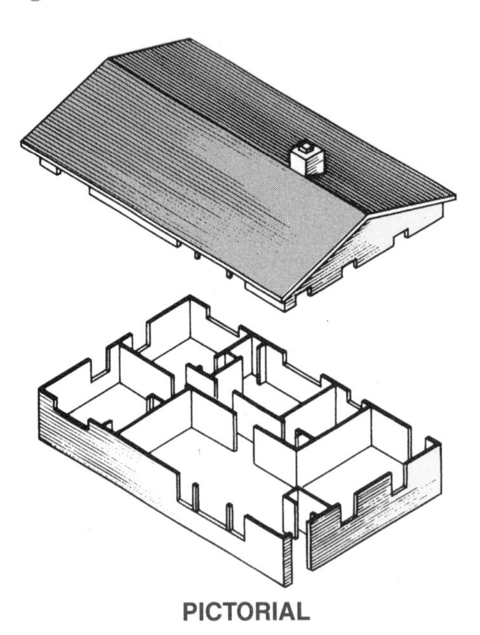

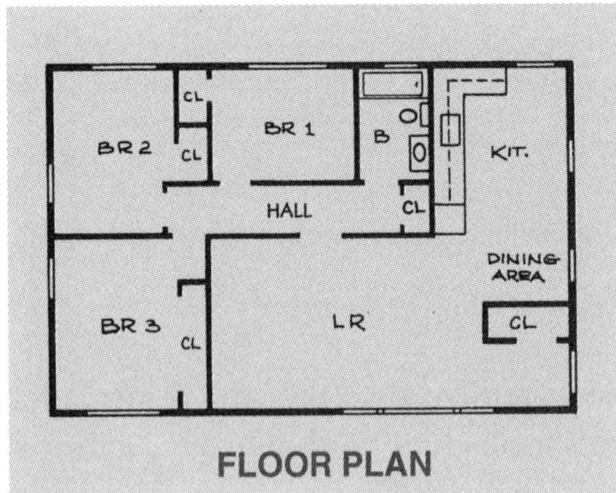

Figure 6-4. A one-story house on a concrete slab has one floor plan.

Three bedrooms and a bath are located off the hallway. The hallway is entered from the living room. The entry to the hallway is located so that it is not directly across from the Bathroom door or the door to Bedroom 1. A linen closet is located at the end of the hallway near the Bathroom. This space could be devoted to a furnace if required. (Note that a chimney for the furnace is shown on the roof in the pictorial sketch.)

All bedrooms are well lighted with large windows. Bedrooms 2 and 3 have cross ventilation. Closets for Bedrooms 1 and 2 are located between the two bedrooms. The closet for Bedroom 3 is larger. Bifold or sliding doors are required for this closet.

All Bathroom fixtures have been arranged along a common partition with the Kitchen. This fixture arrangement facilitates installation and reduces material and labor costs.

Two-Story House

A two-story house with a basement has three floor plans—one plan for the second floor, one plan for the first floor, and one plan for the basement. See Figure 6-5. The three cutting planes that produce the first floor, second floor, and basement floor plans are indicated by cutting plane lines designated A-A, B-B, and C-C, respectively. A-A is the projection for the first floor plan. Cutting plane line A-A passes through the house at the upper sash of the windows and the upper part of the doors on the main floor. B-B is the projection for the second floor plan. Cutting plane line B-B passes through the house at the upper sash of the windows on the upper floor. C-C is the projection for the basement plan. Cutting plane line C-C passes through the basement windows.

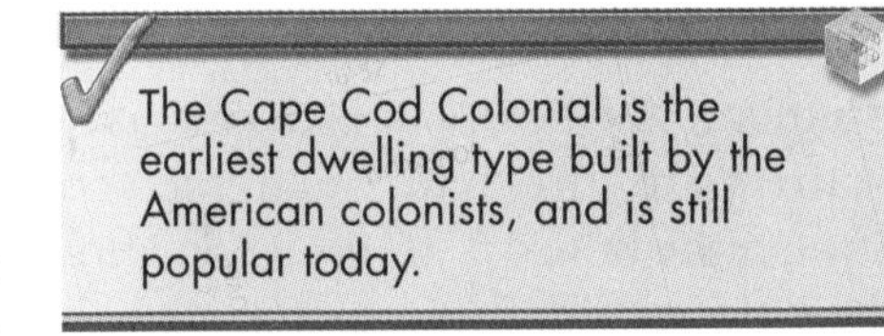

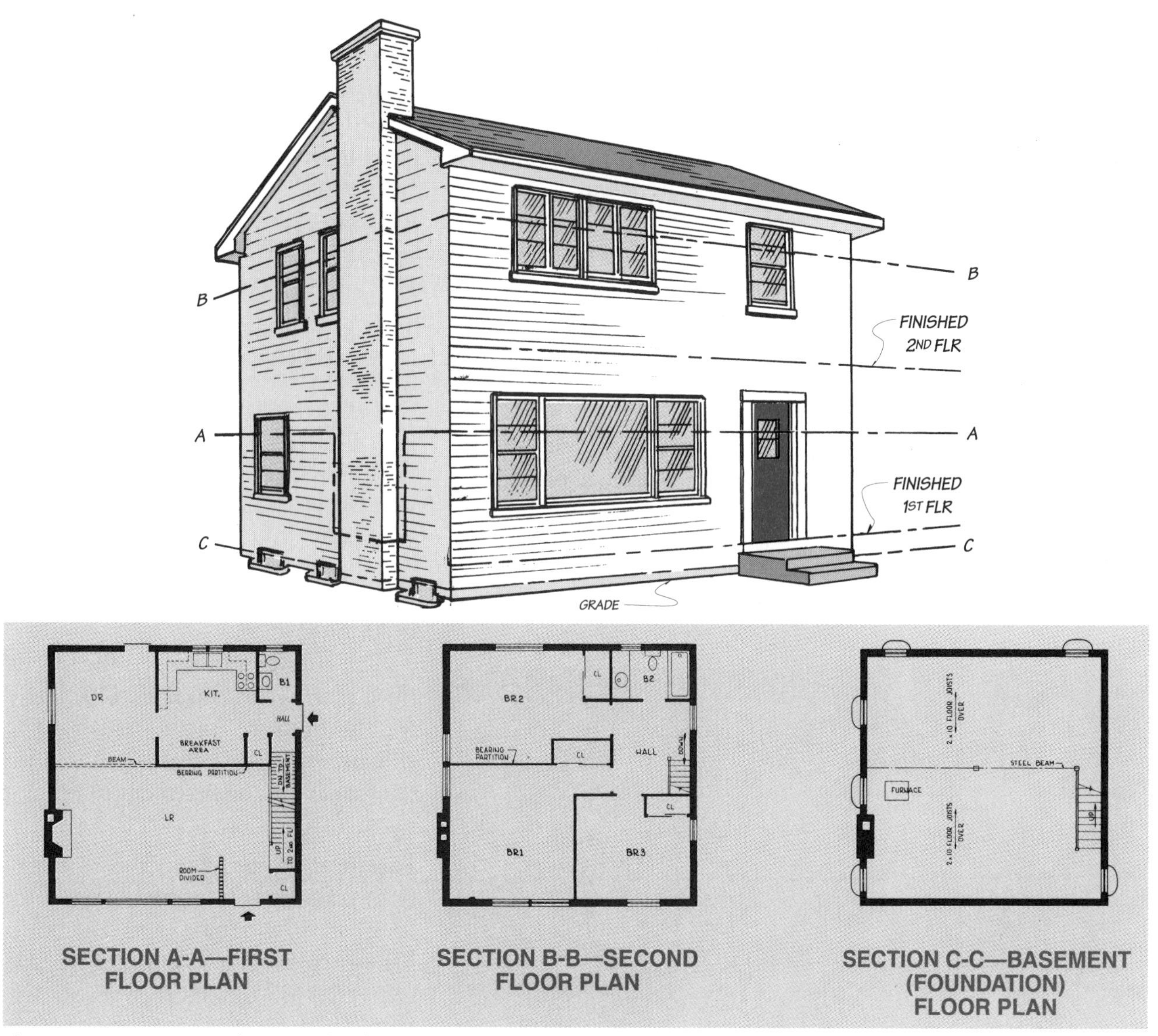

Figure 6-5. A two-story house with a basement has three floor plans.

First Floor Plan. The first floor plan shows that the Living and Dining Rooms are one large L-shaped room. The Kitchen, Bathroom, Breakfast Area, and rear exit occupy the remaining portion of the first floor. A stairway to the second floor starts at a landing one riser above the floor level. The stairway is open with a railing on the Living Room side. Stairs to the basement descend from the rear hallway and are directly below the stairs to the second floor. A landing is provided as a safety measure inside the doorway before the stairs begin. Both stairways stop against break lines.

The construction of the building requires support for second floor joists across the center of the house. A *joist* is a horizontal framing member that supports a floor. A load-bearing partition and a built-up beam support the floor joists. A *built-up beam* is a structural member made of laminated wood members and is designed to carry heavy loads.

The Kitchen has U-shaped base and wall cabinets. Base cabinets are shown with solid lines. Wall cabinets are shown with dashed lines. Note that the wall cabinet near the opening to the Dining Room extends beyond the base cabinet and provides additional storage space above the refrigerator. Additional information about cabinets is usually shown in detail views. Space is provided for a breakfast table along the wall opposite the sink. The appliances are arranged for efficient work flow from the storage area (refrigerator) through the preparation area (sink) to the cooking area (stove). The serving areas (either the Breakfast Area or Dining Room) are conveniently located. The window over the Kitchen sink provides light and ventilation.

The fireplace in the Living Room is drawn with the cutting plane passing 1′-0″ above the floor to show details of its shape on the floor plan. The flue shown is for the basement furnace.

Floor plans provide dimensions for the proper placement of windows.

Second Floor Plan. The second floor plan shows the layout of rooms on the upper level. The top riser of the stairway is shown in its exact location. Treads descend until they stop against a break line. The stairway is open to the hallway and is protected by a railing. The window at the head of the stairs provides light.

Bathroom 2 is located above Bathroom 1 on the first floor. All plumbing fixtures in the house are arranged to be plumbed to a common partition. A small window is located in Bathroom 2.

Bedrooms 1, 2, and 3 all have windows in both exterior walls to provide cross ventilation. All bedroom closets have sliding doors to conserve space.

Door openings are located in walls to provide efficient access for building inhabitants.

A load-bearing partition, directly over the beam shown on the first floor plan, and other necessary structural members (not shown) continue to the wall over the stairs to support the overhead joists.

The chimney shown on the second floor plan shows two flues. One flue is for the fireplace on the first floor to exhaust smoke. The other flue is for the furnace in the basement to exhaust burnt furnace gases.

Basement Floor Plan. The basement floor plan shows information about the foundation, windows, stairs, and building structure. The location of the bottom riser of the stairway is shown. Some of the treads are drawn and terminate against a break line.

The steel beam directly below the load-bearing partition on the first floor supports the joists that run 90° to the beam. The beam terminates on a post at the edge of the stairway and does not continue to the wall, as it would pass through the stairway if it did. The notation 2 × 10 FLOOR JOISTS OVER and the directional symbol indicate the size and direction of the floor joists.

The fireplace foundation is carried down to the footing with only the furnace flue showing. A *footing* is a support base for a foundation wall. Basement windows are located in areaways to provide light and ventilation.

One-and-One-Half-Story House

A one-story house with a steeply sloped roof provides useful attic space that may be converted into living space. For example, the attic space in a traditional Cape Cod house can be used as living space. Dormers can be added to provide additional floor area, ventilation, and architectural effect. A *dormer* is a projection from a sloping roof that provides additional interior area. Three common types of dormers are the gable, hip, and shed dormer. See Figure 6-6.

Two cutting planes are required for a Cape Cod house—one for the first floor and one for the second floor. Each cutting plane is taken 5′-0″ above its finished floor. The cutting plane for the second floor plan reveals the irregular outline of walls with dormers. See Figure 6-7.

READING FLOOR PLANS

The Wayne Residence is a contemporary house with a full basement. (Refer to Wayne Residence, Sheets 1 and 2.) Sheet 1 shows the Foundation/Basement Plan. Sheet 2 shows the Floor Plan.

Foundation/Basement Plan

The title block shows that the Foundation/Basement Plan is drawing number 1 OF 7. This indicates that seven sheets comprise the complete set of plans for the Wayne Residence. The plan was drawn by Pam Hulen (initials PMH) of Hulen & Hulen Designs at the scale of 1/4″ = 1′-0″. The Wayne Residence is to be built on Lot 12 of Country Club Fairways in Columbia, Missouri.

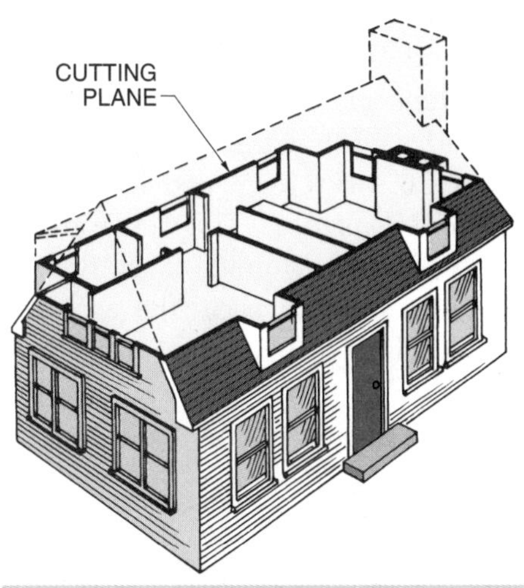

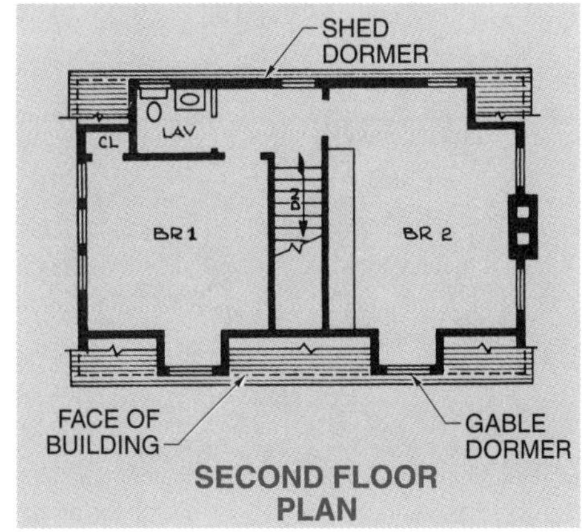

Figure 6-7. Cutting planes on the upper floor of a house with dormers reveal an irregular wall line.

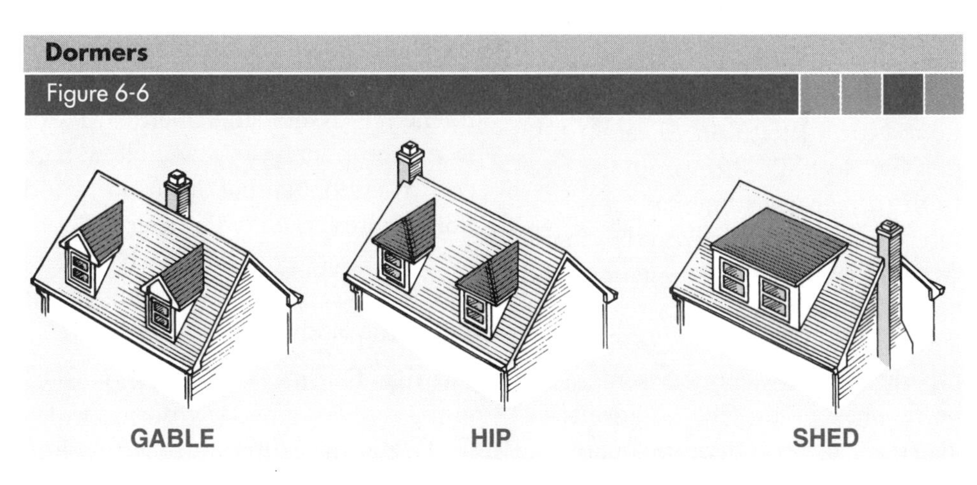

Figure 6-6. Gable, hip, and shed dormers provide additional area for the upper floor.

The title of the drawing—FOUNDATION/BASEMENT PLAN—is shown below the plan and the scale (1/4″ = 1′-0″) is repeated below the title. An arrow showing North indicates the orientation of the house and is used when referring to exterior elevation views of the house.

Foundation. The foundation walls are 8″ and 10″ wide while the footings are 16″ and 20″ wide. (Refer to Basement Wall and Walkout Details, Sheet 5.) Brick veneer is shown on the south and east sides of the foundation wall. A stepped 6″ brick ledge provides a base for the brick veneer. The west foundation wall is below grade level. Retaining walls are shown on the east and south walls to compensate for the drop in grade due to the slope of the lot. (Refer to Elevations, Sheets 3 and 4.)

Foundation walls can be formed using conventional wood or metal forms or by using insulating concrete forms (ICFs).

The north wall is the rear wall. (Refer to North Elevation, Sheet 4.) A lot that has an 8′-0″ drop in grade from front to rear gives builders an opportunity to provide a walkout basement. A *walkout basement* is a basement with standard-sized windows to provide light and standard-sized doors for entry and exit. A walkout basement can be finished to provide additional living space at a competitive square foot cost. The lot slope and use of retaining walls allows for a walkout basement in the Wayne Residence. A window well provides clearance for an awning window in the Future Workshop. (Refer to Window Schedule, Sheet 5.)

> Steel or plastic reinforcement is used in a building foundation to provide additional strength. Rebar, welded wire reinforcement, or plastic fibers are embedded in the concrete.

Steel columns 4″ in diameter are placed in foundation footings to carry the beams that support the floor joists. The W (wide flange) beams are described by their nominal measurement over the flanges and their weight per running foot. A W8 × 10 beam is 8″ high and weighs 10 pounds per running foot.

Slab. The typical note 4″ CONC SLAB W/ 6 × 6—W1.4 × W1.4 WWR ON GRAVEL BASE is shown for the basement floor. The portion of the note 6 × 6—W1.4 × W1.4 WWR refers to a specific size of welded wire reinforcement. Welded wire reinforcement provides additional strength to the concrete. This note is repeated for the unfinished portion of the basement and for the Garage slab.

Stairway. Entry to the basement from the main floor of the house is by the stairs. A 2′-6″ × 6′-8″ × 1⅜″ hollow-core, six panel wood door is at the foot of the stairs. Panels in these doors are formed from moldable fiberboard with wood components used in hollow-core construction. Natural or various wood finishes are available. The stairway contains 14 risers at 7.392″ each. Three-way switches control a ceiling outlet above the landing.

Family Room. A 4′-9″ × 4′-0½″ vinyl-clad casement window and a 6′-0″ × 6′-8″ aluminum-clad sliding glass door provide natural light to this room. Electrical receptacles are located in all walls. The ceiling outlet is shown with its switch leg. A telephone outlet, represented by a triangle, is shown on the west wall.

Bathroom. A 2′-4″ × 6′-8″ × 1⅜″ hollow-core, six panel wood door leads from the Family Room to the Bathroom. The Bathroom contains a water closet, vanity with lavatory, and bathtub. A ground-fault circuit interrupter (GFCI or GFI) receptacle is located near the lavatory, and wall switches control a combination ceiling outlet and fan.

Study. A 2′-6″ × 6′-8″ × 1⅜″ hollow-core, six panel wood door leads to the Study. Casement windows provide light and ventilation. Electrical receptacles are located in framed walls. A switch located near the door controls the ceiling light. Future plans of the owner include enclosing the column and steel beam and adding a floor-to-ceiling combination storage and bookcase unit in the recessed area on the east wall. Alternate plans include using the Study as a fourth Bedroom.

Unfinished Basement. The Storage Area and Future Workshop are left unfinished to reduce the initial cost of the house. The water heater, furnace, and fireplace foundation with flue are completed. A floor drain is located near the water heater. The power panel is located on the rear wall.

Registers for supply and return air are not shown on this set of plans. The mechanical subcontractor will design and install the heating and cooling system. The contractor will coordinate the mechanical subcontractor's work with that of other trades to assure that proper electrical and structural requirements are met.

Floor Plan

Title block information given on the Foundation/Basement Plan is repeated in the title block of the Floor Plan. The Floor Plan is drawn to the scale of ¼″ = 1′-0″. The arrow shows that the house faces south.

Overall dimensions of the house are 44′-4″ × 72′-8″. Dimensions to the centers of the openings are given for locating doors and windows. A 24′-0″ × 12′-0″ Deck is located on the north side of the house. Doors from Bedroom 1 and the Living Room open to the Deck.

The drawing scale shown with a plan view, elevation, or section takes precedence over the scale shown in the title block.

Southern Forest Products Association

Locations of interior load-bearing and non-load-bearing partitions are indicated on floor plans.

The main floor of the Wayne Residence contains the following rooms:

- Entry
- Hallway
- Bedroom 1
- Bedroom 2
- Bedroom 3
- Bathroom
- Master Bathroom
- Living Room
- Dining Room
- Kitchen
- Breakfast Area
- Laundry Room
- Screened Porch
- Garage

Entry and Hallway. The house is entered from a covered stoop through a 3′-0″ × 6′-8″ × 1¾″ exterior door. The door is metal insulated and includes a deadbolt. Light is provided to the entry by 15″ sidelights. Wall switches control exterior and interior ceiling outlets and the ceiling outlet in the coat closet. A rod and shelf are shown in the coat closet.

The hallway provides access to the basement, bathroom, and bedrooms. Dashed lines show a 30″ × 30″ scuttle in the ceiling. A *scuttle* is an access opening in a ceiling or roof with a removable or movable cover. Dashed lines in the linen closet indicate shelf rods. Two ceiling lights controlled by three-way switches provide lighting for the hallway.

Wood or steel framing members may be used to construct the framework of a building.

Bedrooms. Three 2′-6″ × 6′-8″ × 1⅜″ hollow-core, six panel wood doors lead to the bedrooms. Bedroom 1 is the largest bedroom. It measures 11′-9½″ × 17′-0″. Bedrooms 2 and 3 are 11′-2″ × 12′-3½″. Electrical receptacles, ceiling lights with switch legs, and telephone outlets are shown in each bedroom. The closet in Bedroom 2 has bifold doors. Bedrooms 1 and 3 have walk-in closets. A small linen closet is also shown for Bedroom 1.

Bathrooms. The Master Bathroom contains a water closet, bathtub, and shower separated from the double-bowl lavatory. Wall cabinets, mirror, and soffit are shown in Elevations 2-5 and 3-5. Wall switches control ceiling lights in the Master Bathroom. A combination ceiling light and fan is also controlled by a wall switch. Ground-fault circuit interrupter receptacles are placed near the lavatories.

The hallway Bathroom contains a lavatory, water closet, and bathtub/shower combination. A wall switch controls the combination ceiling light and fan. A GFCI receptacle is placed near the lavatory.

> A ground-fault circuit interrupter (abbreviated as GFI or GFCI) is a safety device that automatically de-energizes a circuit when the grounded circuit exceeds a predetermined value.

Living Room. The 19′-4½″ × 21′-6″ Living Room is located at the rear of the house. Four aluminum-framed fixed windows allow light to enter while providing a view of the backyard. The Deck, which is adjacent to the Living Room, is reached by a full glass door.

A masonry fireplace defines the south portion of the Living Room. Cutting planes refer to Fireplace Details 1-6 and 2-6, which contain additional information.

Southern Forest Products Association
Interior load-bearing partitions run perpendicular to ceiling joists, while non-load-bearing partitions run parallel to ceiling joists.

Recessed ceiling lights are controlled by three-way switches located on the east wall near the stair landing and near the patio door. Electrical receptacles are spaced around walls and wired so as not to interfere with the location of a future door to the Breakfast Area. A header is provided for the future door. Dashed lines show a gypsum board wrap on the header over the opening leading to the Kitchen. A telephone outlet is shown on the west wall.

Dining Room. The panned ceiling with recessed lighting provides a formal look to the Dining Room. A *panned ceiling* is a ceiling consisting of two ceiling levels connected by sloped surfaces. (Refer to the Panned Ceiling Detail, Sheet 6.) Vinyl-clad casement windows provide light and ventilation.

Kitchen and Breakfast Area. The Kitchen is arranged with L-shaped, straight-run, and island base cabinets. Wall cabinets are indicated by dashed lines. Sheet 6 includes several Kitchen elevations showing the cabinet arrangement. The arrangement of storage, preparation, and cooking areas is compact and efficient. A desk on the east wall provides planning space. The telephone outlet is located in the wall above the desktop.

Ground-fault circuit interrupter receptacles are located to serve the countertop. A wall switch controls the disposal in the sink. A range receptacle serves the drop-in stove in the island cabinet. The stove has a downdraft vent system to remove cooking odors and smoke. Recessed ceiling fixtures over the bar cabinet are controlled by a single-pole switch. A three-way switch also controls the ceiling outlet in the Breakfast Area. A clock outlet is shown on the south wall. A vinyl-clad casement window is above the Kitchen sink.

The Laundry, Screened Porch, and Garage are entered from the Kitchen. (Refer to Door Schedule, Sheet 5 for types and sizes of doors.)

Laundry. The Laundry contains space for the washer and dryer with wall cabinets above. A 240 V receptacle supplies the dryer. The Laundry includes a pantry for additional storage. A pull chain lighting outlet is shown. A casement window is centered in the west wall of the Laundry.

Dashed horizontal lines in base cabinets indicate shelves.

Clothes washer outlet boxes provide water supply and waste connections for a clothes washer.

A weatherproof GFCI receptacle includes a cover to prevent moisture from entering the receptacle.

Screened Porch. The 10′-4″ × 14′-0″ Screened Porch has a 4″ concrete slab that is sloped away from the house. (Refer to Foundation/Basement Plan, Sheet 1.) A screened door leads to the backyard, and a wood center-hinged patio door leads into the Breakfast Area. A weatherproof GFCI receptacle is shown on the east wall.

Garage. The 23′-4″ × 21′-2½″ Garage has a 4″ concrete slab that slopes 2″ from the north wall to the Garage door. (Refer to Foundation/Basement Plan, Sheet 1.) A steel beam lintel provides a base for brick veneer above the overhead door.

A frostproof hose bibb (FP HB), or sillcock, is located on the north wall, and GFCI-protected electrical receptacles are located on the north, east, and west walls. Switches located at the two doors control lighting. The overhead Garage door will be operated by a garage door opener connected to the switched receptacle in the ceiling.

Name ______________________________ Date ______________

Sketch the following on Warner Residence—Floor Plan. For example, the ceiling outlet (13) in the Dining Room is controlled by two three-way switches (15). *Note:* All exterior walls are brick.

1. Brick exterior wall
2. Double-hung window
3. Casement window
4. Exterior door
5. Interior door
6. Pocket door
7. Accordion door
8. Sliding door
9. Bifold doors
10. Cased opening
11. Water closet
12. Telephone
13. Ceiling outlet
14. Recessed ceiling outlet
15. Three-way switch
16. 120 V receptacle
17. 240 V receptacle
18. 120 V GFI receptacle
19. Pull chain
20. Weatherproof GFI receptacle
21. Ceiling exhaust
22. Kitchen sink
23. Bathtub
24. Hose bibb
25. Shelf and rod

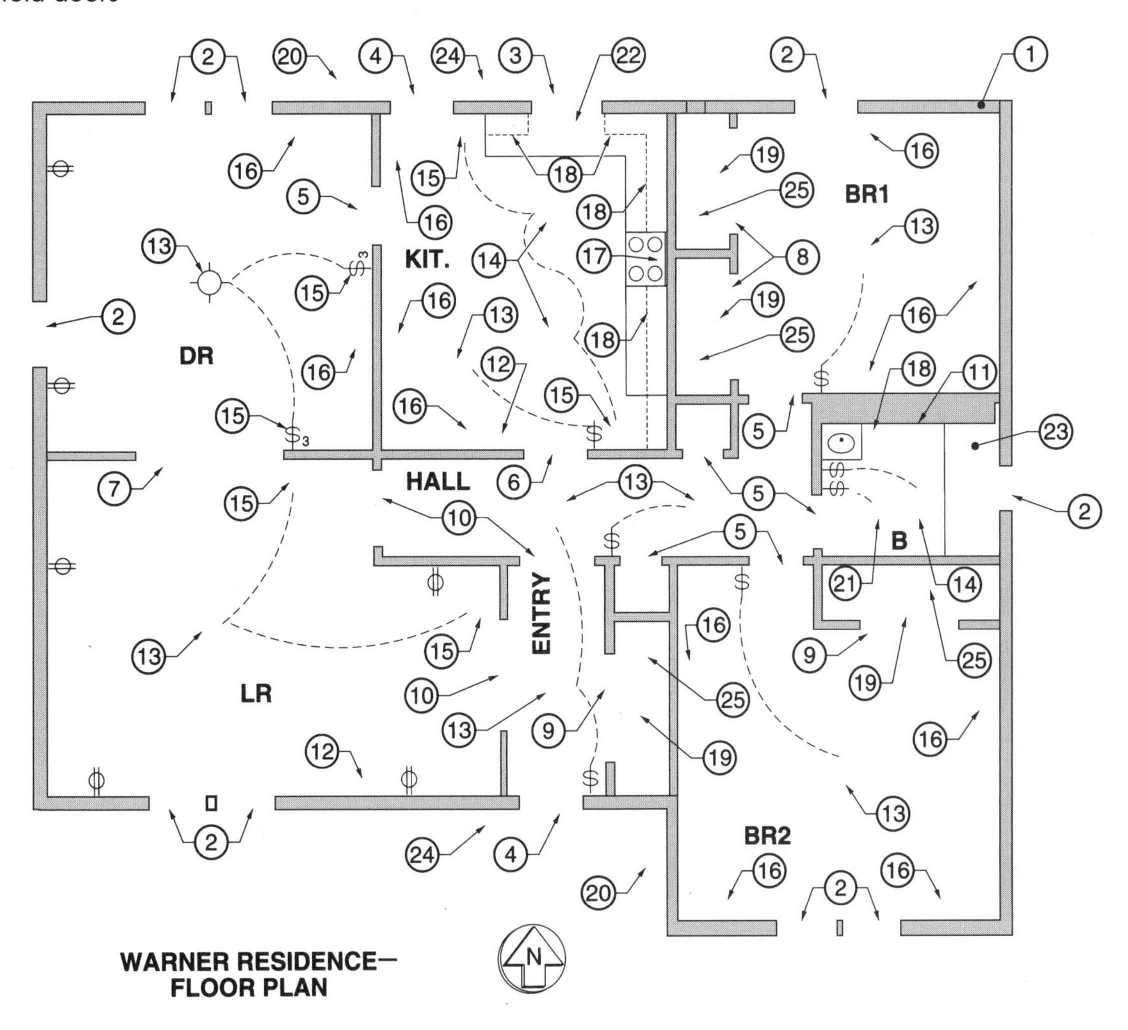

**WARNER RESIDENCE—
FLOOR PLAN**

Floor Plans 6

Review Questions

Name ______________________ Date ____________

True-False

T F **1.** A portion of the stairway from the first to the second floor is shown on both floor plans.

T F **2.** Exterior elevations are generally drawn to a smaller scale than floor plans.

T F **3.** The broad aspects of shape, size, and relationship of rooms are shown on floor plans.

T F **4.** Layering on CAD drawings refers to the total area of the house.

T F **5.** Electrical fixtures and outlets on floor plans are shown with abbreviations only.

T F **6.** Cutting planes for floor plans are taken 3′-0″ above the finished floor.

T F **7.** Floor plans may be oriented to fit the print sheet.

T F **8.** Stairways in plan views are terminated on a break line.

T F **9.** The note 2 × 10 JOISTS OVER on a first floor plan indicates that the joists are overhead and support the second floor.

T F **10.** Floor plans are generally drawn so that the front view of the house is toward the bottom of the sheet.

T F **11.** Abbreviations are not used on floor plans.

T F **12.** The cutting plane for second floor plans passes through the house at the upper sash of the windows on the main floor.

T F **13.** A two-story house with basement requires three floor plans.

T F **14.** A joist is a horizontal framing member that supports the floor.

T F **15.** The tread is the vertical portion of a stair step.

T F **16.** Floor plans are generally drawn to the scale of ¼″ = 1′-0″.

T F **17.** A typical note gives information pertaining to all similar items on the floor plan.

T F **18.** Dashed lines on floor plans show features above the cutting plane.

T F **19.** The scale for each room of a floor plan varies to fit the size of the room.

T F **20.** The first floor plan is generally studied before second floor or basement floor plan.

Printreading 6-1

Refer to Torrance Residence on page 139.

_______________ **1.** A ___ house is shown.
A. one-story
B. one-story with basement
C. one-and-one-half story
D. two-story with basement

_______________ **2.** The Living Room measures ___ × ___.

_______________ **3.** The arch between the Entry and the Living Room is shown with ___ lines.

_______________ **4.** The upstairs Bathroom contains a ___.
A. shower, tub, and water closet
B. vanity, tub, and water closet
C. vanity, shower, and water closet
D. vanity and water closet only

T F **5.** Part of the ceiling in Bedrooms 1 and 2 is sloped.

T F **6.** A dormer is shown on the rear of the house.

T F **7.** A dormer is shown on the front of the house.

T F **8.** The face of the rear dormer is directly above the face of the first floor wall line.

_______________ **9.** The house is ___ wide across the front.

_______________ **10.** The Patio is laid with ___.
A. brick
B. stone
C. stone with brick details
D. brick with stone details

_______________ **11.** Bedroom ___ is the largest bedroom.

_______________ **12.** Two built-in ___ are shown on the upper floor.

T F **13.** Two risers are shown from the grade to the front door.

T F **14.** Stairs to the upper floor begin in the Living Room.

T F **15.** Four exterior doorways are shown on the first floor plan.

_______________ **16.** The Dining Area is ___.
A. part of the Living Room
B. adjacent to the Patio
C. adjacent to the Kitchen
D. all of the above

_______________ **17.** The Family Room has ___.
A. two windows on the front wall
B. two windows on the side wall
C. one window on each wall
D. no windows

_______________ **18.** The house measures ___ from front to back.

T F **19.** The Kitchen sink is centered on the Kitchen windows.

T F **20.** The upstairs Bathroom measures 5′-6″ × 7′-0″.

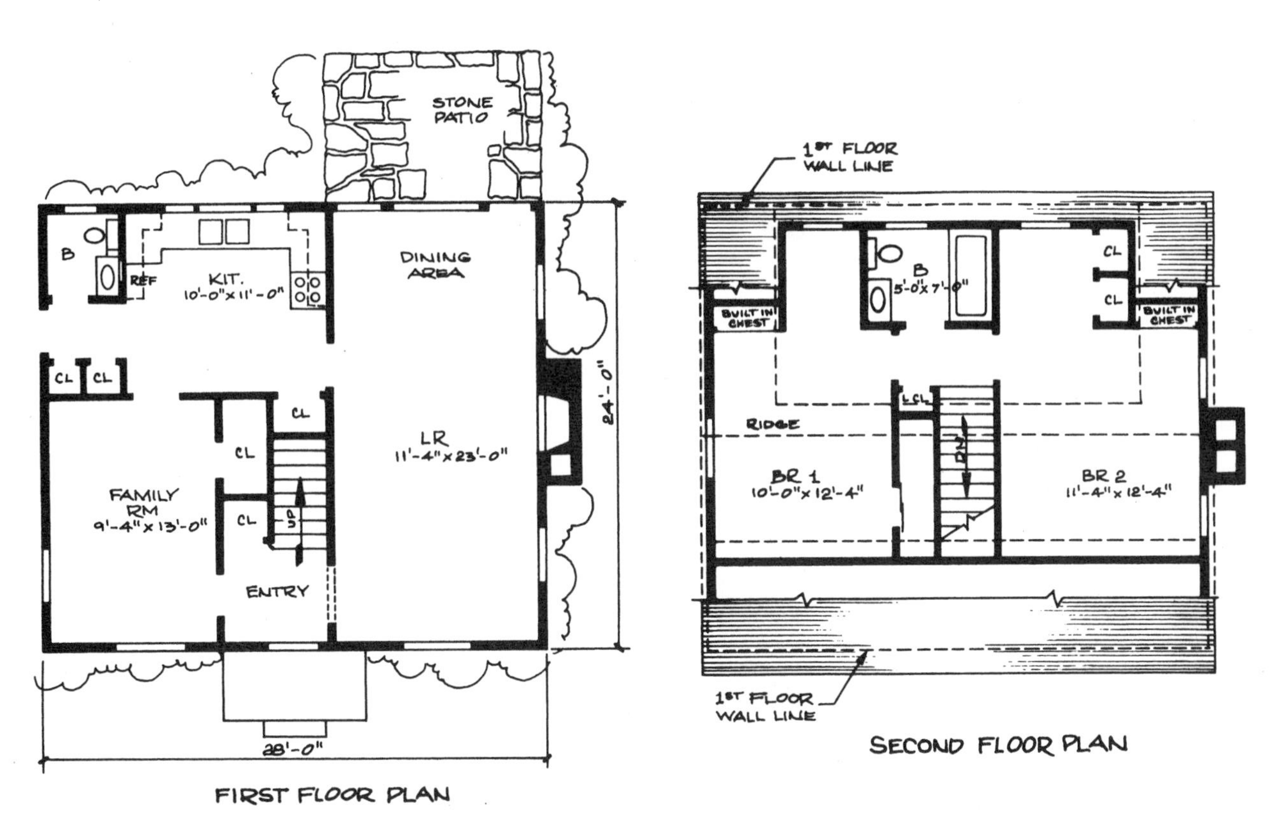

TORRANCE RESIDENCE

Printreading 6-2

Refer to Chapman Residence on page 140.

_______________ **1.** The house has ___.

A. a full basement
B. two fireplaces
C. a two-car garage
D. two levels

_______________ **2.** The Living Room ___.

A. is separated from the Dining Room by a room divider
B. has direct access to Bedroom 1
C. has a fireplace on the west wall
D. is part of a Living-Dining Room area

_______________ **3.** The Dining Area has ___.

A. one awning window
B. a swinging door to the Kitchen
C. sliding glass doors to the Patio
D. two casement windows

_____________ **4.** The Kitchen ___.

A. has wall cabinets above the sink
B. has two flush exterior doors
C. is U-shaped
D. also serves as a Laundry

_____________ **5.** The hallway ___.

A. opens to all bedrooms
B. has no door that swings out into it
C. is L-shaped
D. has two linen closets

_____________ **6.** Regarding closets, ___.

A. Bedroom 1 has two closets
B. linen closets have sliding doors
C. Bedrooms 2 and 3 have closets of equal size
D. Bathroom 2 has one linen closet

_____________ **7.** Regarding plumbing, ___.

A. all fixtures in the bathrooms are located along the same wall
B. two tubs are shown
C. the Kitchen sink is located under the Kitchen window
D. the washing machine is located on the east wall of the Kitchen

_____________ **8.** The South Elevation has ___ windows.

A. double-hung
B. hopper
C. casement
D. awning and horizontal sliding

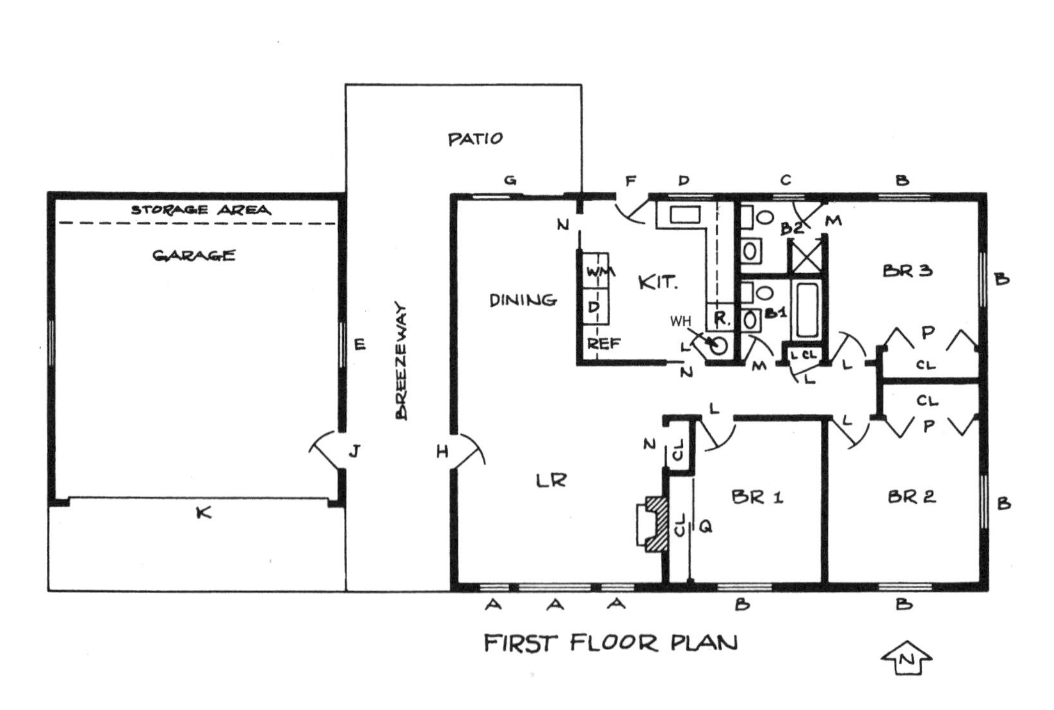

WINDOW & DOOR SCHEDULE	
A	AWNING
B	HOR SLIDING
C	DH
D	DH
E	DH
F	FL DOOR 1 LT
G	SL GL DOORS
H	FL DOOR 2 LT
J	PANEL DOOR
K	GARAGE DOOR
L	FL DOOR
M	FL DOOR
N	SL DOOR
P	BI-FOLD DOOR
Q	SL DOOR

ABBREVIATIONS	
W/M	WASHING MACHINE
D	DRYER
REF	REFRIGERATOR
WH	WATER HEATER
R	RANGE
CL	CLOSET
LCL	LINEN CLOSET
LAV	LAVATORY
HOR	HORIZONTAL
DH	DOUBLE HUNG
FL	FLUSH
SL	SLIDING
GL	GLASS

CHAPMAN RESIDENCE

_______________ **9.** The East Elevation has ___ windows.

A. awning
B. horizontal sliding
C. double-hung and fixed sash
D. double-hung and casement

_______________ **10.** The North Elevation has ___ and ___ windows.

A. horizontal sliding; double-hung
B. horizontal sliding; casement
C. double-hung; casement
D. awning; double-hung

Identification

Refer to Appendix.

_______________ **1.** Ceiling outlet

_______________ **2.** Bifold doors

_______________ **3.** Double-hung window

_______________ **4.** 240 V receptacle outlet

_______________ **5.** Three-way switch

_______________ **6.** Common brick

_______________ **7.** Telephone

_______________ **8.** Weatherproof receptacle

_______________ **9.** 120 V receptacle outlet

_______________ **10.** Double-acting door

Ⓐ Ⓑ Ⓒ Ⓓ

Ⓔ Ⓕ S_3 Ⓖ

WP Ⓗ Ⓘ Ⓙ

Floor Plans 6

Trade Competency Test

Name ______________________________ Date ______________

PRINTREADING 6-1

Refer to Wayne Residence—Foundation/Basement Plan, Sheet 1.

True-False

T F **1.** The Porch slab is sloped away from the house.

T F **2.** The power panel is located directly beside the furnace.

T F **3.** Five 120 V receptacles are shown in the Study.

T F **4.** An areaway is shown on the east wall.

T F **5.** All exterior walls are brick veneer.

T F **6.** A frostproof hose bibb is located on the north wall.

T F **7.** Three retaining walls are shown on the exterior of the house.

T F **8.** The Garage slab is 4″ thick.

T F **9.** The basement contains a total of 4 doors.

T F **10.** The Foundation/Basement Plan is drawn to the scale ¼″ = 1′-0″.

T F **11.** All steel beams run east to west.

T F **12.** A three-way switch controls the ceiling outlet above the stair landing.

T F **13.** The overall length of the house is 72′-8″.

T F **14.** Casement windows are shown in the Study.

T F **15.** The water heater is located in the utility closet.

T F **16.** A #6 door is located at the foot of the stairway.

T F **17.** The foundation wall at the front of the Garage is notched 6″ for floor slope.

T F **18.** The Garage slab slopes 2″ to the door.

T F **19.** Steel posts in concrete footing pads in the Storage Area are 9′-2″ OC.

T F **20.** Three recessed ceiling fixtures in the family room are controlled by three-way wall switches.

PRINTREADING 6-2

Refer to Wayne Residence—Floor Plan, Sheet 2.

Multiple Choice

________________ **1.** Regarding the Dining Room, ___.

A. six electrical receptacles are spaced on four walls
B. awning windows are paired on the south wall
C. a panned ceiling is shown
D. wall openings are arched

________________ **2.** Regarding the stairway, ___.

A. 16 R DN is shown
B. the stairs are straight run
C. an interior door is located at the top of the stairs
D. tread height is 7.392″

________________ **3.** Regarding the Master Bathroom, ___.

A. a combination tub-shower is shown
B. the overall size is 7′-3½″ × 8′-8″
C. lighting is provided by a switch-chain outlet
D. a 3′-0″ interior door is shown

________________ **4.** Regarding the Kitchen, ___.

A. a partition separates the Kitchen and Breakfast Area
B. a double-bowl sink is shown in the island cabinet
C. a clock outlet is shown on the west wall
D. three-way switches control a light fixture above the island

________________ **5.** Regarding the Kitchen, ___.

A. GFI receptacles serve the countertop
B. a telephone outlet is located in the west wall
C. all of the above
D. none of the above

________________ **6.** Regarding the Laundry, a ___.

A. casement window is shown in the south wall
B. casement window is shown in the west wall
C. double-hung window is shown in the south wall
D. double-hung window is shown in the west wall

________________ **7.** Regarding the Laundry, ___.

A. lighting is provided by a recessed fixture
B. a floor drain is centered in the room
C. entry is from the Screened Porch
D. the pantry has a pull chain light fixture

________________ **8.** Regarding the bedrooms, ___.

A. all bedrooms are the same size
B. Bedroom 2 and Bedroom 3 are the same size
C. Bedroom 1 is smaller than Bedroom 2
D. none of the above

_______________ **9.** Regarding the bedrooms, ___.

A. Bedroom 1 is located in the northwest corner of the house
B. telephone outlets are shown in each bedroom
C. sliding windows are shown in each bedroom
D. 28″ interior doors lead to the hallway

_______________ **10.** Regarding the bedrooms, ___.

A. Bedroom 2 has a walk-in closet
B. Bedroom 3 has a walk-in closet
C. Bedroom 1 has two linen closets
D. all closets are the same size

_______________ **11.** Regarding the bedrooms, the Deck may be entered directly from ___.

A. Bedroom 1
B. Bedroom 2
C. Bedroom 3
D. none of the above

_______________ **12.** Regarding the Living Room, ___.

A. a cased opening separates the Living Room from the Dining Room
B. casement windows provide a view of the backyard
C. the handrail for the stairway is on the west wall
D. recessed ceiling fixtures are controlled by three-way switches

_______________ **13.** Regarding the Garage, ___.

A. a frostproof hose bibb is located on the north wall
B. the interior light is controlled by three-way switches
C. a steel beam lintel is placed over the Garage door
D. all of the above

_______________ **14.** Regarding the Screened Porch, ___.

A. two casement windows are shown in the west wall
B. a screen door is shown on the east wall
C. its overall size is 10′-4″ × 14′-0″
D. entry to the house is through the Laundry

_______________ **15.** Regarding the Entry, ___.

A. a bank of wall switches controls the Porch light, ceiling light, and Entry closet light
B. sidelights are shown to the left of the exterior door
C. a telephone outlet is located near the Entry closet
D. ceramic tile is shown on the floor

PRINTREADING 6-3

Refer to Wayne Residence—Foundation/Basement Plan and Floor Plan, Sheets 1 and 2.

Completion

_______________ **1.** Floor plans for the Wayne Residence are drawn to the scale of ___.

_______________ **2.** The front of the house faces ___.

_______________ **3.** The Garage is 21′-2½″ × ___.

_______________ **4.** Access to the attic is through a(n) ___ scuttle.

_______________ **5.** The concrete footing pad for the fireplace measures ___.

_______________ **6.** Seven ___″ columns support the steel beams.

_______________ **7.** Two ___ outlets provide lighting in the Future Workshop.

_______________ **8.** The front foundation wall is stepped down to provide a(n) ___ ledge.

_______________ **9.** The Entry is ___ wide.

_______________ **10.** The hall bath on the main floor has ___ GFI receptacle(s).

_______________ **11.** ___ lavatories are shown in the vanity for the Master Bedroom.

_______________ **12.** ___ linen closets are shown on the main floor.

_______________ **13.** Bedroom windows on the south wall measure ___ C to C.

_______________ **14.** The Dining Room measures 13′-4½″ × ___.

_______________ **15.** A(n) ___″ concrete pier is shown below the front stoop.

_______________ **16.** The linen closet at the head of the stairs is ___ deep.

_______________ **17.** Lighting for the Deck is controlled by wall switches located in Bedroom 1 and the ___.

_______________ **18.** The Garage contains ___ GFI receptacles.

_______________ **19.** The east foundation wall is ___″ thick.

_______________ **20.** A microwave oven is located on the ___ wall of the Kitchen.

_______________ **21.** The light fixture in the walk-in closet in Bedroom 3 is operated by a(n) ___.

_______________ **22.** Bedroom 1 measures 11′-9½″ × ___.

_______________ **23.** The Wayne Residence measures ___ across the front.

_______________ **24.** The fireplace opening faces ___.

_______________ **25.** The ___ bath is the only bathroom with a separate shower.

Elevations 7

Elevations (or elevation views) show the exterior of a house or other building. A minimum of four exterior elevations is required to show all sides of a house. The basic design of the house and the type of roof are shown. Elevations are directly related to floor plans and are generally drawn to the same scale. Symbols and notations are used to simplify elevations. In some cases, interior elevations may also be included in a set of plans to show interior wall details.

ELEVATIONS

An elevation is a scaled orthographic projection showing the walls of a building. Elevations may be drawn as exterior or interior elevations and may also show features of the building. Four exterior elevations, each showing a different side of the building, are part of the working drawings typically prepared by the architect. Exterior elevations show the house design, the location of openings, and the materials to be used.

North, East, South, and West are the four major directions of the compass. Compass directions are used to designate exterior elevations. The North Elevation shows the side of the building facing north, not the direction a person faces to see that side of the house. The East Elevation shows the side of the house facing east, the South Elevation shows the side of the house facing south, and the West Elevation shows the side of the house facing west. See Figure 7-1.

The prints in a set of working drawings all relate to each other. It is often necessary to refer to several print sheets to find all the information on one subject. A complete and comprehensive set of prints is drawn so that the foundation plan, floor plans, and elevations exactly match regarding the location of windows, doors, and other details. Exterior elevations provide a significant amount of information on completing the exterior work on a construction project to tradesworkers.

Elevations provide a general sense of the building design by showing the location of offsets, patios, stairways, bays, dormers, chimneys, and other features. Elevations may also indicate the dimensions and materials to use for footings and foundations.

James Hardie Building Products

Elevations provide a general sense of the building design and show the relationship between building features.

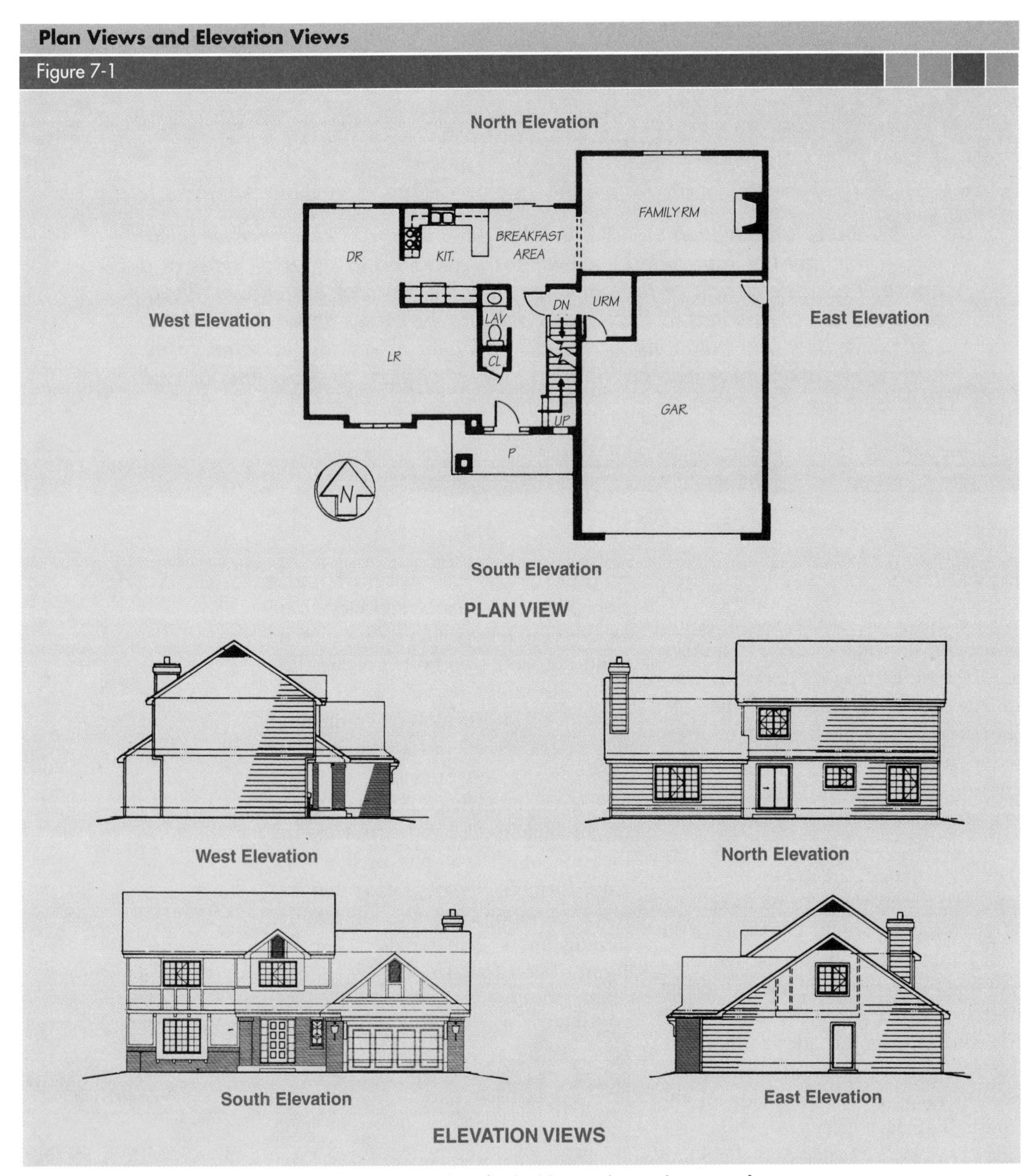

Figure 7-1. Exterior elevations show the exterior sides of a building without adjustments for perspective.

Roof information on elevations includes type of roof design (gable, hip, shed, gambrel, flat, or a combination of these styles), slopes, materials to be used, vents, gravel stops, projection of eaves, and soffits.

Most residential roofs in North America are covered with asphalt shingles. Asphalt shingles are easy to apply, are economical, and have a life expectancy of 15 to 25 years.

Window and door openings are shown on elevations. For windows, various symbols show window types, sizes, swings, and locations. In a similar manner, standard symbols for doors indicate the types of doors, sizes, number of lights in each door, and locations.

Typically, elevations contain very few dimensions. Dimensions that are shown on elevations are usually vertical dimensions that include established grade to finished basement floor, established grade to finished first floor, heights from floor to floor, heights of special windows above a floor, and heights from roof ridges to tops of chimneys. These dimensions are usually provided in feet and inches.

Exterior finish materials appear on exterior elevations and are noted with symbols and abbreviations. Finish materials include types of siding, concrete, concrete masonry units, brick, stone, stucco, exterior insulation and finish systems (EIFS), or other materials. Treatment of exterior trim around windows, entrance doorways, columns, posts, balustrades, cornices, and other features are also shown on exterior elevations. Other miscellaneous details shown on exterior elevations include electrical fixtures, utility outlets, hose bibbs, gutters and downspouts, flashing, and waterproofing.

Symbols

Symbols showing building materials and fixtures save space on elevations. Most elevations for residential construction are drawn to the scale of ¼″ = 1′-0″, which is 1/48 the size of the actual building. A scale such as ¼″ = 1′-0″ dictates the use of symbols to clearly show the materials to be used. See Figure 7-2.

Window and door schedules supplement information pertaining to windows and doors shown on elevations.

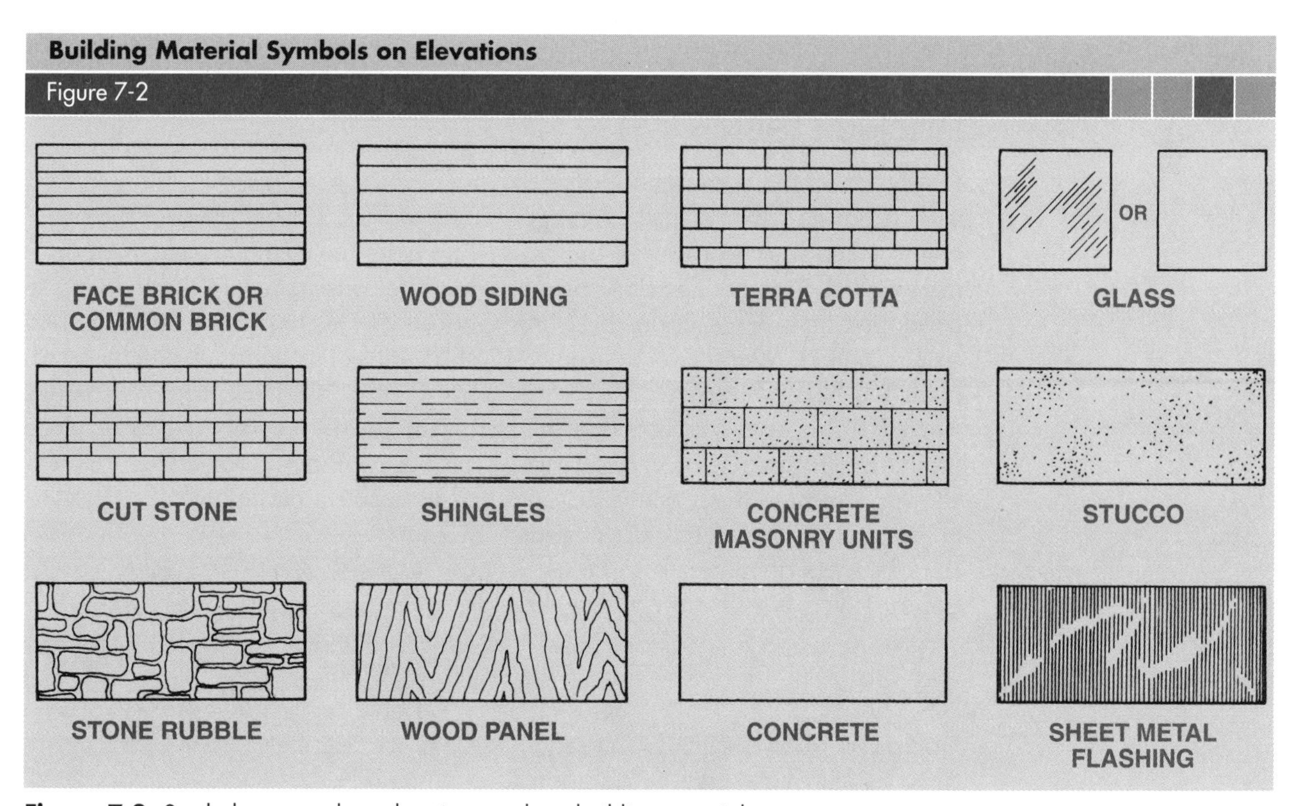

Figure 7-2. Symbols are used on elevations to show building materials.

Abbreviations

Abbreviations are also used on elevations to conserve space. Abbreviations may be used alone, in notations, or with symbols to describe materials. Standard abbreviations are typically used to avoid misinterpretation. Only uppercase letters are used in abbreviations. Abbreviations that form a word are followed by a period. For example, IN. is the abbreviation for inch.

California Redwood Association

A contemporary design includes more complex rooflines, geometric windows, and curved or angled windows and walls.

BUILDING DESIGN

Houses are most commonly designed with room and living space arrangements taken into account as the primary design element. The floor plan is developed first, based upon the living space required. After the room arrangement has been determined, the exterior of the house is planned. Local ordinances regarding lot size, setback, allowed exterior finish materials, and house size are considered. Client preferences, types of materials to be used, and overall costs are carefully reviewed.

Styles

Houses are often designed in a particular style. This style may be traditional or contemporary. A *traditional design* is an architectural design that reflects long-standing design elements. For example, Cape Cod, Colonial, and Old English designs remain popular today because of their design elements. Traditional design elements commonly include horizontal and vertical window alignment and straight walls with few angular offsets. A *contemporary design* is an architectural design that reflects current trends and may include more complex rooflines, geometric windows, and curved or angled walls.

Details such as the general proportions, roof type, windows, and trim must be consistent with the design. Elevations show the elements of style and design. Whenever a change is shown in a wall or roofline, a modification of the basic rectangular shape of the house is indicated on the elevation.

All visible lines on elevations are drawn as object (solid) lines. Any part of the building below grade level is shown with hidden (dashed) lines. For example, foundation footings, walls below grade, and areaways (window wells) for below-grade windows are shown with hidden lines.

Roofs

The roof of a house protects the structure from the elements. The roof style must be consistent with the style of the house. The six basic roof styles are flat, shed, gable, hip, gambrel, and mansard. See Figure 7-3. These roof styles may be modified or combined to produce various rooflines. Roof styles are most easily seen on elevations.

Roof slope is typically shown on a drawing using a pitch symbol. However, fractional pitch may be used on drawings to indicate roof slope. Fractional pitch is calculated by dividing the total rise by the total span.

A flat roof must slope at least ⅛″ per foot of run for proper water runoff. Shed roofs slope in one direction (indicated with arrow) with opposing walls constructed to different heights. Gable roofs slope in two directions and are the most common roof style. Hip roofs slope in four directions. All walls of buildings with hip roofs are the same height. Gambrel roofs have a double slope in two directions. Gambrel roofs are a popular roof style on barns and country-style houses. Mansard roofs have a double slope in four directions and are used for multistory dwellings.

Slope. Roofs must slope to provide proper water runoff. *Slope* is the relationship of unit rise to unit run of a roof. *Unit rise* is the vertical increase in height per foot of run. *Unit run* is a unit of the total run based on 12″. Roof slope is shown on elevations with the pitch symbol. See Figure 7-4.

Roof slope may also be expressed as pitch. *Pitch* is the angle a roof slopes from the roof ridge to the outside walls of the building. Pitch is expressed as a fraction.

Elevations provide information about finish roofing materials.

Roof Styles

Figure 7-3

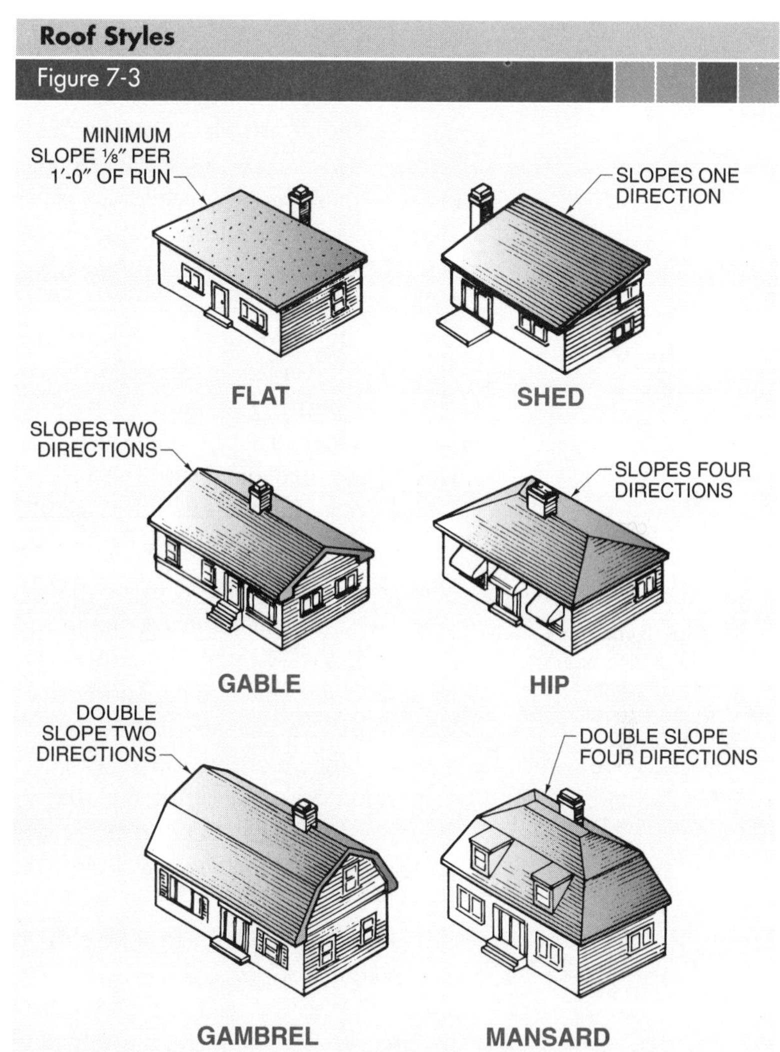

Figure 7-3. Six basic roof styles are used in residential construction. Roof styles are shown on elevations.

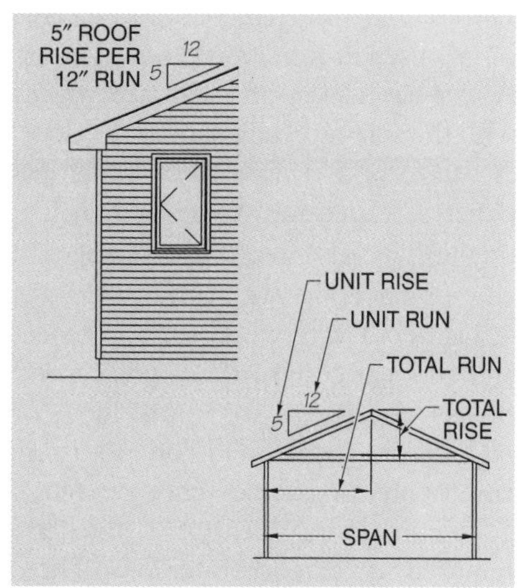

Figure 7-4. Roof slope is shown on elevations with a pitch symbol.

Wall Openings

Windows, doors, and other wall openings are shown on elevations in their exact locations. The floor plan, details, and door and window schedules must be referred to for additional information. For example, the elevation shows the general exterior appearance of a window and its vertical location in the wall. The floor plan, however, provides dimensions for the location of the window from the corner of the house. The window size may be given on an elevation or on a window schedule, typically located on another sheet. See Figure 7-5.

High-performance windows commonly consist of a multipane unit that is coated with a low-emittance (low-E) coating with argon or krypton gas between the panes.

Windows. The most common window types used in residential construction are fixed sash, double-hung, horizontal sliding-sash, casement, awning, and hopper windows. Elevations show the window sash and hinge locations on a swinging-sash window. Dashed lines are used to show window swing. The apex of the triangle drawn on the window designates the side to be hinged. See Figure 7-6.

The size of the light may be shown on elevations. A *light* is a pane of glass. For example, the sash of a window designated 28/24 indicates that the light size is 28″ × 24″. Note that the width is given first when designating window size, followed by the height.

Doors. Flush and panel doors are commonly used in residential construction. A *flush door* is a door with flat surfaces with the stiles and rails within the door. A *panel door* is a door with individual panels between the stiles and rails.

Flush wood doors have either a solid core or a hollow core. A *solid-core door* is a flush door with wood surface veneer and an inner core made of solid wood blocks, engineered wood products, or high-density foam. Solid-core doors are commonly used for exterior openings. Solid-core doors at least 1¾″ thick are required for exterior openings. Three hinges are used to support the weight of solid-core doors. A *hollow-core door* is a flush door with wood surface veneers, a solid wood frame, and a mesh core. Hollow-core doors are commonly used for interior openings. A minimum of two hinges is required for interior doors.

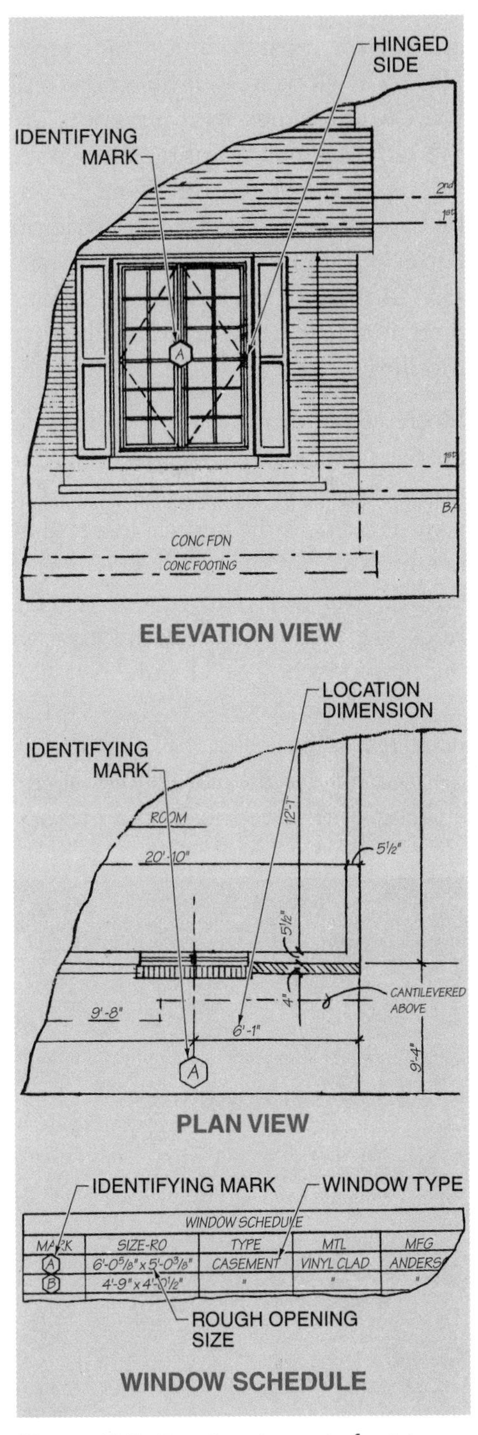

Figure 7-5. Drawings in a set of prints are related. A window shown on an elevation is also shown on the floor plan. The rough opening size is commonly included on the window schedule.

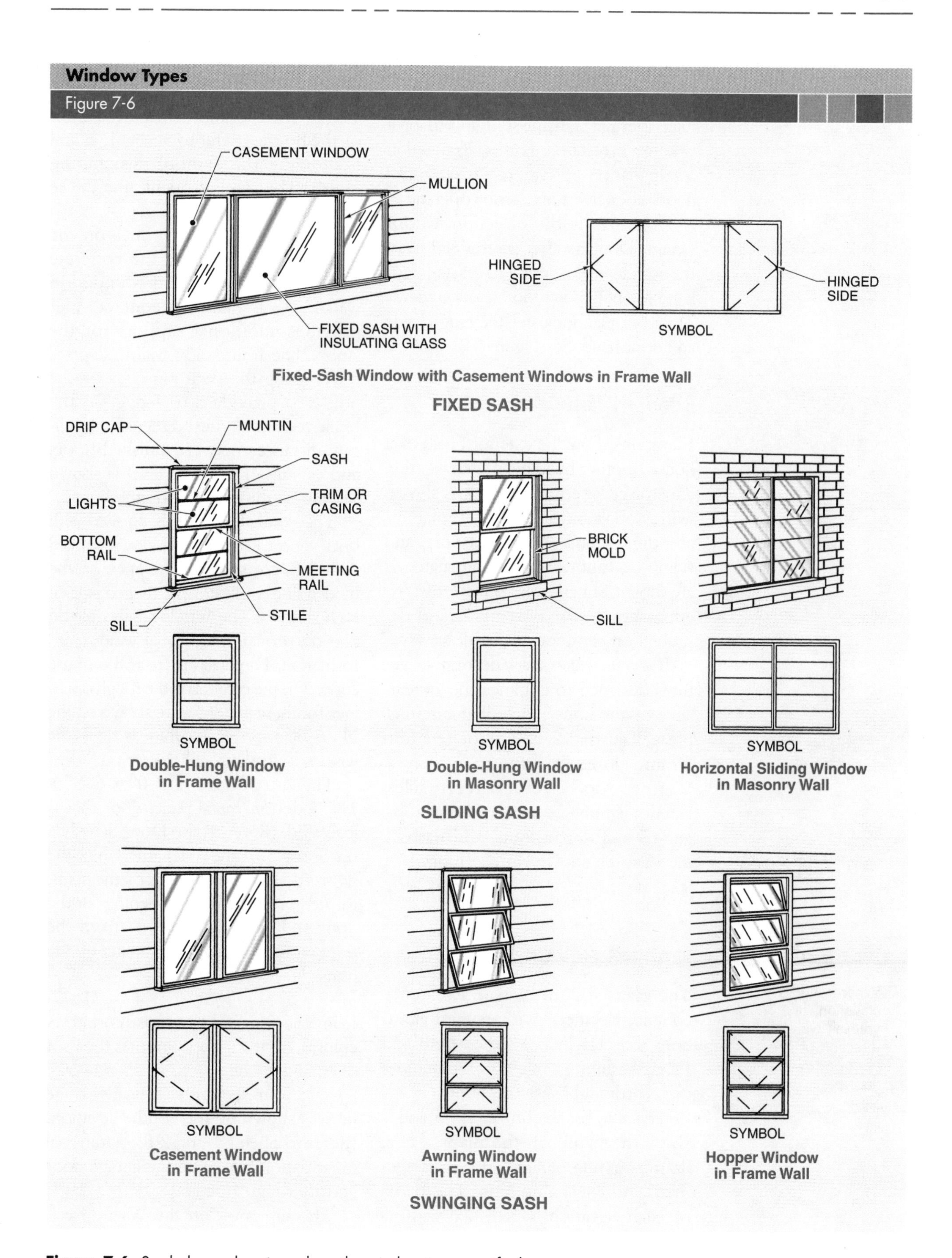

Figure 7-6. Symbols on elevations show the window type specified.

Metal doors may be flush or panel style. Metal doors are insulated to minimize thermal transmission and to provide fire protection. Lights, arranged in a wide variety of patterns, make metal doors attractive for exterior openings.

Door hand is the direction a door swings. Door hand is determined from the outside of the door. Hinges and door hand are not shown on elevation views. The floor plan indicates the hinged side and door hand. See Figure 7-7.

Exterior Finish

Elevations show the exterior finish of a house or other building. Materials, trim, and miscellaneous details are shown with symbols and described by notations and dimensions. Wood, masonry, and siding of various shapes and materials are shown. Shingles and other roof coverings are drawn as symbols and described on elevations using notes.

Elevations also show decorative features designed to enhance the appearance of the house, including columns, posts, balustrades, and decorative trim around doors, windows, and eaves. Weatherproof electrical receptacles, exterior lighting, frostproof hose bibbs, gutters and downspouts, and flashing may also be shown on elevations.

When reading an elevation, first orient yourself with the direction of the North arrow on the floor plan.

READING ELEVATIONS

The plans for the Wayne Residence contain two sheets with exterior elevations. Sheet 3 includes the South and East Elevations while Sheet 4 includes the North and West Elevations.

The title blocks on Sheets 3 and 4 show the name of the plans, scale, drafter, lot number and address, design firm, and sheet numbers. The name of each elevation is provided with its scale. The scale for all exterior elevations of the Wayne Residence is ¼″ = 1′-0″.

South Elevation

The South Elevation shows the front of the house. (Refer to Sheets 1, 2, and 6 showing the symbol designating North. The symbol orients the house and identifies the elevations.)

The hidden (dashed) lines on the South Elevation show the concrete foundation footing and foundation wall. A 1′-0″ diameter concrete pier provides additional support for the stoop. The foundation wall is stepped down along the Bedroom area of the house to provide a 6″ ledge for the brick veneer. Vertical dimensions indicate distances between finished floors and ceilings. The grade level is shown sloping to the east end of the house.

The South Elevation is finished with brick veneer. Casement windows, marked A and B, are shown. The apex of the hidden lines indicates the hinged side of each window. The Window Schedule on Sheet 5 provides additional window information. Dimensions from the house corners to the centers of the rough openings for these windows are shown on the Floor Plan, Sheet 1. Shutters flank the windows on the South Elevation.

The entry door is a 3′-0″ × 6′-8″ × 1¾″ exterior, metal panel door that is insulated. (Refer to the Door Schedule on Sheet 5.) One sidelight flanks the door. As shown on Sheet 2 of the plans, the center of the rough opening for the door and sidelight is 31′-3″ from the framing for the southeast corner of the house (7′-8″ + 13′-2″ + 7′-8″ + 2′-9″ = 31′-3″) or 31′-7″ (31′-3″ + 4″ = 31′-7″) from the finished southeast corner. A column on the stoop supports the roof overhang at the stoop.

The four-section overhead garage door is shown on the South Elevation. (Refer to Sheets 1 and 2 for additional information regarding the garage door and its rough opening size.)

The hip roof on the Wayne Residence has a 6 on 12 slope. The roof is covered with cedar shakes and has a

1′-6″ overhang. A chimney is shown rising above the roof. Notes specify brick bonds to be used on the chimney. (Refer to the Roof Plan on Sheet 5 for additional information regarding the roof and chimney.)

Doors
Figure 7-7

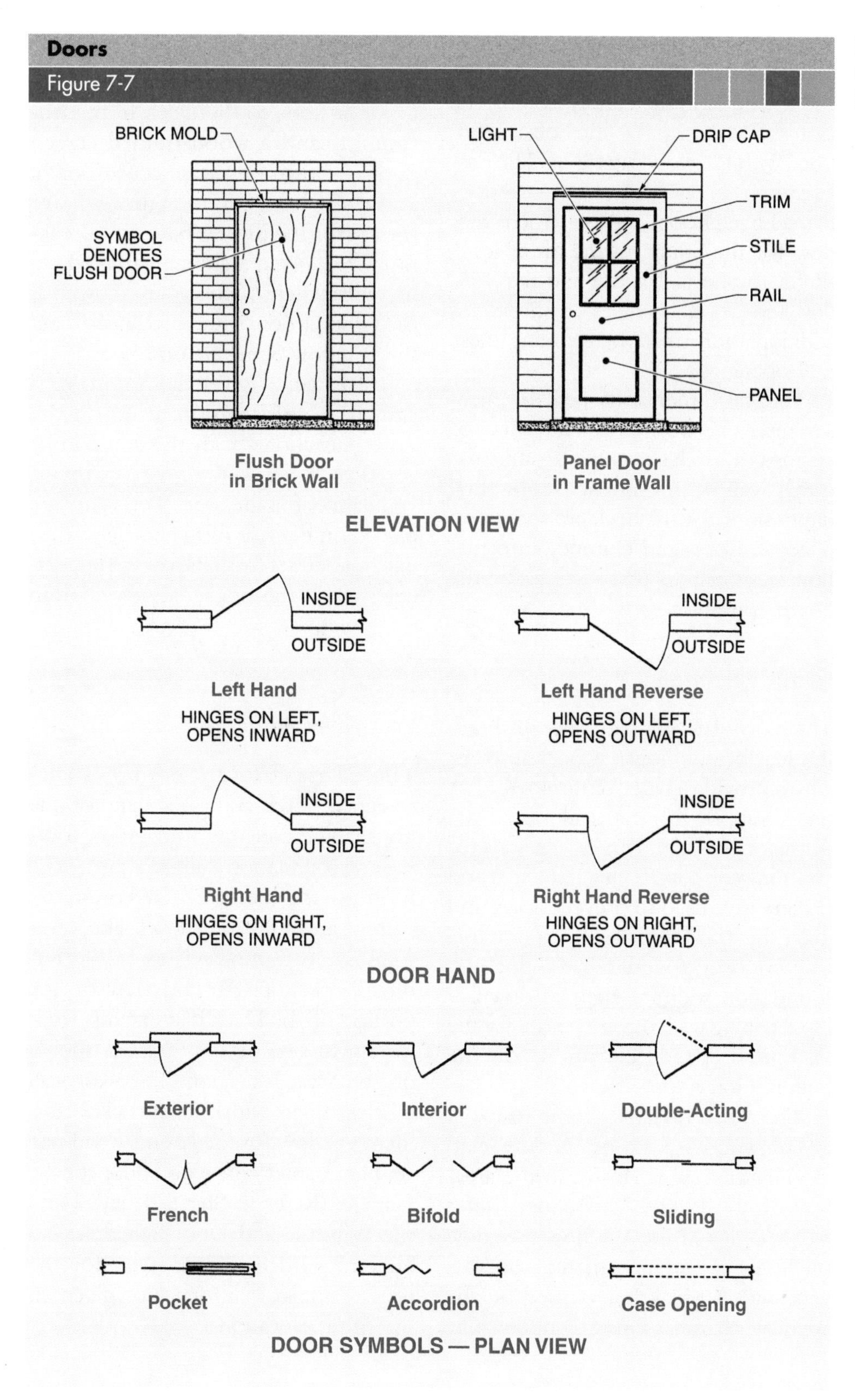

Figure 7-7. Door hand and symbols are shown on plan views.

East Elevation

Foundation footings and walls are stepped down at the retaining wall to compensate for the drop in the lot grade. The east wall is brick veneer with two casement windows, one in Bedroom 1 and the other in the Family Room. The east wall at the rear of the house is finished with 12″ wood lap siding. The East Elevation shows a full-glass wood door in the Living Room and a casement window for the Study on the lower level. (Refer to Sheets 1 and 2 and the Window and Door Schedules on Sheet 5 for additional information regarding these wall openings.)

Vertical dimensions provide measurements between finished floors and ceilings. The deck wraps around the north wall of the house. Horizontal dimensions for the deck are shown on Sheet 2. Roof and chimney information is similar to that shown for the South Elevation.

Inhalation of sawdust from treated wood and panel products should be avoided. Gloves, an approved particulates mask, and goggles should be worn.

North Elevation

The North Elevation on Sheet 4 shows the rear of the house. Concrete piers, which provide a base for the 4 × 4 ACA posts supporting the deck, are 1′-0″ diameter. The acronym ACA stands for ammoniacal copper arsenate, a wood preservative used to prevent decay and make the wood suitable for exterior use. The concrete foundation footing and foundation wall are stepped down at the retaining wall to compensate for the changes in grade elevations on the building lot.

The finish material on the rear wall is 12″ wood lap siding. A vinyl-clad awning window is shown in the areaway of the Future Workshop. (Refer to Sheet 1 and the Window Schedule on Sheet 5 for information on this window.) A break line is used on the areaway so the awning window may be shown with object (solid) lines. A sliding glass door is shown for the Family Room. (Refer to Sheet 1 and the Door Schedule on Sheet 5 for additional information on the sliding glass door.)

A wood-framed, center-hinged patio door leads to the Deck from Bedroom 1, and a wood-framed screen door is shown on the screened Porch. Vinyl-clad casement windows are shown for the Master Bath and Breakfast Area. Fixed aluminum-framed patio replacement windows are shown for the Living Room. (Refer to Sheet 1 and the Window Schedule on Sheet 5.)

Information given for the roof is similar to information given on the other elevations, with the addition of seven roof vents. Roof vents provide ventilation for the attic. The vents are placed on the rear of the house so they will not detract from the overall appearance when the house is viewed from the street.

West Elevation

The West Elevation shows the screened porch and garage end of the house. The concrete foundation footing and foundation wall are shown. The footings for the screened Porch and Garage are at the same level. The lower foundation footing relates to the Future Workshop. An end view of the portion of the Deck that extends from the north side of the house is shown on the West Elevation. The west wall is brick veneer with one vinyl-clad casement window for the Laundry and one six-panel, solid-core wood door for the Garage. (Refer to Sheets 1 and 2 and the Window and Door Schedules on Sheet 5.) Information given for the roof is similar to information given on the other elevations.

Elevations 7 Sketching

Name ______________________________ Date ______________

Sketching 7-1

Sketch elevation symbols in the spaces provided. Refer to Appendix.

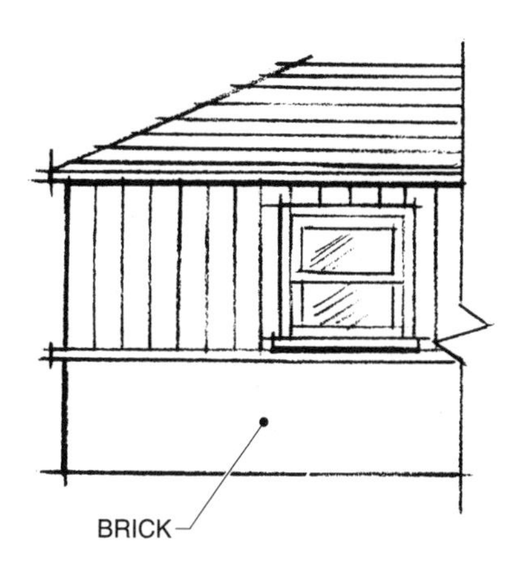

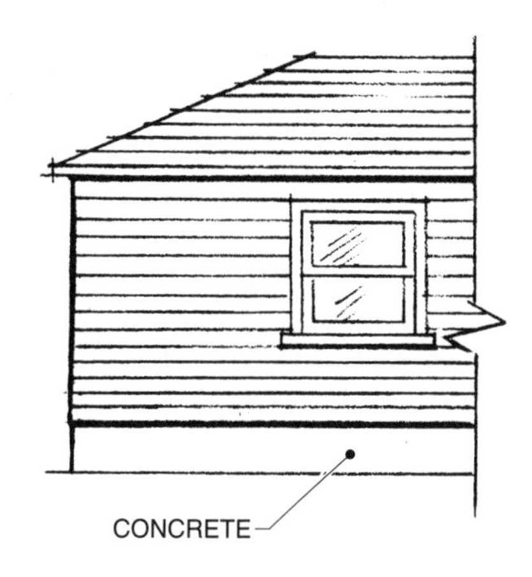

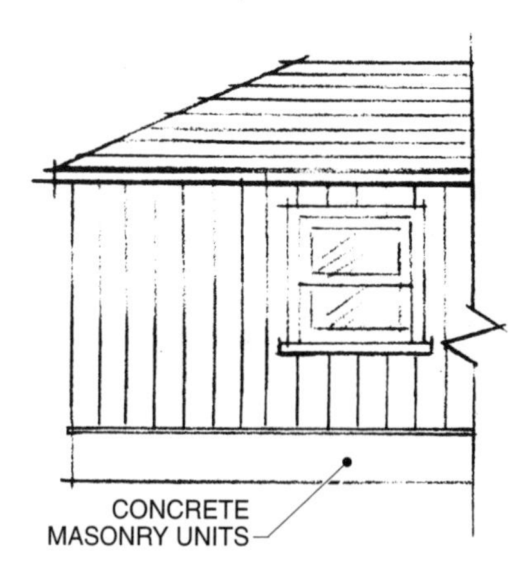

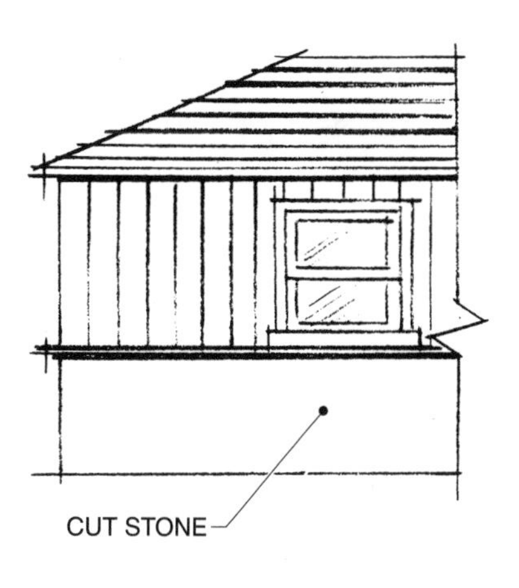

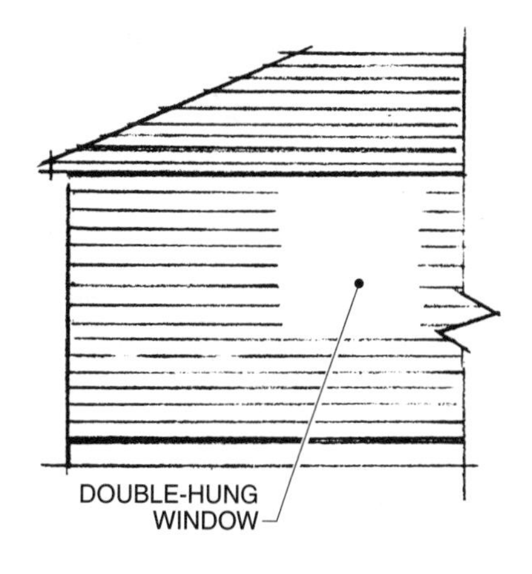

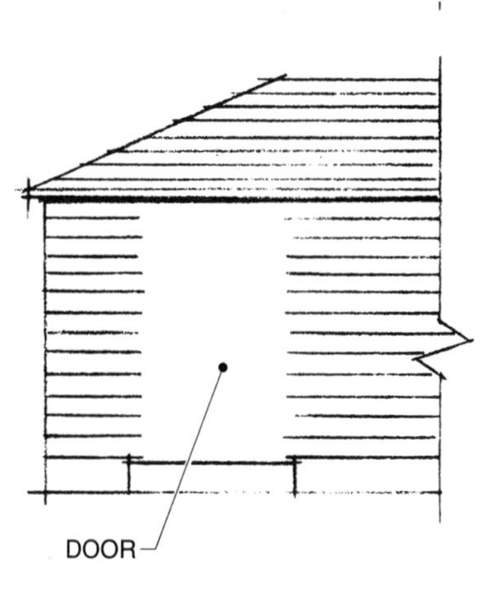

Sketching 7-2

Complete the sketch to show the windows hinged on the sides indicated by checkmarks.

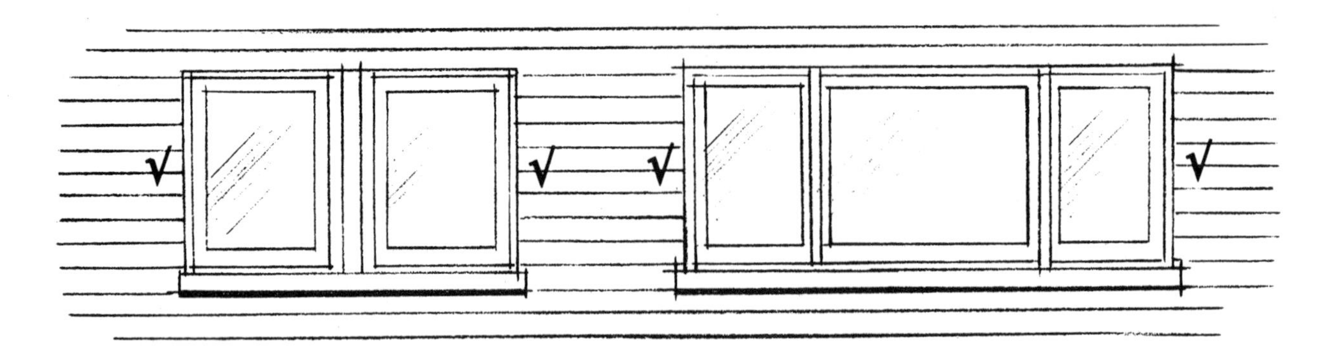

Sketching 7-3

Add symbols and notations as required to complete the sketch of the South Elevation.

1. Roof slope 5 on 12
2. Cedar shake roof
3. Full-length shutters on both sides of windows and doors
4. Brick veneer to 4'-0" above finish grade with wood lap siding above
5. Casement window with mullion between the door and window shown

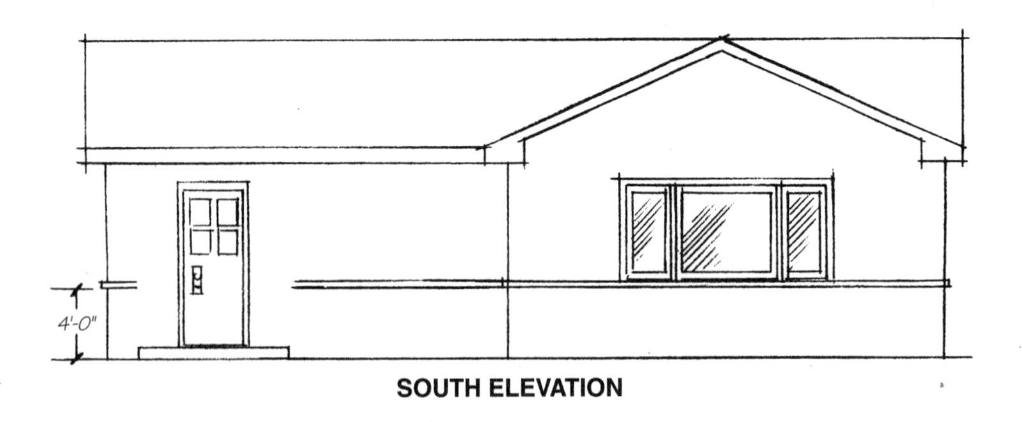

SOUTH ELEVATION

Sketching 7-4

On separate sheets of paper, sketch North, East, South, and West Elevations of the Warner Residence. Add symbols and notations as required.

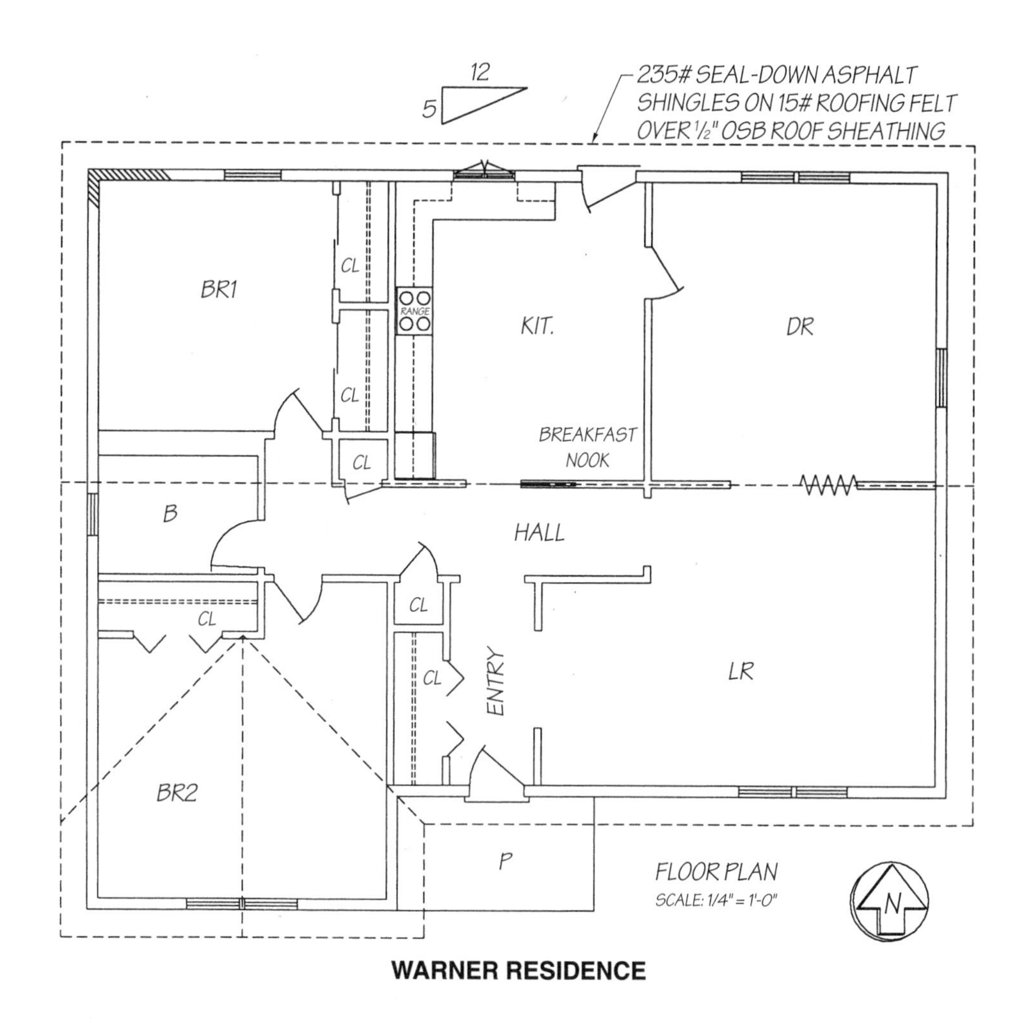

WARNER RESIDENCE

Elevations 7

Review Questions

Name ______________________ Date ______________

True-False

T	F	**1.** Elevations are pictorial drawings showing exterior views of a building.
T	F	**2.** Elevations show where exterior openings are located.
T	F	**3.** Major points of the compass may be used to designate elevations.
T	F	**4.** Various prints in a set of plans are related to each other.
T	F	**5.** Most elevations for residential construction are drawn to the scale of ⅛″ = 1′-0″.
T	F	**6.** Abbreviations in notations on elevations may be used with symbols to show building materials.
T	F	**7.** Abbreviations that form a word are followed by a period to distinguish them from the actual word.
T	F	**8.** A Cape Cod house has a traditional design.
T	F	**9.** A flat roof must slope at least ⅛″ per foot to provide for water runoff.
T	F	**10.** The slope of a roof is the relationship of unit rise to unit run.
T	F	**11.** Elevations show horizontal dimensions for door and window openings.
T	F	**12.** The apex of the triangle drawn on a window on an elevation points to the hinged side.
T	F	**13.** A window designated 32/48 indicates that the glass is 48″ wide.
T	F	**14.** Flush doors may be either solid core or hollow core.
T	F	**15.** Three hinges are required for exterior doors in residential construction.

Matching

______________	**1.** Gable roof	A. Double slope in two directions
______________	**2.** Hip roof	B. Double slope in four directions
______________	**3.** Flat roof	C. Minimum slope of ⅛″ per foot
______________	**4.** Gambrel roof	D. Single slope in two directions
______________	**5.** Mansard roof	E. Single slope in four directions

Multiple Choice

_______________ **1.** The North Elevation is the ___.

A. elevation facing north
B. elevation facing south
C. direction a person faces to see the north side of the house
D. none of the above

_______________ **2.** The scale of ¼″ = 1′-0″ indicates the drawing is ___ the size of the actual building.

A. ¼
B. 1⁄12
C. 1⁄48
D. 1⁄96

_______________ **3.** ___ roofs are popular on barns and country-style houses.

A. Flat
B. Hip
C. Gambrel
D. Mansard

_______________ **4.** In relation to roofs, unit run is the ___.

A. vertical increase in roof height
B. unit of total run
C. overall distance between building corners
D. overall distance, less space for brick veneer, between corners

_______________ **5.** The hinged side of a door can be determined from the ___.

A. plot plan
B. floor plan
C. elevation
D. none of the above

_______________ **6.** Doors for exterior openings are required to be ___ and at least ___″ thick.

A. hollow core; 1⅜
B. hollow core; 1¾
C. solid core; 1⅜
D. solid core; 1¾

_______________ **7.** The size of a window may be given ___.

A. on the elevations
B. on a window schedule
C. with the width stated first, followed by the height
D. all of the above

_______________ **8.** Information commonly found on elevations includes ___.

A. dimensions for locating rough openings for doors and windows
B. overall size and shape of the lot
C. dimensions from established grade to finished first floor
D. size and location of joist spacing

Name ______________________________ Date ______________

Refer to Wayne Residence—Elevations, Sheets 3 and 4.

Multiple Choice

______________ **1.** The Wayne Residence is a ___.
A. one-story house with living space in the attic
B. two-story house with attached garage
C. one-story house with full basement
D. one-story, L-shaped house

______________ **2.** The elevations show the ___.
A. garage on the South and East Elevations
B. deck on the North, East, and South Elevations
C. porch on the South and West Elevations
D. none of the above

______________ **3.** ___ roof vents are shown on the North Elevation.
A. Four
B. Five
C. Six
D. Seven

______________ **4.** Brick veneer walls are shown on the ___, ___, and ___ Elevations.
A. North; West; South
B. East; West; North
C. West; South; East
D. South; East; North

______________ **5.** The main level ceiling is ___ above the main level floor.
A. 8′-0⅛″
B. 8′-1⅛″
C. 9′-0⅛″
D. 9′-1⅛″

______________ **6.** A total of ___ E windows is shown on all elevations.
A. two
B. four
C. six
D. eight

__________ **7.** The slope of the roof is ___ on 12.

A. 4
B. 5
C. 6
D. 7

__________ **8.** A total of ___ retaining wall(s) is shown for the Wayne Residence.

A. one
B. two
C. three
D. four

__________ **9.** A brick ledge is required in the foundation walls shown on the ___ and ___ Elevations.

A. East; West
B. North; South
C. North; West
D. South; East

__________ **10.** Lap siding is shown on the ___ and ___ Elevations.

A. South; East
B. North; West
C. North; East
D. North; South

__________ **11.** Elevations are drawn to the scale of ___″ = 1′-0″.

A. ⅛
B. ¼
C. ⅜
D. none of the above

__________ **12.** Floor-to-ceiling height in the basement is ___.

A. 7′-9½″
B. 8′-1⅛″
C. 8′-7½″
D. none of the above

__________ **13.** The screened porch is shown on the ___ and ___ Elevations.

A. South; East
B. South; West
C. North; East
D. North; West

__________ **14.** The front entry door is ___.

A. a flush door
B. shown on the North Elevation
C. a panel door
D. none of the above

__________ **15.** Shutters are shown ___.

A. with all B windows
B. with all A and B windows
C. on the North Elevation
D. none of the above

True-False

T F **1.** The basement window for the Future Workshop is below finished grade level.

T F **2.** The South and East Elevations are shown on Sheet 3.

T F **3.** Brick quoins project 1″ and are 24″ wide.

T F **4.** A 1′-6″ diameter concrete pier is shown supporting the column on the front porch.

T F **5.** Cedar shakes are specified for the roof.

T F **6.** The garage floor is above finished grade level.

T F **7.** The plans were drawn by PLH.

T F **8.** All doors opening to the deck are the same type and size.

T F **9.** Paired windows are hinged toward the center.

T F **10.** East and West Elevations are drawn to a larger scale than North and South Elevations.

T F **11.** The foundation wall on the front of the house is stepped down to provide for the sloping lot.

T F **12.** The deck has grade-level access.

T F **13.** The retaining wall on the back of the house is sloped to follow the lot slope in the backyard.

T F **14.** The lower windows on the East Elevation are aligned vertically.

T F **15.** The top of the retaining wall on the East Elevation is flat.

Completion 7-1

______________ **1.** The Wayne Residence is located in ___, Missouri.

______________ **2.** One window and one door are shown on the ___ Elevation.

______________ **3.** A sidelight is shown to the right of the exterior door designated # ___.

______________ **4.** The ___ Elevation shows the front of the house.

______________ **5.** The front yard slopes to the ___ end of the house.

______________ **6.** Finish material on the rear wall is ___″ lap siding.

______________ **7.** ___ 1′-0″ diameter concrete piers support 4 × 4 deck posts.

______________ **8.** A handrail is shown on ___ sides of the deck.

______________ **9.** Masonry information related to the chimney is specified by dimensions and ___.

______________ **10.** ___ brick quoins are located on each front corner of the garage.

Completion 7-2

Place checkmarks in the appropriate boxes to show the windows and doors designated on each elevation view. For example, the checkmark shown indicates that a G window is shown on the North Elevation.

ELEVATION	WINDOWS						DOORS					
	A	B	C	E	F	G	1	2	3	4	5	10
SOUTH												
EAST												
NORTH						✓						
WEST												

Completion 7-3

______ **1.** The 4 × 4 posts supporting the deck are treated with ___.

______ **2.** The East Elevation has brick veneer and ___″ lap siding.

______ **3.** A total of ___ shutters is required for the South Elevation.

______ **4.** Brick quoins are set out ___″ from the face of the brick veneer.

______ **5.** The basement ceiling is ___ above the basement floor.

______ **6.** A(n) ___ door is shown on the West Elevation.

______ **7.** The roof extends ___ beyond the face of the brick veneer.

______ **8.** A total of ___ E windows is shown on the North Elevation.

______ **9.** Roof vents are shown on the ___ Elevation.

______ **10.** A total of ___ porch column(s) is required at the front entrance.

Sections 8

Sections are created by passing a cutting plane through a portion of a building to show construction details. Longitudinal and transverse sections show the relationships of rooms and floors. Details of platform, balloon, and post-and-beam framing, as well as brick veneer and masonry construction, are shown using sections.

SECTIONS

Floor plans and elevations provide overall and location dimensions; show the materials to be used for walls, partitions, and floors; and indicate locations of rough openings for doors and windows. Sections provide additional detailed information about the building. A *section* is a scaled view created by passing a cutting plane vertically through a portion of a building. Sections drawn to a larger scale than the plan from which the cutting plane is taken are known as details. Details provide greater clarity of features due to their larger size.

Typical Details

A typical detail is an enlarged section that shows the materials and construction details common throughout the building. For example, a typical foundation footing and wall section shows, in detail, materials required and necessary dimensions for constructing foundation footings and walls wherever they appear on the set of plans.

A typical wall detail is created by passing a cutting plane vertically through the foundation, footing, and wall shown on the plan view. See Figure 8-1. Arrows indicate the direction of sight for the section. For example, the detail of the section is identified as Detail 1 of Sheet A10. The Typical Wall Detail indicates the size of the footing and foundation wall and the size and spacing of rebar. The footing is 10″ × 16″. The foundation wall is 8″ thick. Two #4 rebar are placed horizontally throughout the length of the footing. The foundation wall contains #4 rebar placed horizontally and vertically 4′-0″ OC (on center). The 4″ concrete slab contains 6 × 6—W1.4 × W1.4 welded wire reinforcement.

ITW Ramset/Redhead

Sections provide information about roof and wall sheathing, insulation, and underlayment.

Figure 8-1. Sections are created by passing a cutting plane through a building feature to show its interior details. Details are sections that are drawn to a larger scale.

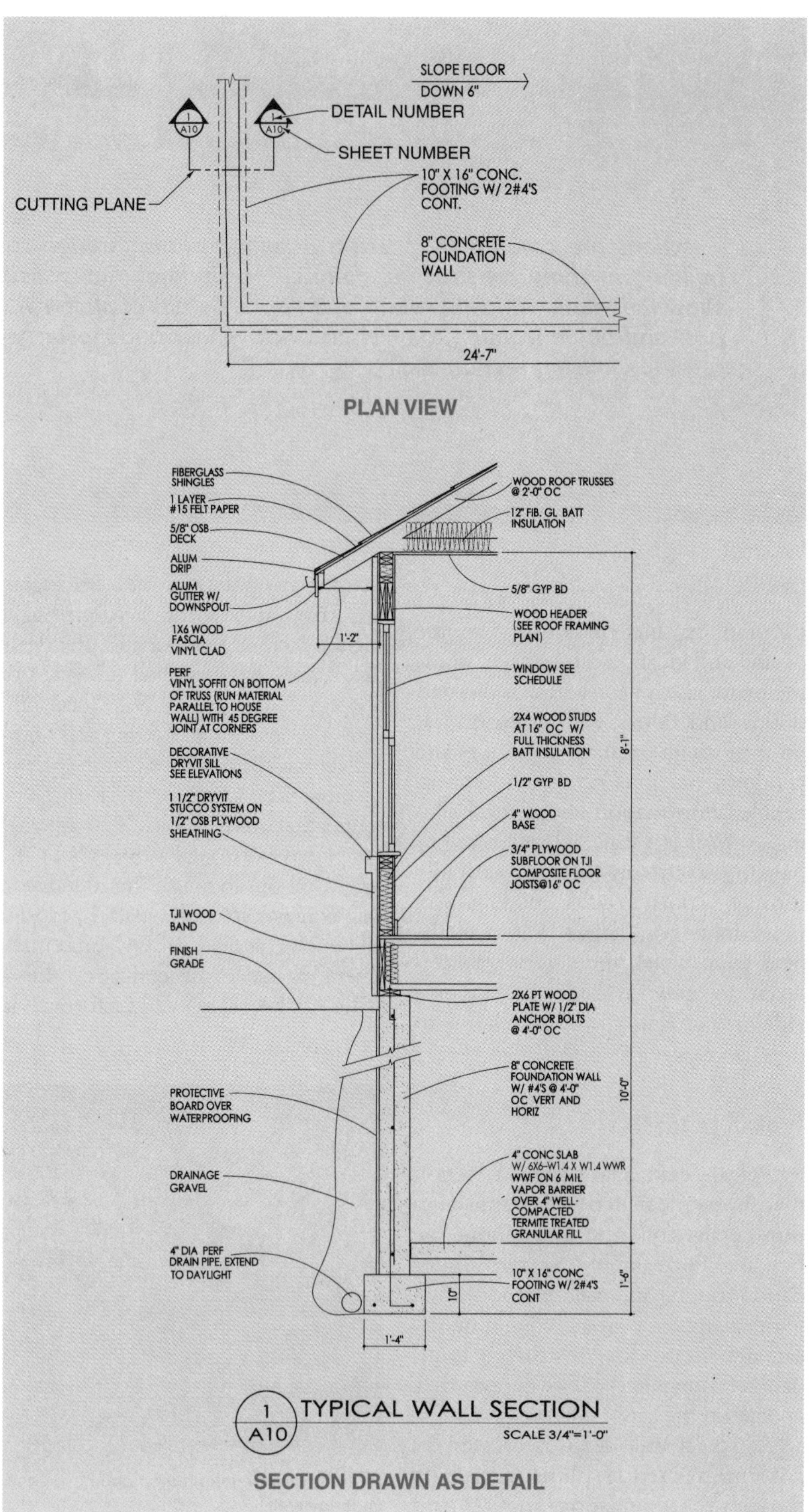

Plan Ahead, Inc.

A 2 × 6 pressure-treated sill is secured to the foundation wall with ½″ diameter anchor bolts. Composite floor joists are placed 16″ OC and covered with ¾″ plywood subfloor. Interior wall surfaces are covered with ½″ gypsum board and the ceiling is covered with ⅝″ gypsum board. Exterior walls are sheathed with ½″ oriented strand board (OSB), which is then finished with 1½″ Dryvit® system. The Dryvit system is a type of exterior insulation and finish system (EIFS). An *exterior insulation and finish system (EIFS)* is a system used to provide exterior building protection through application of exterior insulation, insulation board, reinforcing mesh, a base coat of acrylic copolymers and portland cement, and a finish coat of acrylic resins. EIFSs provide an uninterrupted layer of rigid insulation that is attached directly to the building sheathing using adhesives and/or mechanical fasteners.

Longitudinal and Transverse Sections

Passing vertical cutting planes completely through a house to show the relationship of rooms and floors creates longitudinal and transverse sections. A *longitudinal section* is a section created by passing a cutting plane through the long dimension of a house. A *transverse section* is a section created by passing a cutting plane through the short dimension of a house. See Figure 8-2. The cutting planes for longitudinal and transverse sections are shown on the floor plan.

Joists and rafters usually run perpendicular to the long dimension of a house. Therefore, the ends of joists and rafters are shown in longitudinal sections and the faces of joists and rafters are shown in transverse sections.

In the example, the longitudinal section—Section A-A—is created by passing a cutting plane through the Bathroom, down the Hall, and through the Living Room. The subfloor is ½″ plywood and 1 × 2 cross bridging reinforces the 2 × 10 floor joists placed 16″ OC. Beam pockets in the foundation walls and a 4″ diameter steel post support a W8 × 15 steel beam. The ceiling is covered with ⅜″ gypsum board.

The transverse section—Section B-B—is created by passing a cutting plane through the front door and Entry, across the Hall, and through the Kitchen. Interior walls are covered with ½″ gypsum board. Doors are referenced to a door schedule, shown on another sheet of the prints. Kitchen cabinets and appliances are shown on the west wall of the house. The countertop and 4″ backsplash are made from Micarta®. A 12″ soffit is above the wall cabinets. Batt insulation is placed between the 2 × 8 ceiling joists that are 16″ OC. Rafters are also 16″ OC. The exterior of the house is face brick, and the roofing is 235 lb asphalt shingles on 15 lb roofing felt over ½″ OSB sheathing.

Steel beams and posts are usually coated with a paint primer to prevent corrosion.

RESIDENTIAL CONSTRUCTION

Three common types of residential construction are frame construction, brick veneer construction, and solid masonry construction. Frame construction is divided into platform (western), balloon, and post-and-beam framing. A brick veneer building may also be considered to be frame construction since the supporting structure is wood or steel framing.

In solid masonry construction, the exterior walls are concrete, brick, concrete masonry units, stone, structural clay tile, or a combination of masonry products. The other parts of the building, including the floor structure, its supports, and the interior partitions, are conventional frame construction.

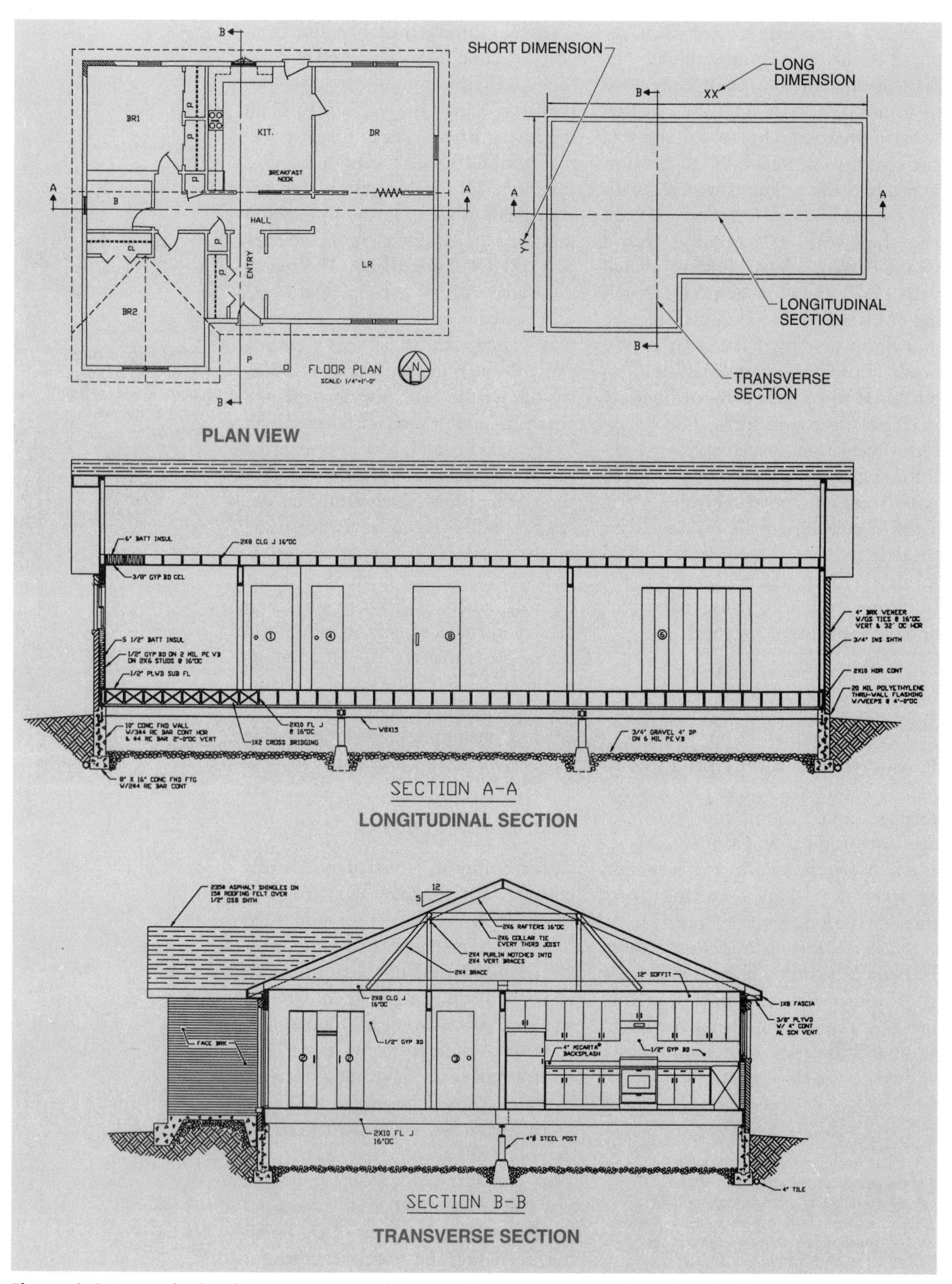

Figure 8-2. Longitudinal and transverse sections are created by passing cutting planes through plan views.

Platform Framing

The most common type of residential frame construction is platform framing. *Platform framing* is a system of frame construction in which each story of a building is framed as a unit, with studs being one story in height. Exterior walls in a platform-framed building consist of bottom plates, studs, and double top plates. The floor unit consists of joists, subfloor, and solid or cross bridging. The main characteristic of platform framing is that a complete floor system is built as a platform at each level.

> Balloon framing has largely been replaced by platform framing due to decreased availability of longer lengths of lumber.

In platform framing, studs in exterior walls, load-bearing interior partitions, and beams through the center of a building support floor members. See Figure 8-3. Exterior wall studs provide rough-framed openings for the doors and windows. Horizontal members, such as joists or rafters, support floor or roof members. An OSB or plywood subfloor and exterior wall sheathing tie the building members together and provide the stiffness needed to resist strong winds and other lateral forces.

A sill plate is the lowest member of platform framing. Sill plates are pressure-treated lumber, typically 2 × 6s or 2 × 8s, that are bolted to the foundation wall to act as a base for the floor joists. Floor joists rest on the sill plates and are tied together by a header joist, which maintains floor joist alignment. Floor joists are commonly spaced 16″ OC so that subfloor panel edges fall directly over the joists. Longer floor joists are supported by glued laminated (glulam) or steel beams on steel posts set on concrete footings. Solid or cross bridging installed between the floor joists strengthens and stiffens the floor and prevents the joists from twisting.

OSB or plywood subfloor panels are fastened to the floor joists in a staggered pattern. The exterior walls and partitions are then laid out and built on the subfloor, squared, raised into position, plumbed, and properly braced.

Second floor joists are placed on top of the double top plates of the first floor walls and bearing partitions and are fastened together. The second floor OSB or plywood subfloor panels are then fastened to the tops of the joists to cover the platform. Exterior walls and interior partitions are laid out and built on the subfloor, squared, raised into position, plumbed, and properly braced. Prefabricated trusses may be installed to provide a nailing surface for roof sheathing as well as a surface for attaching ceiling finish materials. If trusses are not used, ceiling joists are laid on the second floor double top plates and rafters are cut and placed on the wall plates to transfer the load of the roof to the walls. The roof is then sheathed and covered with the roof finish materials.

> Oriented strand board (OSB) is a panel product made of layers of wood strands bonded together with a phenolic resin.

Studs in a platform-framed building are one story in length. Each story is framed as an individual unit.

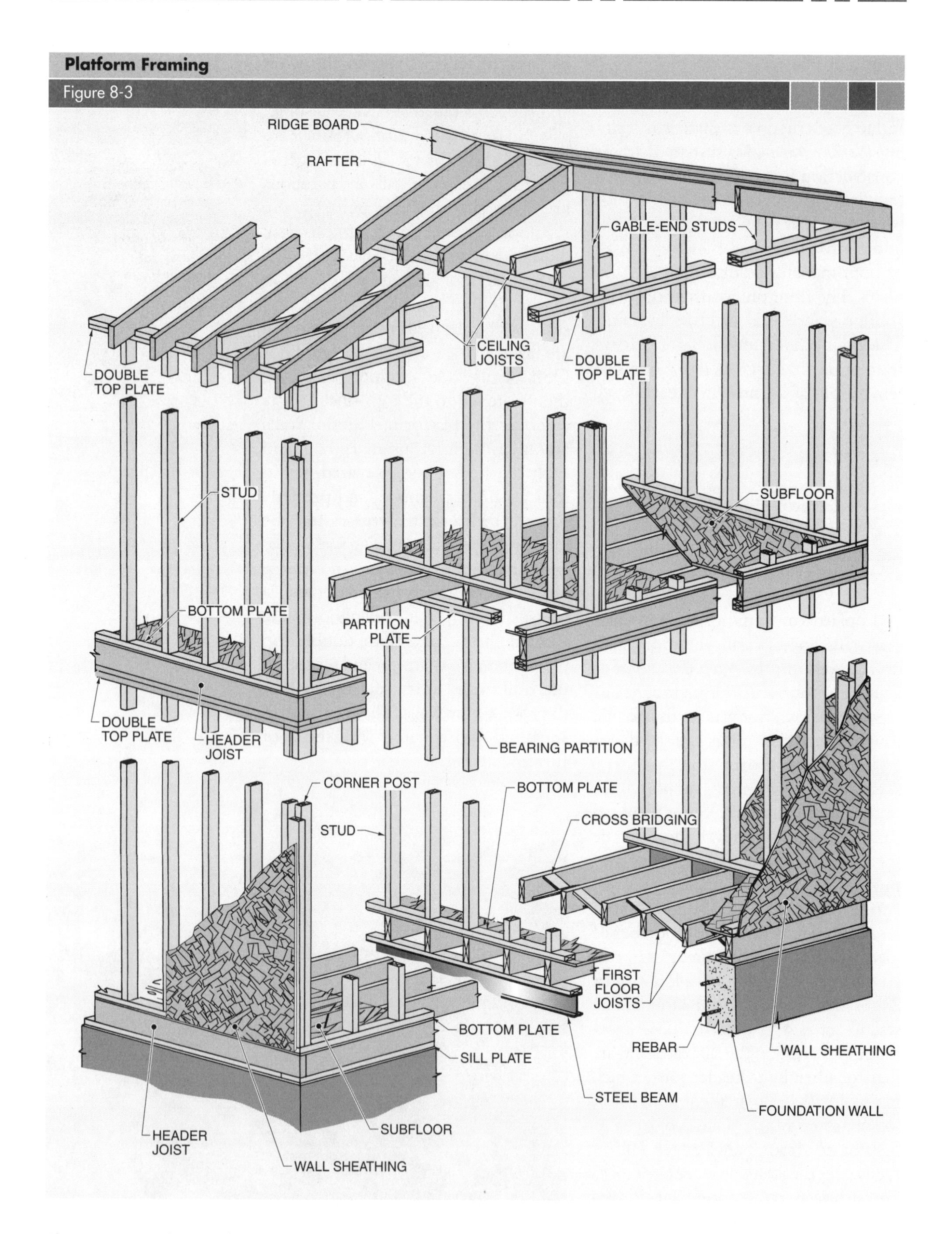

Figure 8-3. Studs in platform-framed buildings are one story high.

Balloon Framing

Balloon framing is a system of frame construction in which one-piece exterior wall studs extend from the first floor line or sill plate to the double top plate. Balloon framing is not commonly used in new construction but may be encountered in remodeling work. Studs at bearing partitions are as long as conveniently possible to reduce shrinkage. Minimal shrinkage is especially valuable when brick veneer, EIFS, or stucco is used on a building over one floor in height.

Pressure-treated sill plates are required by most model building codes.

As in platform framing, the sill plate in a balloon-framed building is also the lowest member and is bolted to the foundation wall. See Figure 8-4. Floor joists and studs are fastened together on the sill, and fire-stop blocks are fastened to the floor joists and studs. The floor joists are stiffened with cross bridging, and the subfloor is fastened to the tops of the floor joists. Let-in braces, metal straps, or plywood panels are used to stiffen the structure.

A 1 × 4 ribbon is notched into the studs or a horizontal ledger is fastened to the studs at the second floor. Floor joists rest on the ribbon and are fastened to the studs. Fire-stop blocks and fireblocking are installed. Fireblocking impedes the spread of a fire through a wall cavity by reducing airflow. Second floor bridging is installed between the floor joists. The subfloor is fastened to the tops of the second floor joists. Prefabricated trusses are set in place or ceiling joists and rafters are fastened to the double top plate. The roof is sheathed and covered with roof finish materials.

Kolbe & Kolbe Milwork Co., Inc.

Wide expanses of glass in exterior walls, plank ceilings, and heavy exposed beams are key characteristics of post-and-beam framing.

Post-and-Beam Framing

Post-and-beam framing is a system of frame construction in which posts and beams provide the primary structural support. Post-and-beam framing provides for wide expanses of glass in exterior walls and allows for large unobstructed areas inside the building. Plank ceilings and heavy exposed beams are key characteristics of post-and-beam structures. Solid or built-up beams used to support the floor and roof are spaced at wide intervals.

Planks of 2″ nominal thickness are used for floors and roofs to span the wider distances and provide structural support. The *nominal size* of a piece of wood is its size before it is planed to finished size. For example, a plank with a 2 × 8 nominal size is 1½″ × 7½″ when planed to finished size. Tongue-and-groove planks are commonly used in post-and-beam framing.

Structural insulated panels may be used to cover roofs and walls. A *structural insulated panel* is a structural member consisting of a thick layer of rigid foam insulation pressed between two OSB or plywood panels. Structural insulated panels require less time to install than planks.

Structural insulated panels (SIPs) are available in many standard sizes ranging from 4′ × 6′ to 8′ × 24′.

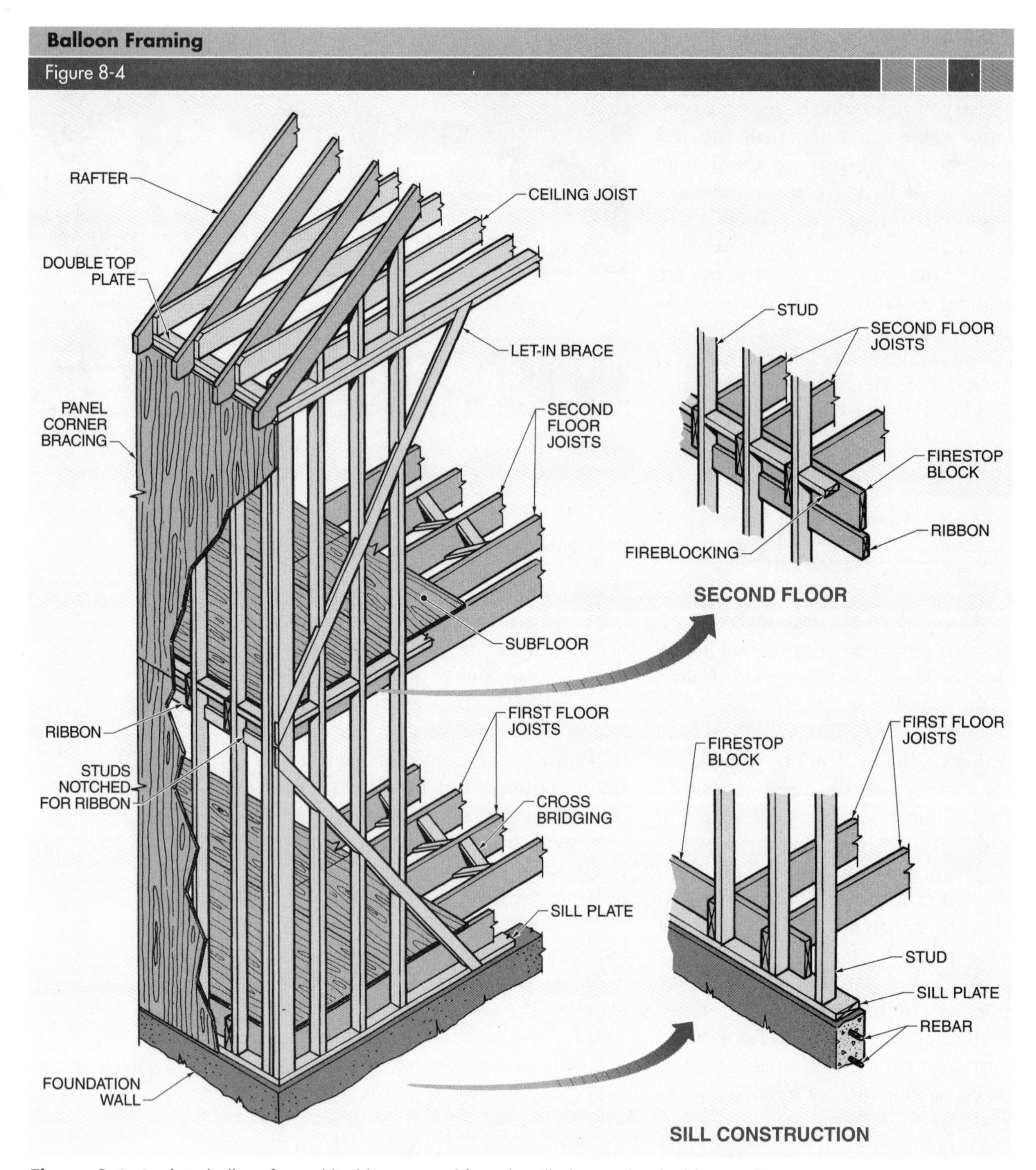

Figure 8-4. Studs in balloon-framed buildings extend from the sill plate to the double top plate.

Posts are usually spaced at equal intervals in exterior walls to support roof or ceiling beams. Posts may also be used at the center of a building or a load-bearing partition may be used to support the ridge beam.

In post-and-beam framing, the sill plate is also the lowest wood member. The sill plate is fastened to the foundation wall using anchor bolts. See Figure 8-5. Header joists and floor beams are fastened to the sill plate and the plank floor is laid. A bottom plate is installed on the plank floor above the header joists to support the posts. The posts are 4 × 4 or larger and extend up from the bottom plate to a 4 × 6 or larger girder. Roof beams are cut to fit the girder and the ridge beam. Structural insulated panels or 2″ planks are then applied as the roof.

Brick Veneer Construction

Brick veneer construction is frame construction with a brick exterior facing. Brick veneer walls may be constructed with platform-framed or balloon-framed structures. An air space is maintained between the brick and wall sheathing to permit movement between the two walls. Metal ties are used to attach the brick veneer wall to the frame construction. Weep holes at the base of the brick veneer wall allow moisture to escape.

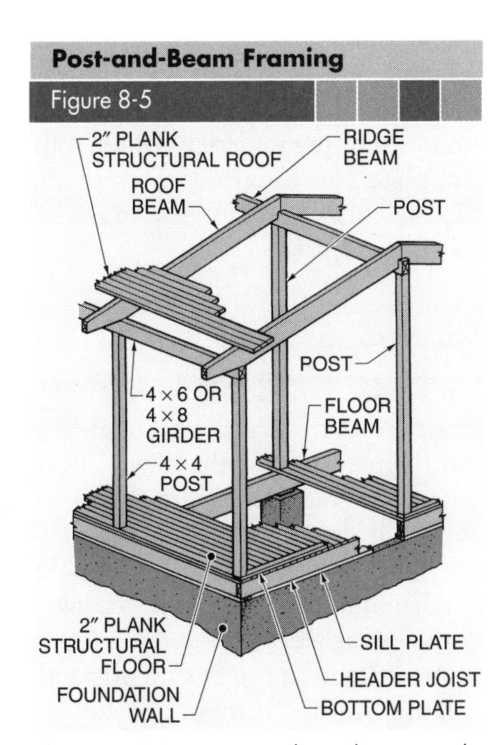

Figure 8-5. Large wood members provide structural support in post-and-beam construction.

The foundation wall for brick veneer construction is offset to provide a base for the brick veneer. See Figure 8-6. Metal or plastic base flashing extends up at least 6″ behind the housewrap to prevent moisture from penetrating into the structure. Metal ties are nailed to the studs and embedded in the mortar when the brick is applied.

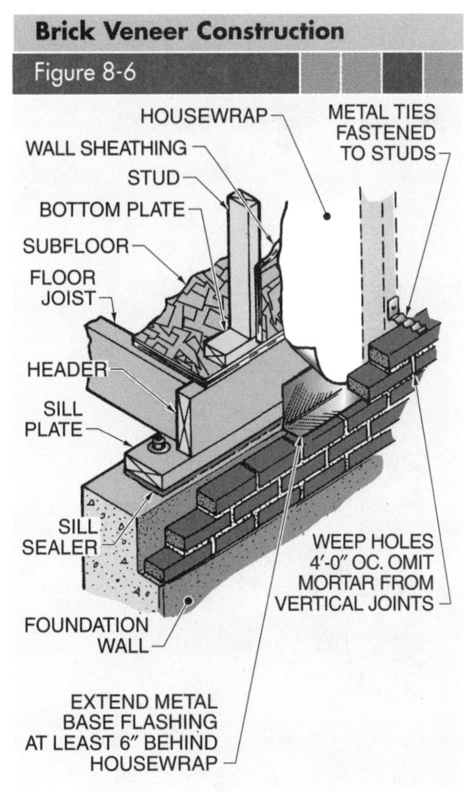

Figure 8-6. Brick veneer construction is frame construction with brick exterior facing.

Masonry Construction

Masonry construction is a system of construction in which masonry units such as brick, concrete masonry units, stone, or structural clay tile are formed into walls to carry the load of floor and roof joists. Fire-cut wood joists are secured to a masonry wall with joist anchors designed to pull out of the wall in case of fire. A *fire cut* is an angled cut in the end of a wood joist that allows a burnt joist to fall out without disturbing the solid brick wall.

Foundation footings and walls sufficient to carry the load serve as the base for solid masonry walls. See Figure 8-7. The first floor joists rest on a wythe of concrete masonry units or brick. A *wythe* is a single continuous masonry wall, one unit thick. The balance of the solid masonry wall is two units thick. Upper-story wood ceiling and roof joists are fire cut. Load-bearing partitions through the center of the building are platform framed.

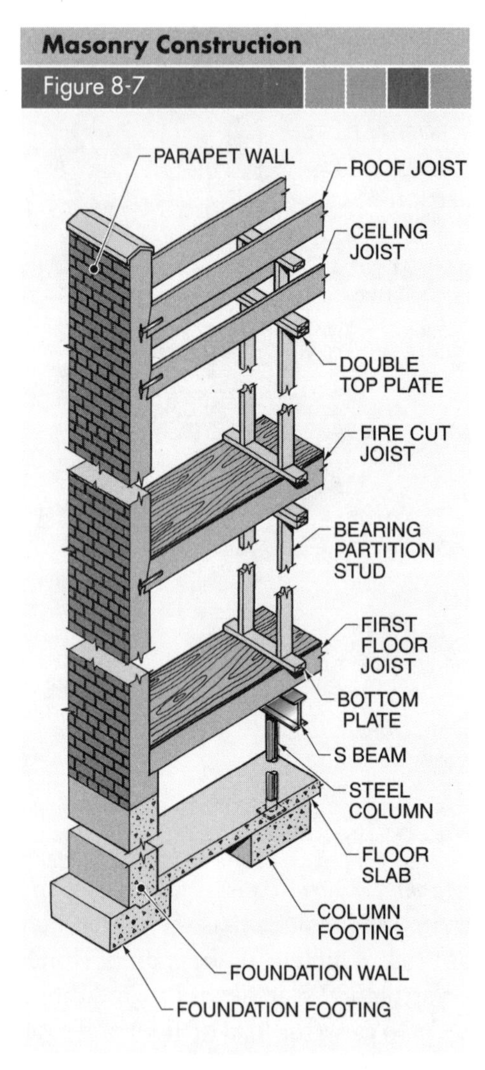

Figure 8-7. Masonry construction uses solid masonry walls to carry the load of floor and roof joists.

READING SECTIONS

Interior spaces of the Wayne Residence are shown on the sections on Sheet 7. Cutting planes for the section views are shown on Sheet 2. Longitudinal Section A-A is created by the cutting plane passing through the Garage, Dining Room, Entry, Hallway, and Walk-in Closet. Transverse Section B-B is created by the cutting plane passing through the Living Room, Entry, and Entry door. Both section views are drawn to the scale of ¼″ = 1′-0″.

Longitudinal Section A-A

Longitudinal Section A-A is created by passing a cutting plane through the long dimension of the Wayne Residence. A W8 × 10 steel beam resting on a 4″ diameter steel post is shown in the Family Room. An additional steel column is shown in the Storage Area. The upper portion of the #6 door at the foot of the stairway is visible. Two shelves in the Walk-in Closet of Bedroom 3 are made of 1 × material. Clothes rods are also shown. Doors to the Master Bedroom, Hall Bath, Linen Closet, and the door leading from the Garage to the Kitchen are referenced on the Door Schedule.

Floor-to-ceiling clearances and sizes of framing materials for the roof complete the longitudinal section. Ceiling joists are 2 × 10s and roof rafters are 2 × 8s.

Transverse Section B-B

Transverse Section B-B is created by passing a cutting plane through the short dimension of the Wayne Residence. The Stair Detail, also shown on Sheet 7, is directly related to the stairway seen on this section. A wrought iron handrail extends from the Hallway to the Living Room.

An 18″ concrete pier supports a 4″ column at the front of the house. Doors and windows are referenced on the schedules, and floor-to-ceiling heights are provided.

Sections 8 Sketching

Name ______________________________ Date ______________

Sketching 8-1

In the space provided, sketch a typical section of a platform-framed house from the concrete foundation footing to 24″ above the finished first floor. Use the ¾″ = 1′-0″ scale. Label all parts of the sketch.

1. 10″ × 20″ foundation footing w/ 2″ × 4″ keyway and 3 #4 rebar horizontal continuously and #4 rebar horizontal at 2′-0″ OC
2. 4″ PVC drain tile
3. 10″ × 36″ foundation wall to 6″ above grade level with #4s horizontal continuously at 1′-0″ OC and #4s vertical at 2′-0″ OC
4. Termite shield
5. 2 × 6 sill plate with ½″ Ø × 8″ anchor bolts
6. 2 × 10 header joists
7. 2 × 10 floor joists with 1 × 2 wood cross bridging
8. ½″ OSB subfloor
9. ¾″ oak tongue-and-groove finished floor
10. 2 × 6 bottom plate
11. 2 × 6 studs
12. ½″ exterior OSB sheathing
13. 8″ vinyl siding
14. ½″ gypsum board
15. R-21 batt insulation in walls

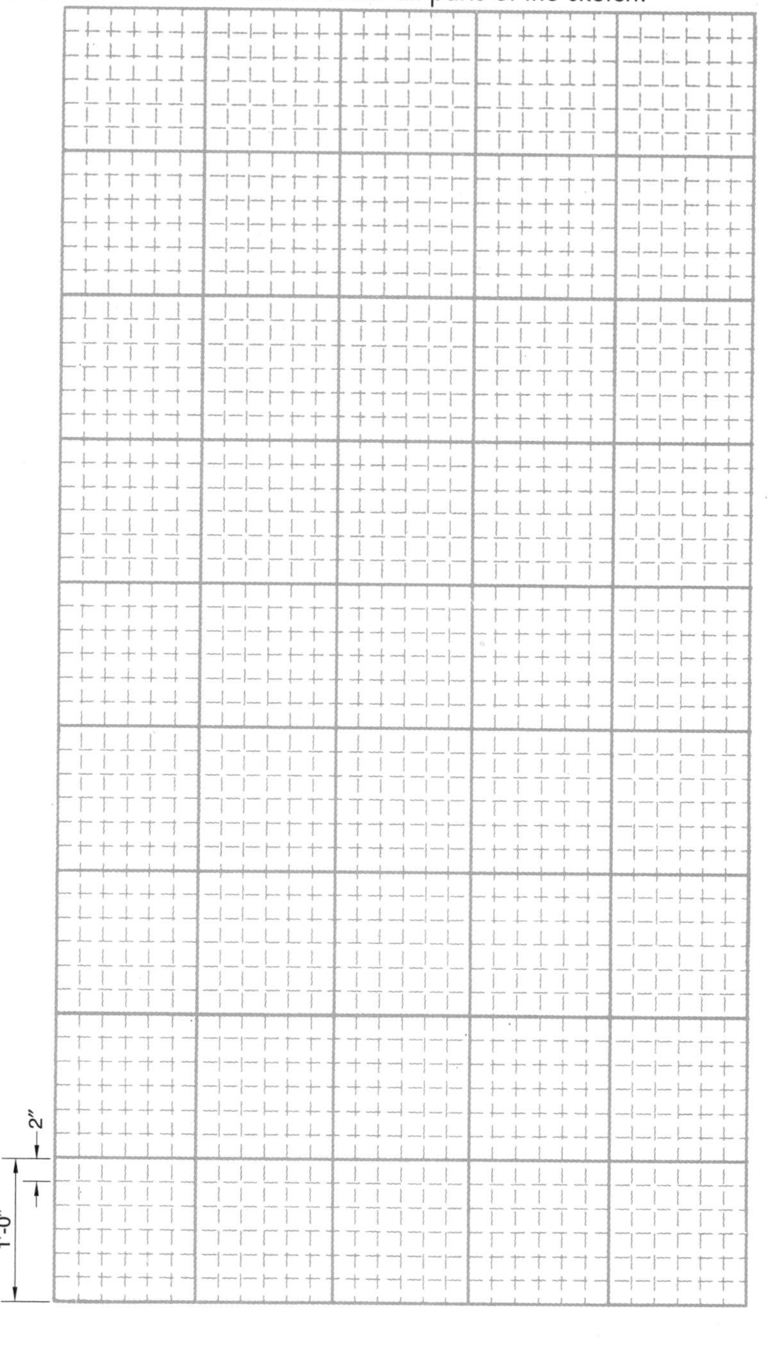

Sketching 8-2

In the space provided, sketch a typical section of a balloon-framed house from the concrete footing to 24″ above the finished first floor. Use the ¾″ = 1′-0″ scale. Label all parts of the sketch.

1. 10″ × 20″ foundation footing with V-shaped keyway and two #4 rebar horizontal continuously and #4 rebar horizontal at 2′-0″ OC
2. 4″ PVC drain tile
3. 10″ × 30″ foundation wall to 4″ above grade level with #4s horizontal continuously at 1′-0″ OC and #4s vertical at 2′-0″ OC
4. Termite shield
5. 2 × 8 sill plate with ½″ Ø × 8″ anchor bolts
6. 2 × 10 floor joists with solid bridging
7. Batt insulation to 30″ in from exterior wall
8. 2 × 10 fire-stop blocking
9. ¾″ plywood subfloor
10. Vinyl tile finished floor
11. 2 × 6 studs
12. ½″ exterior OSB sheathing
13. 4″ wood lap siding
14. ½″ gypsum board
15. R-21 batt insulation in walls

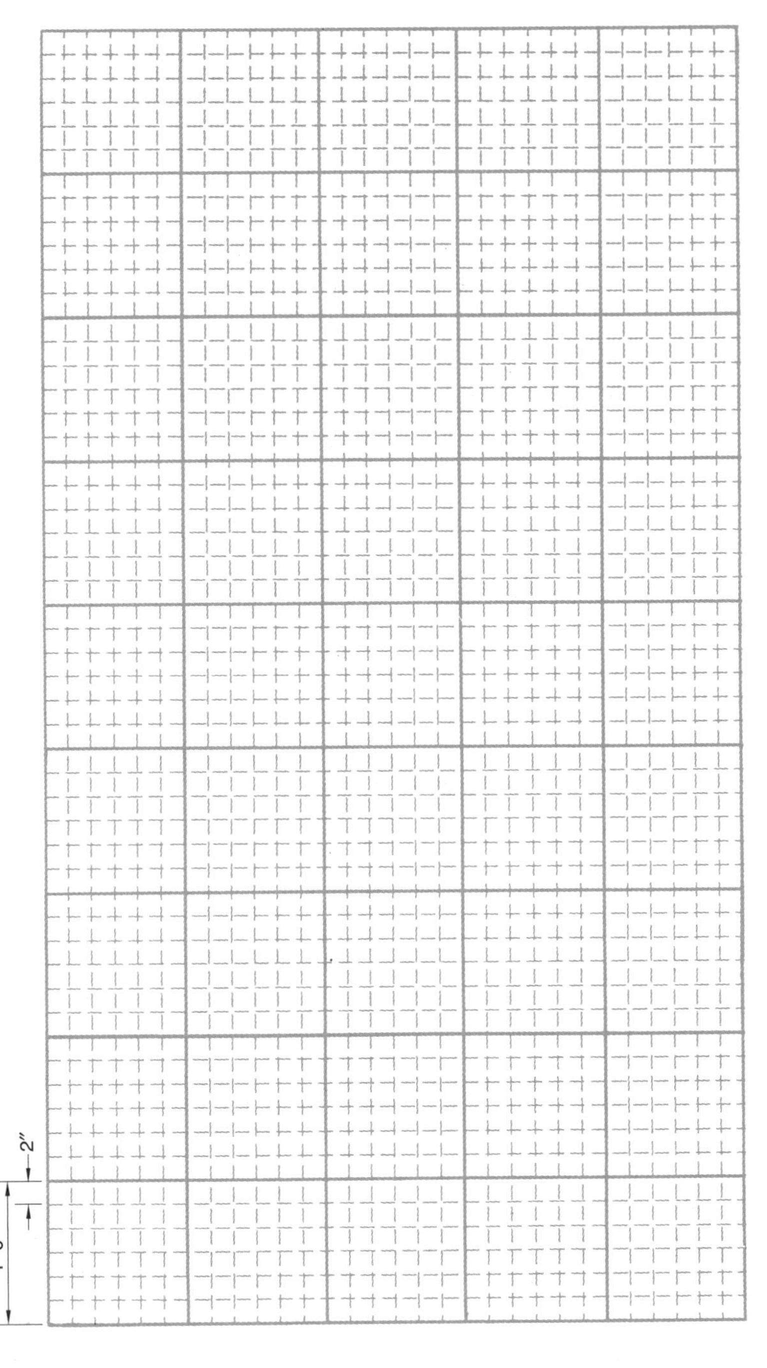

Sections 8 Review Questions

Name ______________________________ Date ______________

Identification 8-1

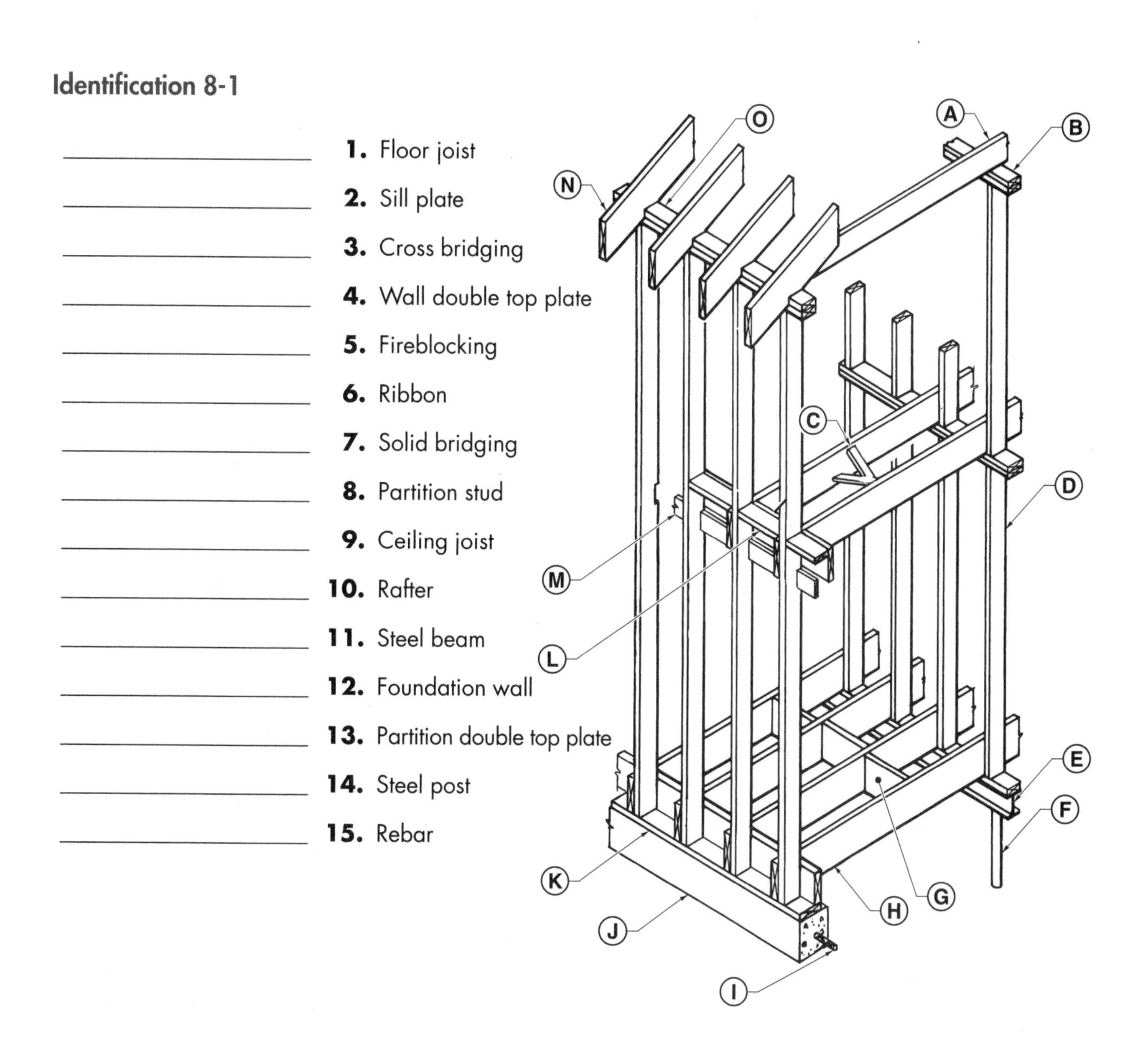

______ **1.** Floor joist

______ **2.** Sill plate

______ **3.** Cross bridging

______ **4.** Wall double top plate

______ **5.** Fireblocking

______ **6.** Ribbon

______ **7.** Solid bridging

______ **8.** Partition stud

______ **9.** Ceiling joist

______ **10.** Rafter

______ **11.** Steel beam

______ **12.** Foundation wall

______ **13.** Partition double top plate

______ **14.** Steel post

______ **15.** Rebar

Completion

Refer to the Typical Wall Section of the Garner Residence on page 181.

_______________ **1.** Ties for brick veneer are spaced ___″ OC vertically.

_______________ **2.** The concrete foundation wall measures ___ from the top of the foundation footing to the bottom of the sill plate.

_______________ **3.** Roof trusses are placed ___″ OC.

_______________ **4.** R-___ batt insulation is placed in exterior walls.

_______________ **5.** Floor support is provided by 12″ top-bearing floor ___.

_______________ **6.** Metal flashing extends ___ below the bottom wythe of brick.

_______________ **7.** The foundation footing contains ___ continuous #4 rebar.

_______________ **8.** The inside of the exterior wall is finished with ___.

_______________ **9.** The eave overhang measures ___ at an 8/12 slope.

_______________ **10.** Medium ___ shingles are applied to the roof.

_______________ **11.** Anchor bolts are spaced ___″ OC in the foundation wall.

_______________ **12.** The foundation wall is ___″ wide.

_______________ **13.** The Typical Wall Section for the Garner Residence is drawn to the scale of ___.

_______________ **14.** Batt or ___ insulation may be used above the ceiling.

_______________ **15.** The subfloor is ___″ plywood.

True-False

T F **1.** Sections are created with cutting planes.

T F **2.** A typical section shows details of material and construction common to that type of construction throughout the building.

T F **3.** Detail 2/6 is found on Sheet 2.

T F **4.** A transverse section is made by passing a cutting plane through the long dimension of a house.

T F **5.** Solid masonry is a type of frame construction.

T F **6.** The most common type of residential construction is platform framing.

T F **7.** The sill plate is the lowest wood member in platform framing.

T F **8.** Studs run full length from the sill plate to the double top plate in balloon framing.

T F **9.** Solid or built-up beams may be used in post-and-beam construction.

T F **10.** The nominal size of a piece of wood is its size after planing.

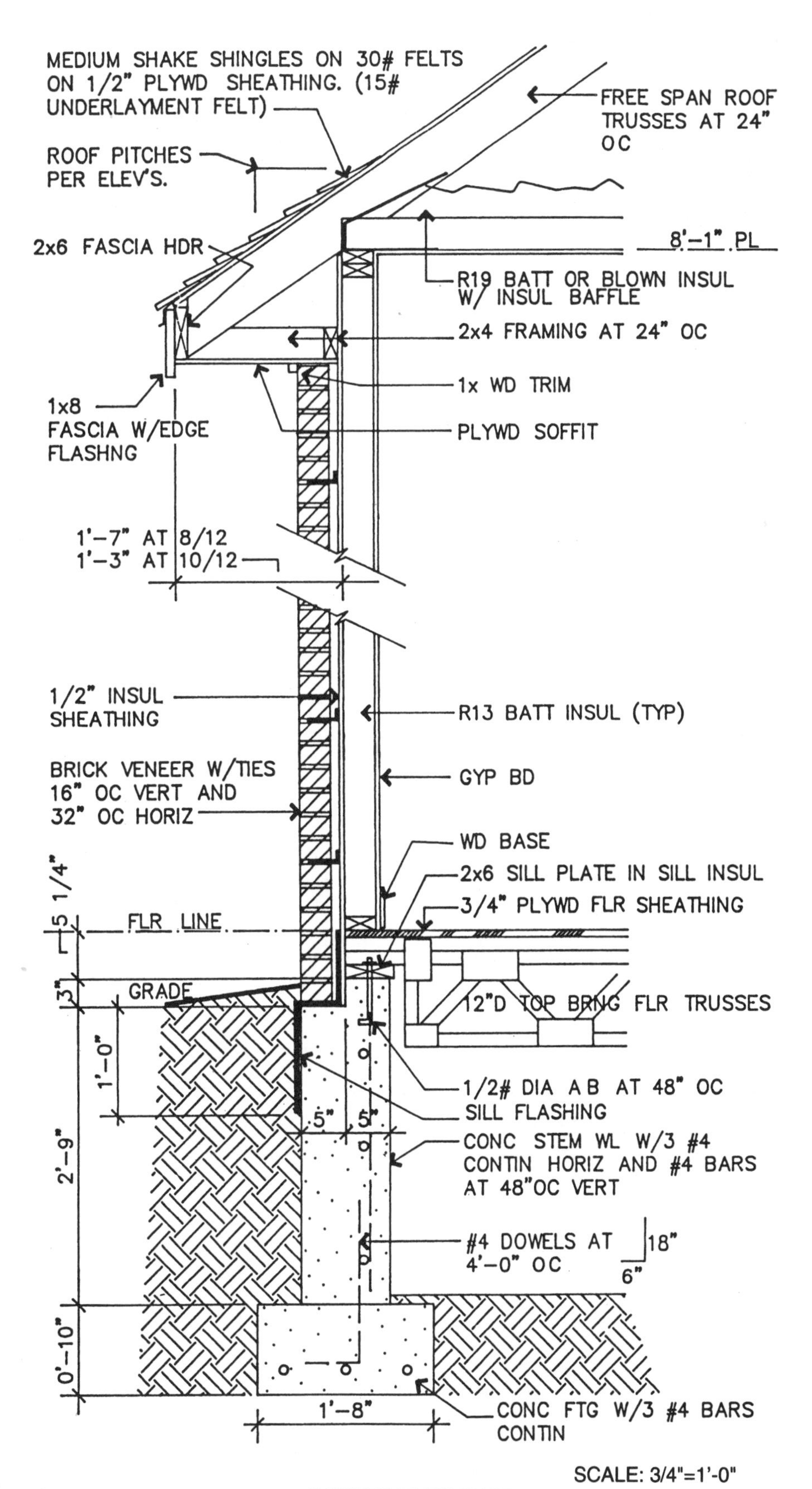

SCALE: 3/4"=1'-0"

GARNER RESIDENCE

Identification 8-2

____________ **1.** Foundation wall	____________ **8.** Plywood subfloor
____________ **2.** Rafter	____________ **9.** Rebar
____________ **3.** First floor joist	____________ **10.** Header
____________ **4.** Sill plate	____________ **11.** Ridge board
____________ **5.** Double top plate	____________ **12.** Corner post
____________ **6.** Partition plate	____________ **13.** Stud
____________ **7.** Ceiling joist	____________ **14.** Cross bridging

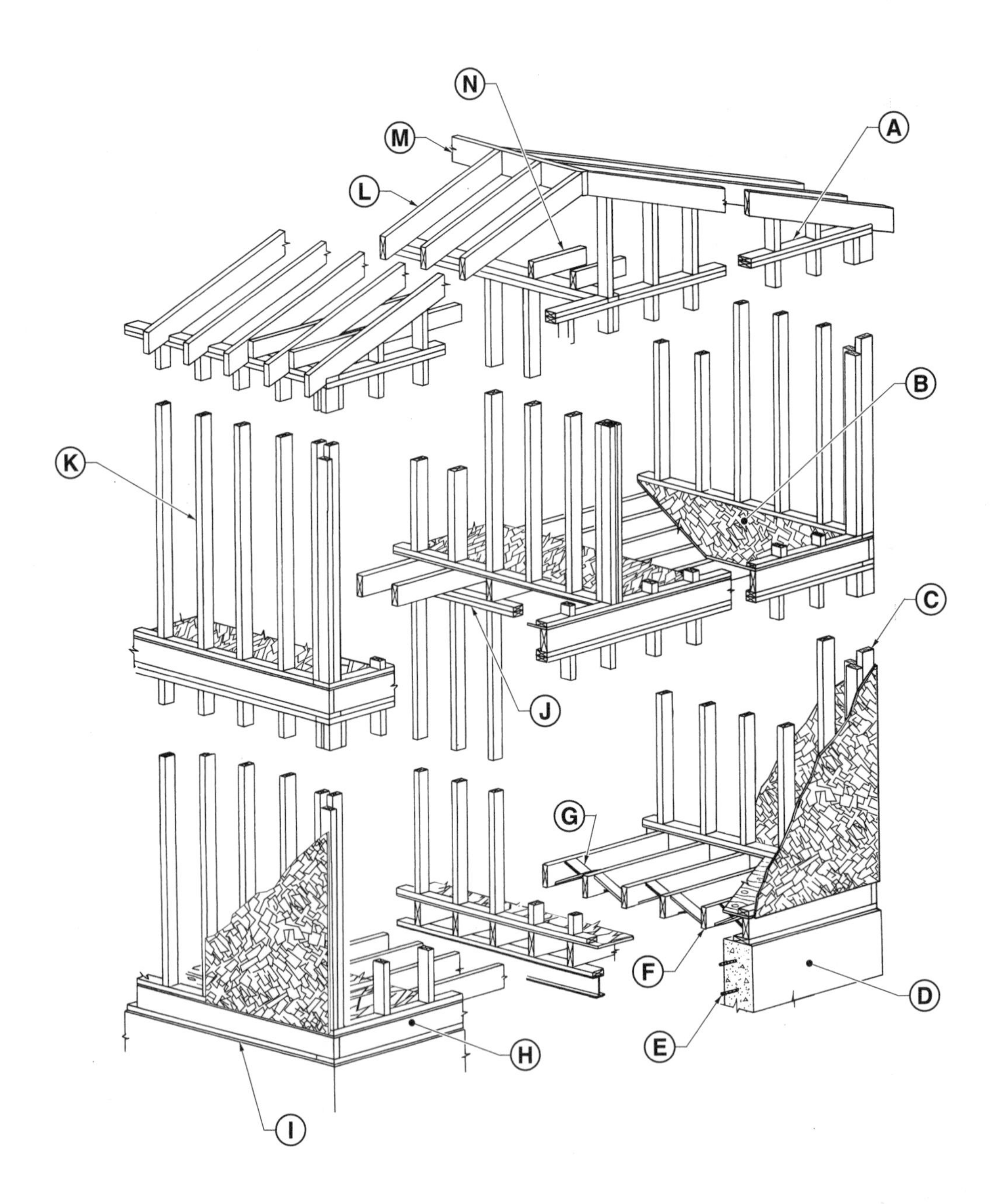

Sections 8

Trade Competency Test

Name ______________________ Date ____________

Refer to Wayne Residence—Sections and Detail, Sheet 7.

Multiple Choice

______ **1.** Regarding the stairway, ___.
A. 12 risers are shown
B. 14 risers are shown
C. treads are 1 × 10 oak
D. a ½″ nosing is required for each tread

______ **2.** Regarding roof framing, ___ are required.
A. 2 × 8 rafters and ceiling joists
B. 2 × 10 rafters and ceiling joists
C. 2 × 8 rafters and 2 × 10 ceiling joists
D. none of the above

______ **3.** Regarding doors, ___.
A. the #7 doors shown in Longitudinal Section A-A are located in the Living Room
B. the #5 door is 30″ wide
C. the #3 door leads to the Deck
D. the #6 door is located at the top of the stairway

______ **4.** Regarding concrete work, ___.
A. an 18″ diameter concrete pier supports the 4″ column at the Front Entry
B. the basement floor is a concrete slab
C. a concrete footing pad is located under the 3″ diameter steel column
D. all of the above

______ **5.** Regarding the roof, ___.
A. the ridge board is shown in the longitudinal section
B. the roof slope is 1 in 5
C. the roof is finished with asphalt shingles
D. none of the above

______ **6.** Regarding ceiling heights, ___.
A. the basement ceiling height is 7′-9½″
B. the main level ceiling height is 8′-1⅛″
C. the distance from the basement floor to the main level ceiling is 16′-8⅝″
D. all of the above

_____ **7.** The panned ceiling is in the ___.

A. Living Room
B. Family Room
C. Breakfast Area
D. none of the above

_____ **8.** Regarding the scale, ___.

A. both sections are drawn to the same scale
B. Section A-A is drawn to a ¼″ = 1′-0″ scale
C. Section B-B is drawn to a ¼″ = 1′-0″ scale
D. all of the above

_____ **9.** Regarding wall finish materials, ___.

A. the Living Room wall is papered
B. the Family Room wall is paneled
C. the bedroom walls are to be finished by the owner
D. wall finish information is not provided in the sections

_____ **10.** Regarding framing, ___ is shown.

A. platform framing
B. balloon framing
C. solid masonry construction
D. none of the above

True-False

T F **1.** The storage area has a concrete slab floor.

T F **2.** A cased opening provides a passageway between the Dining Room and Living Room.

T F **3.** The Wayne Residence roof is covered with cedar shakes.

T F **4.** Two interior stairwells provide a passageway between the basement level and main level.

T F **5.** The stair landing is 4′-0″ deep.

T F **6.** Attic access is provided by a built-in stairway.

T F **7.** A wrought iron handrail separates the Living Room from the stairwell.

T F **8.** The roof is framed with trusses.

T F **9.** Total tread width including the nosing is 10½″.

T F **10.** The Stair Detail is drawn to a ¼″ = 1′-0″ scale.

Completion

_____ **1.** The carriages for the stairway are cut from ___.

_____ **2.** The roof slopes ___″ in 12″.

_____ **3.** The earth symbol is shown only in Section ___.

_____ **4.** The deck handrail is made of ___-treated lumber.

_____ **5.** The stairway landing is constructed with ___″ T&G plywood.

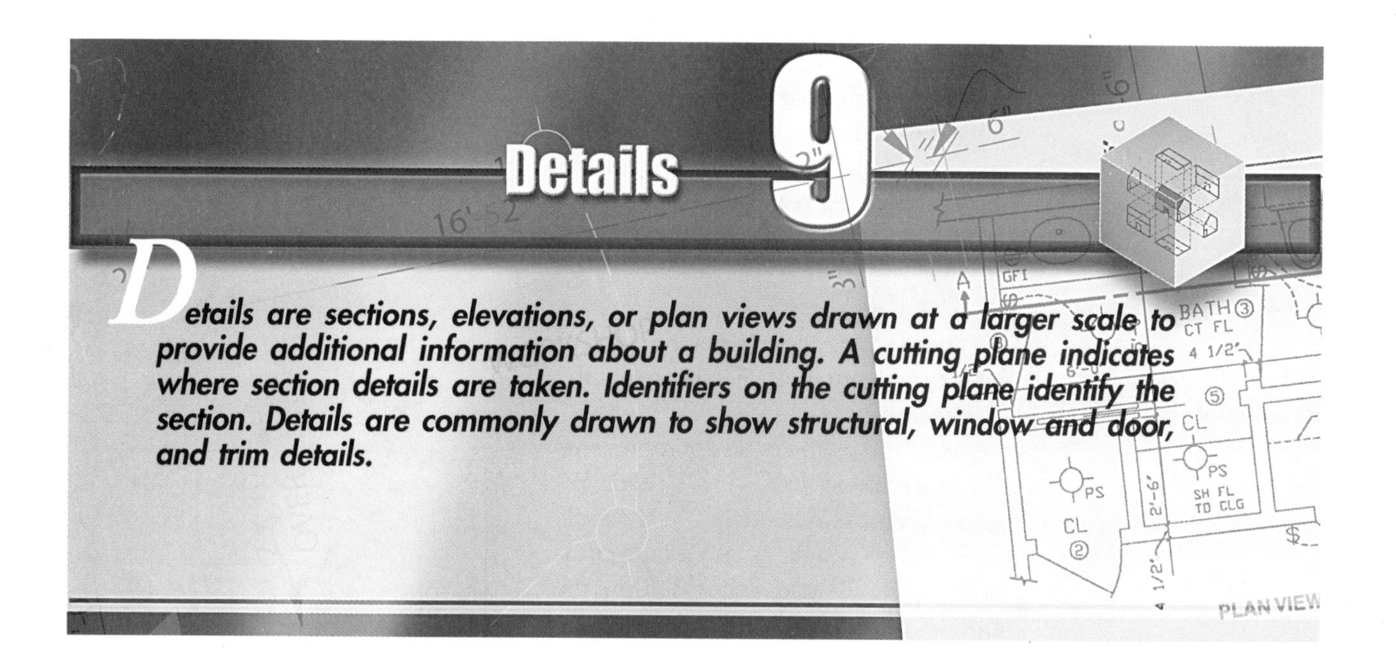

9 Details

Details are sections, elevations, or plan views drawn at a larger scale to provide additional information about a building. A cutting plane indicates where section details are taken. Identifiers on the cutting plane identify the section. Details are commonly drawn to show structural, window and door, and trim details.

DETAILS

Architects try to include all the graphic information required by tradesworkers on plan views and elevations. Specifications provide additional written explanations and descriptions. Some features, however, must be drawn at a larger scale since sufficient information cannot be shown at the smaller scale of plans and elevations.

A *detail* is a scaled plan, elevation, or section drawn to a larger scale to show special features. See Figure 9-1. Prints for residential construction may include details of interior wall construction, structural parts and special features, windows and doors, and exterior and interior trim. Details are usually drawn to a larger scale than the view from which they are taken.

Scales

Details of sections showing structural information such as footings and walls, floor systems, and stairways are drawn at a larger scale than the view from which they were taken for greater clarity. Details of special features such as fireplaces, stair trim, and windows are also drawn at a larger scale. Exterior trim details showing cornices, dormers, and entryways and interior trim such as cabinets, bookcases, mantels, and wall trim are drawn at a larger scale as well.

An elevation detail of a plan view may be drawn at the same scale as the plan view. The use of the same scale allows dimensions to be transferred directly from the plan view to the elevation detail.

American Hardwood Export Council

A detail is required to construct built-in features such as a built-in desk or cabinets.

The scale at which a detail is drawn depends on the complexity of the feature shown and the amount of drawing space available on the sheet. The following are preferred scales for detail views:

- ¾″ = 1′-0″
- 1½″ = 1′-0″
- 3″ = 1′-0″

These scales allow measurements to be worked out with a tape measure. For example, at a scale of ¾″ = 1′-0″, each 1⁄16″ on a tape measure represents 1″. At a scale of 1½″ = 1′-0″, each ⅛″ on a tape measure equals 1″. At a scale of 3″ = 1′-0″, each ¼″ on a tape measure equals 1″ and each 1⁄16″ on the tape measure equals ¼″.

When using a tape measure to calculate dimensions, beginning at the 1″ mark ensures accuracy. See Figure 9-2. When scaling critical dimensions on a drawing, the dimensions must be approved by the engineer or architect assigned to the construction project. Scaling critical dimensions is only done when no other means of obtaining the proper dimensions are available.

Full-size details are rarely drawn for residential construction unless the feature is very small or very complex. For example, full-size details may be shown of small features such as moldings or complex parts such as stair handrails.

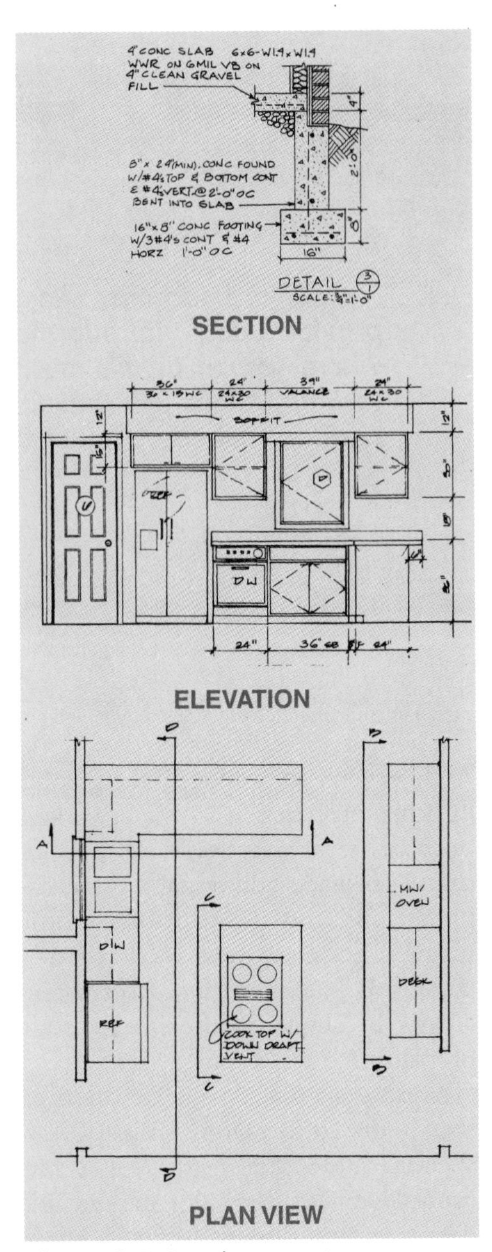

Figure 9-1. Details are sections, elevations, or plan views drawn to a larger scale to show more information.

Dimensions

If dimensions are not shown on plan views and elevations, they will be shown on details. Generally, dimensions are repeated from drawing to drawing only when necessary. Confusion would result if the same feature were unintentionally dimensioned differently on different drawings. When dimensions are shown only once, it may be necessary to refer to several prints to determine the dimension.

Exterior trim details showing dormers or other special features may be drawn to larger scales.

The two basic types of dimensions are location dimensions and size dimensions. A *location dimension* is a dimension that locates a particular feature in relation to another feature. For example, a dimension showing the distance from a building corner to the center of a rough opening for a window is a location dimension. A *size dimension* is a dimension that indicates the size of a particular area or feature. For example, dimensions showing the overall size of a house and dimensions showing the height and width of foundation footings are size dimensions.

Architects place dimensions on the most appropriate view for the feature being shown. It is often necessary to refer to several prints to identify all dimensions related to a particular feature.

References

Details refer to the view from which they are taken through cutting plane lines, direction of sight symbols, and notations. These references indicate that more information is available on a particular sheet of the prints. If possible, a detail is placed on the same sheet as its reference.

Cutting Planes. Cutting planes are imaginary slices passing through a feature to show its cross section and provide additional construction details. A *cutting plane line* is a line that identifies the exact place where the cutting plane passes through the feature. Arrows on a cutting plane line indicate the direction of sight for the section. When several sections are taken through walls, foundations, or other parts of a building, they are designated with a section notation and an alphabetic identifier, for example, SECTION A-A.

A cutting plane line is a thick line consisting of a long dash and two short dashes.

Notations. A notation such as SEE DETAIL is often used to indicate that additional information concerning the referenced feature is shown elsewhere on the prints. Notations with symbols may be used to indicate the direction of sight for interior elevation details. The elevation detail number and the sheet number on which the detail is drawn are shown on the symbol.

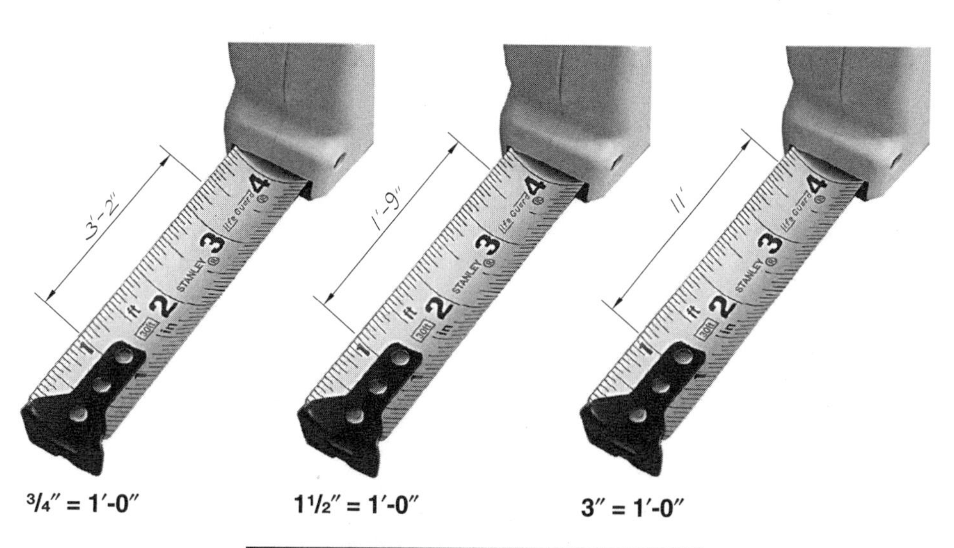

SCALE	TAPE MEASURE
3/4" = 1'-0"	1/16" = 1"
1 1/2" = 1'-0"	1/8" = 1"
3" = 1'-0"	1/4" = 1"

Figure 9-2. A tape measure may be used to scale detail dimensions on a job site. Critical dimensions must be approved by an engineer or architect.

See Figure 9-3. For example, a symbol with the notation 1/5 refers to DETAIL 1, SHEET 5 and a symbol with the notation 2/5 refers to DETAIL 2, SHEET 5. For typical details, no specific reference is required. The detail is noted as a typical detail by a notation such as TYPICAL WALL SECTION.

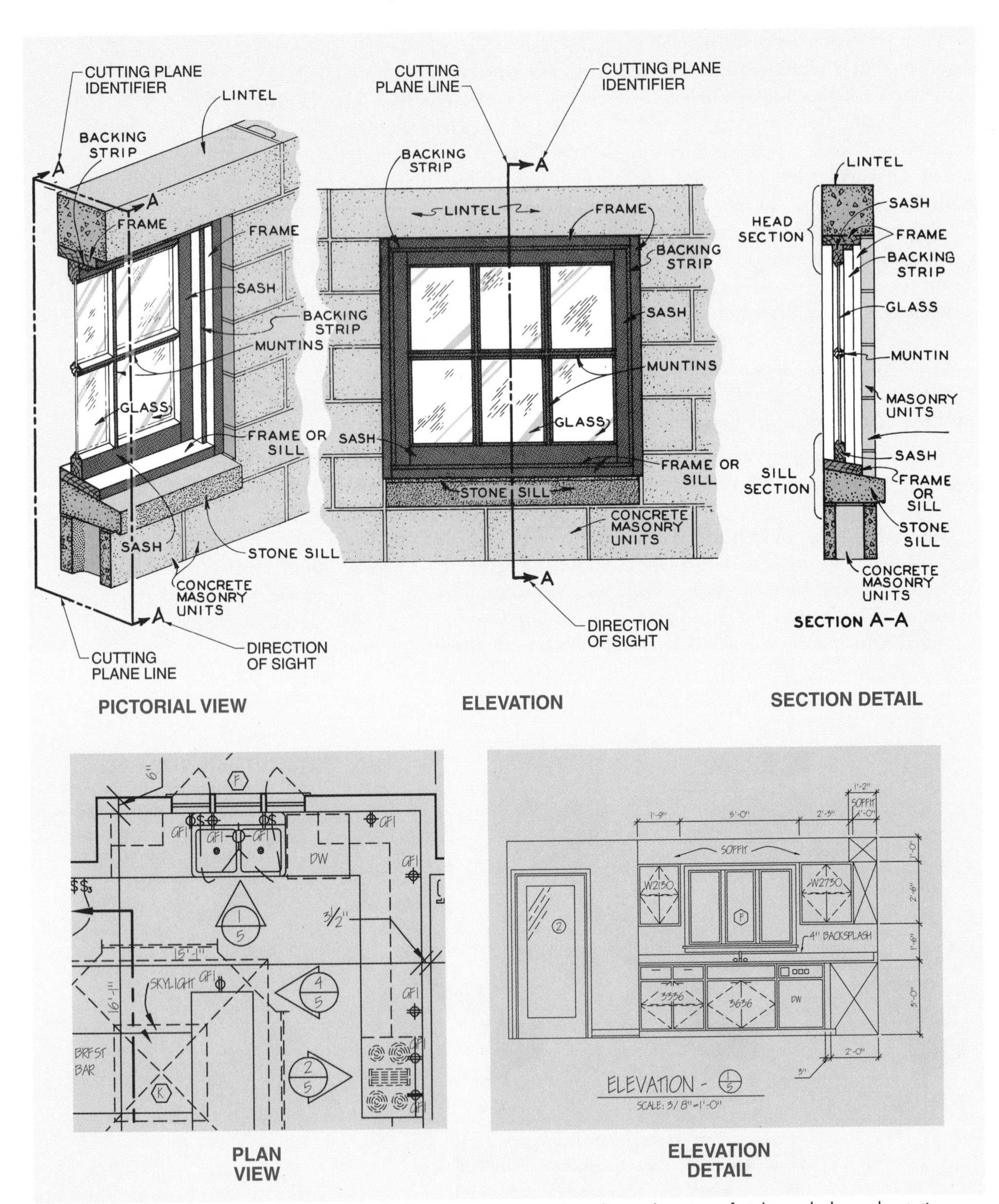

Figure 9-3. References for section details are shown using cutting plane lines, direction of sight symbols, and notations. References for elevation details typically include direction of sight symbols and notations.

Interior Wall Elevations

Details of interior wall elevations provide more information than is shown on the plan view. The plan view of a kitchen shows the location of base and wall cabinets, space for the sink and built-in appliances such as dishwashers, compactors, and wall ovens, and space for movable appliances such as refrigerators and ranges. Other information on a plan view indicates the location of windows, doors, and structural parts of the building along with electrical receptacles and outlets.

However, tradesworkers may need additional information to build and finish soffits, cabinets, and countertops, and to install lighting, ventilation hoods, fans, and other features. Manufacturer specifications and dimensioned drawings may be included as part of the working drawings.

The plan view of a Kitchen and its elevation details are related. See Figure 9-4. The plan view shows that the Kitchen and Breakfast Room measure 11′-8″ × 19′-4″. L-shaped base and wall cabinets are placed on the south and east walls, and an island cabinet containing a range and grill separates the Kitchen and Breakfast Room. The location of appliances, electrical receptacles, telephone outlet, windows, broom closet, and the cased opening to other rooms of the house are also shown on the floor plan. The three elevations of the kitchen provide additional information on features such as cabinet doors, countertop and full backsplash, paneled broom closet doors, and a brick surround on the island cabinet with a copper hood above. Height dimensions for the cabinets are also given.

A kitchen work triangle extends from the refrigerator to the sink to the range. An efficient work triangle should not exceed 21′.

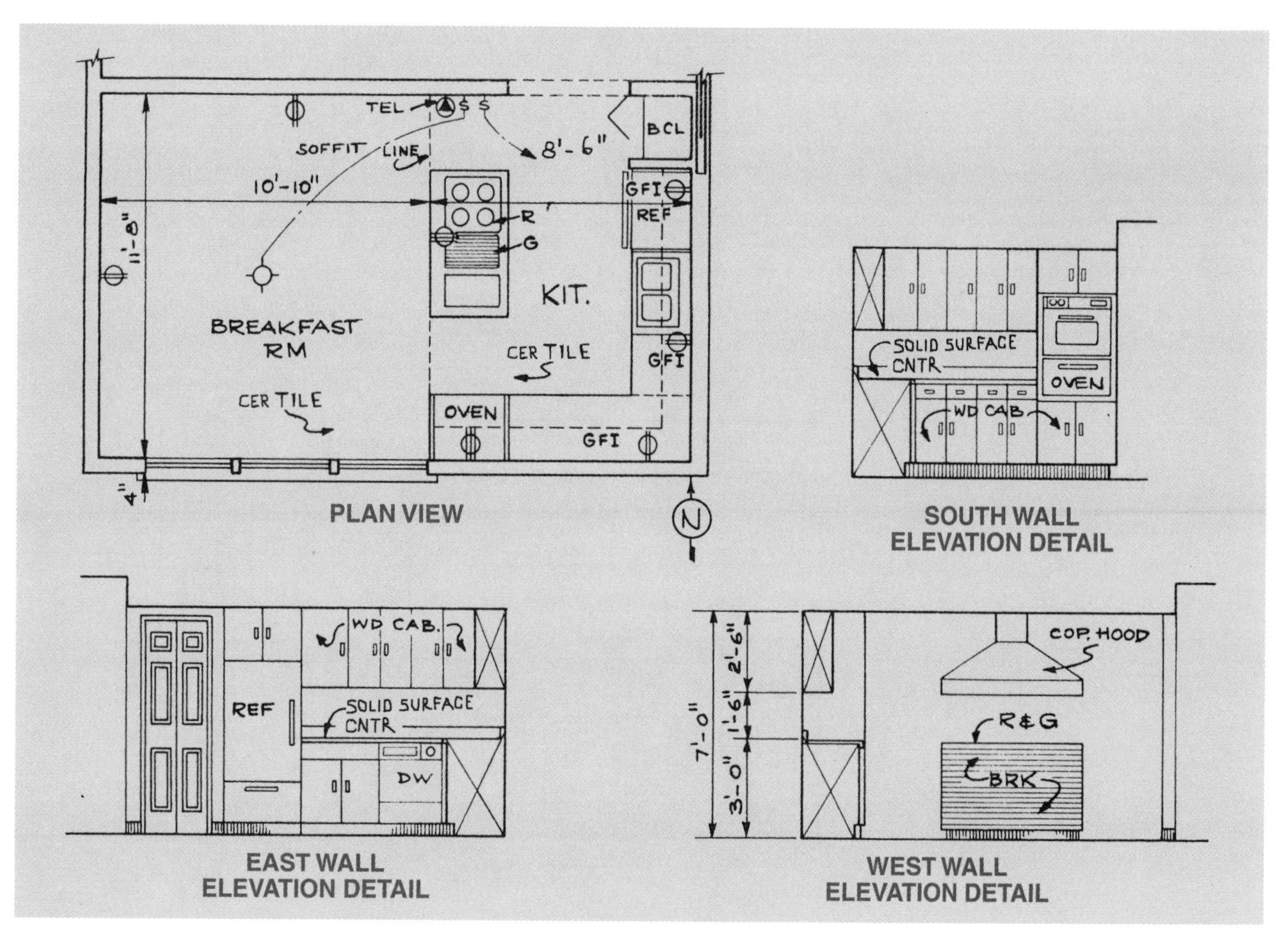

Figure 9-4. Kitchen elevation details supplement information shown on the floor plan.

The plan view of a Bath and its elevation details are related. See Figure 9-5. The plan view shows a Bath containing a water closet and bathtub with a separate 5′-0″ × 6′-0″ Powder Room. By relating the plan view to the elevation detail, the tradesworker can see that the double-bowl vanity measures 30″ × 6′-0″. The vanity has three drawer units and has a shelf running along its length. GFI receptacles are located in walls on both ends of the vanity. A 4′-0″ × 5′-11″ mirror rests on the 6″ vanity backsplash and extends to the 12″ soffit. Two #3 doors are shown. The closet on the Powder Room has a pocket door, identified as a #5 door. Additional information concerning the size of doors is given on a Door Schedule included elsewhere on the prints. The closet is illuminated by a pull chain light fixture. Shelves in the Powder Room closet extend from floor to ceiling.

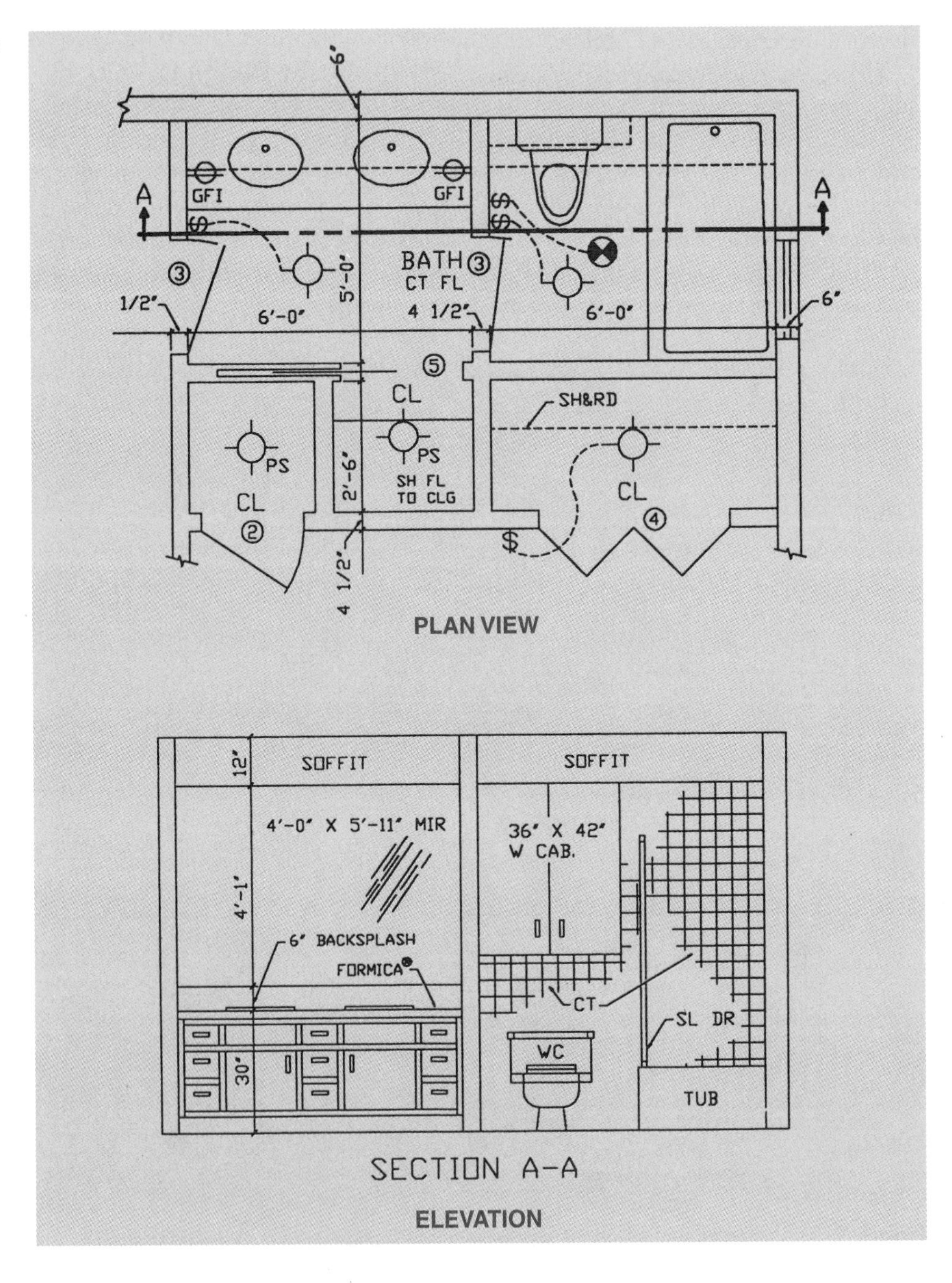

Figure 9-5. Bath elevation details supplement information shown on the floor plan.

The tub and water closet are located in a 5′-0″ × 6′-0″ room with ceramic tile from floor to soffit. The tub has a sliding door. The soffit is the same size as the soffit in the Powder Room. A 36″ × 42″ wall cabinet is hung above the water closet and extends to the soffit.

Illumination for the Bath and Powder Room is provided by ceiling lights controlled by wall switches. A wall switch also controls the ceiling fan above the water closet. The floor for the Bath and Powder Room is ceramic tile.

Other walls throughout a house may have special treatments that also require elevations to show greater detail. For example, fireplaces, bookcases, and special window trim may be shown on elevation details.

Structural Details

A section of the foundation footings and walls provides structural information. Frequently, however, more structural information is required. Additional structural information may be required when the foundation footings and walls need to be formed differently in order to meet special conditions, such as grade slopes or varying load requirements. Details are drawn to provide the additional information. Each detail is labeled with letters that correspond to letters on the plan view. Arrows on a plan view show the direction of sight of the detail. See Figure 9-6.

For many basement walls, a keyway is formed along the top of the footing to improve shear strength.

Stair details provide dimensions and other critical information needed to properly construct a stairway.

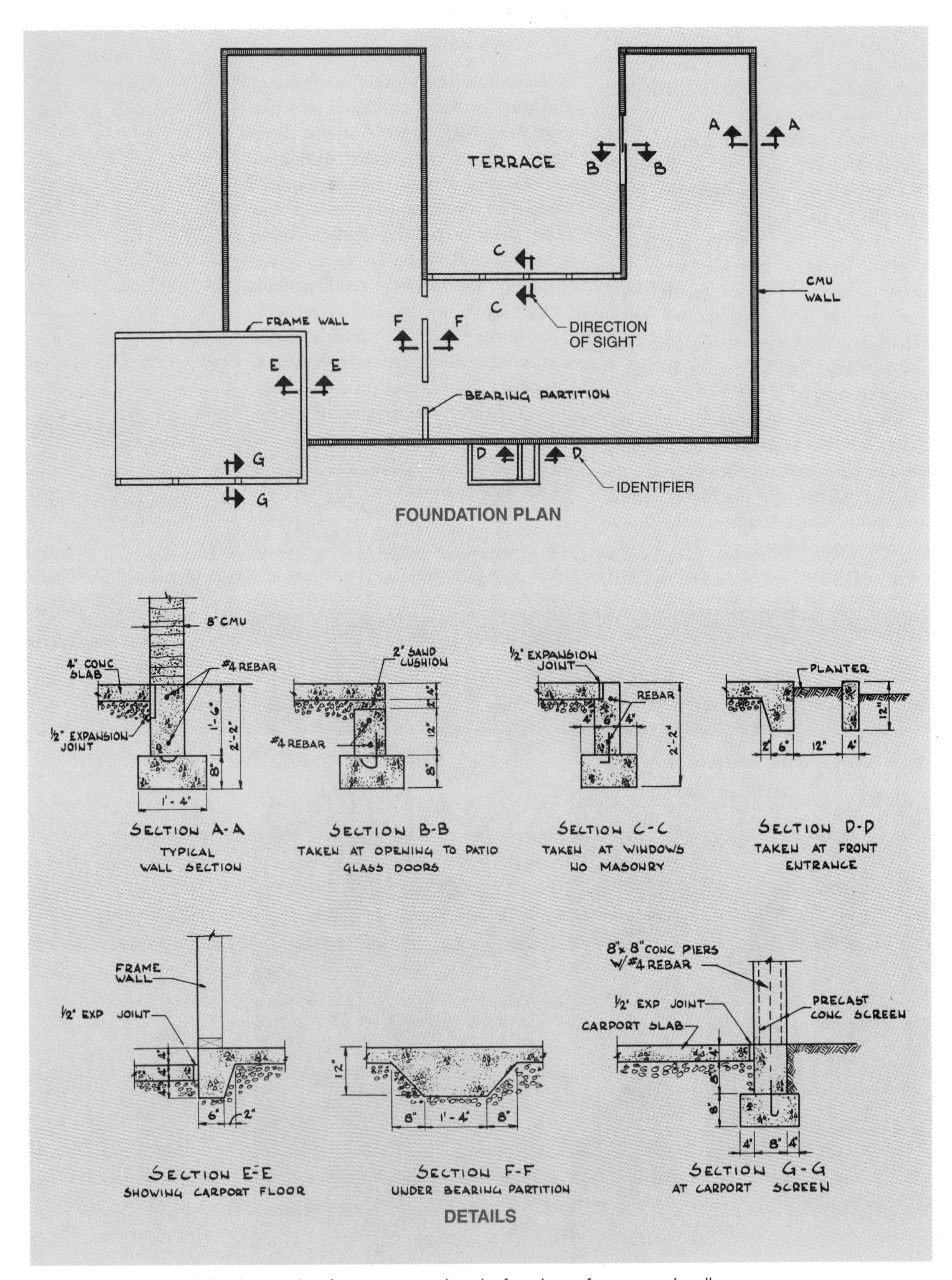

Figure 9-6. Structural details provide information regarding the foundation footings and walls.

Fireplace details cannot be shown adequately on plan views and small-scale elevations. Details regarding the shape and size of a fireplace are required for construction. The arrangement of the damper, shape and size of the transition to the chimney, and size of the firebox must conform to the dimensions on the plans for the fireplace to have efficient combustion and draft and meet building code requirements. The hearth must be designed so that there is no danger of transmitting heat to the floor joists. See Figure 9-7.

Rough framing details for wall and floor openings are generally worked out by the tradesworkers on a construction site. However, an architect will provide details if structural design is involved or if loads must be supported in certain ways. Details showing wall and floor framing or stairway opening framing may be prepared for special situations. Roof framing plans are only shown if the roof has a complex layout. See Figure 9-8. Roof trusses are often detailed for use by the manufacturing facility fabricating the trusses.

Window and Door Details

Window and door details are not included on most prints since residential construction commonly uses prehung window and door units. A rough opening of the proper size must be framed to receive each prehung unit. For special applications, however, details are drawn to show structural elements, installation procedures, and trim. Bay windows or window units composed of several window types may require window details for proper installation.

APA—The Engineered Wood Association

Rough framing details may be provided for wall openings when structural design is involved.

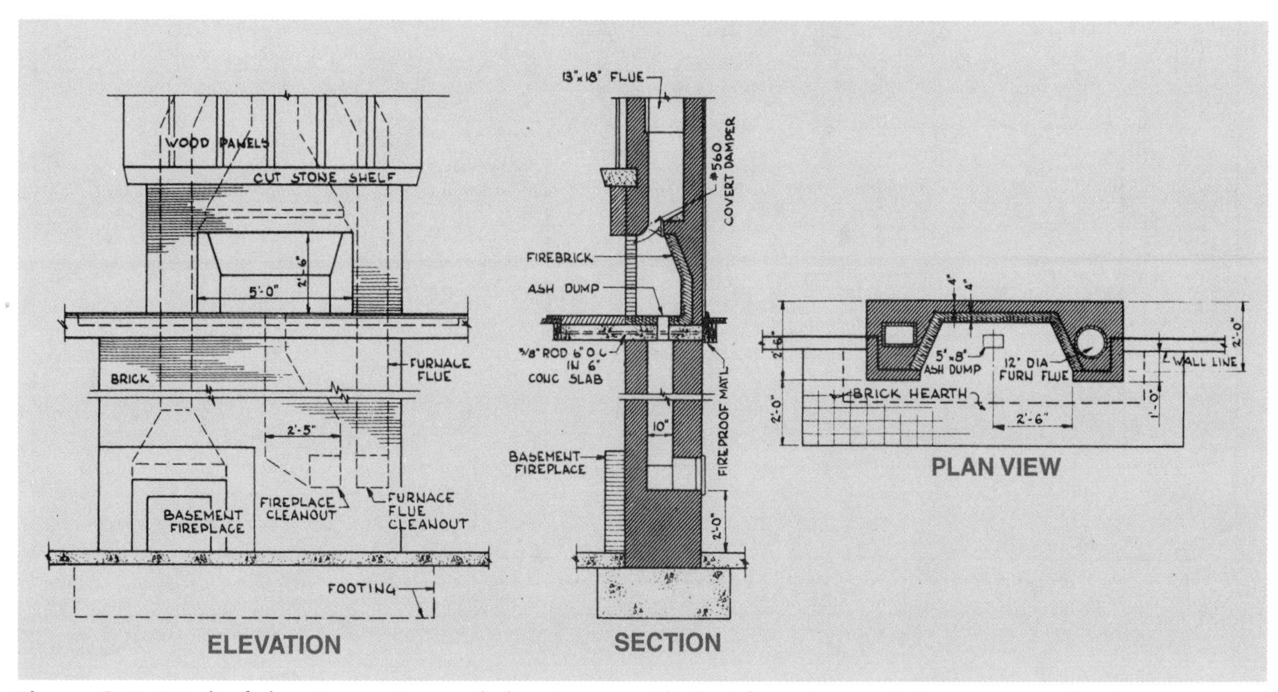

Figure 9-7. Details of elevation, section, and plan views provide the information needed to construct a fireplace.

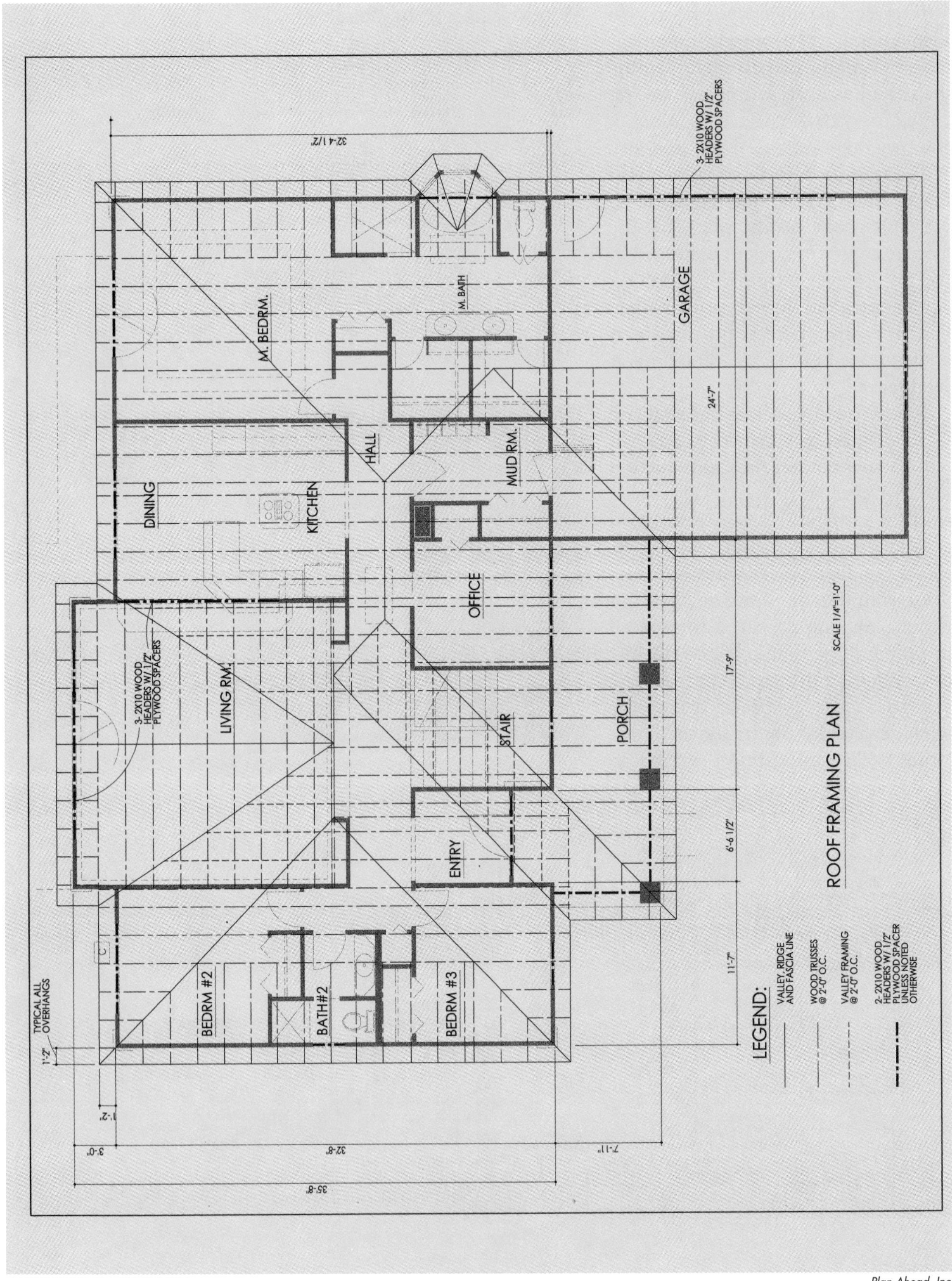

Plan Ahead, Inc.

Figure 9-8. Roof framing plans are provided for roofs with complex layouts.

Window details are drawn as sections through the head, sill, and jamb. The sections through the head and sill are drawn looking directly at the side of the window. A section of a jamb is drawn looking directly down through the window with the cutting plane taken slightly above the sill. Symbols are used to show various members. Rough structural members are shown with an X, and finish members are shown as wood grain. See Figure 9-9.

Most windows are delivered to a job site with the glass in place, window in the frame, and all hardware installed. Dimensions are provided so tradesworkers can construct the proper size rough openings. See Figure 9-10. The section of the jamb is revolved 90° and placed in-line with the sections through the head and sill. The height and width of a prehung window unit are used by the tradesworker to determine rough opening size. The size should also allow for shimming the unit to plumb and level. The unit dimension height and width include exterior trim members such as sills and brick mold. The sections through the head and the jamb show the difference between the top and side pieces of the sash and the top and side pieces of the window frame.

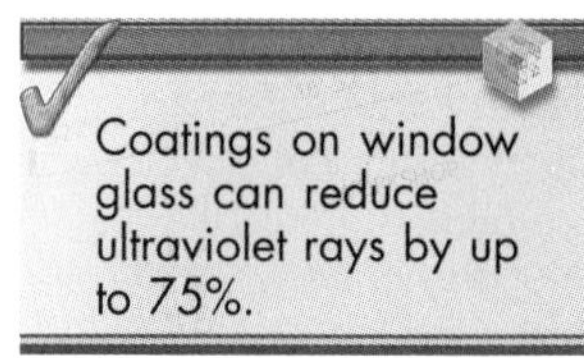

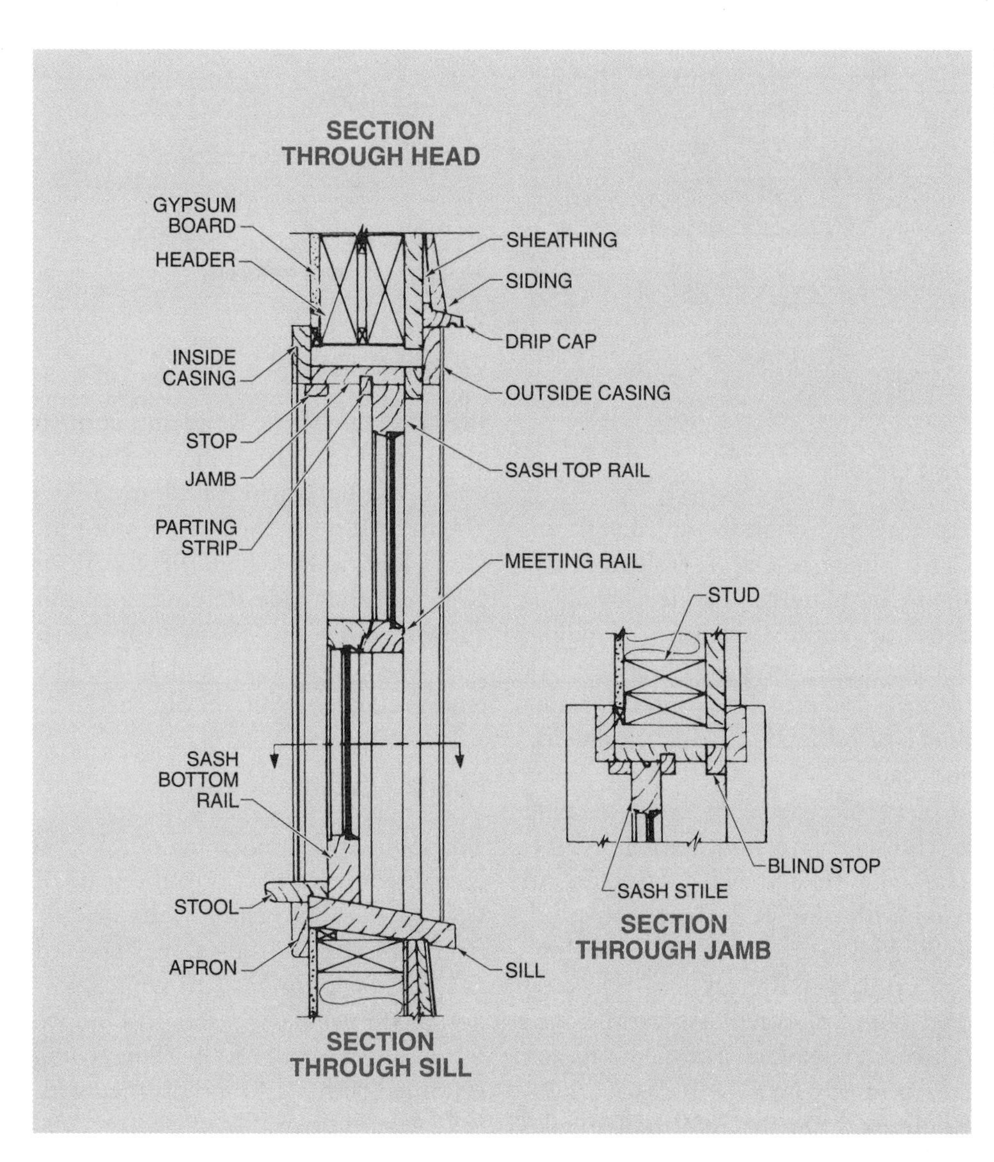

Figure 9-9. Section details through the head, sill, and jamb provide construction information for windows.

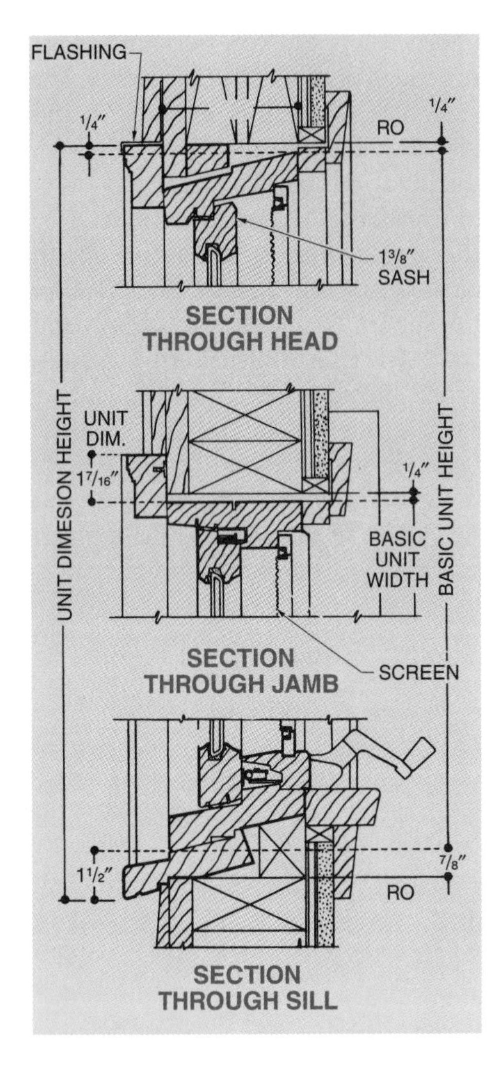

Figure 9-10. Section details through the head, sill, and jamb of a window are related. (Note the relationship between the sill and head.)

Brick details often show the brick bond. The brick bond is the pattern formed by the exposed faces of the brick.

Trim Details

Exterior and interior trim details are usually drawn to a larger scale to clearly show the materials needed. Exterior details are commonly drawn of siding, dormers, doorways, and unique brick patterns. Interior details are commonly drawn of cabinets, stairways, millwork, and trim.

A cornice detail provides framing and finish information. See Figure 9-11. The wall is framed with 2 × 6 studs and filled with 5½″ R-21 batt insulation. Ceiling joists are 2 × 10s filled with 6″ batt insulation. Rafters are 2 × 8s covered with ½″ oriented strand board (OSB) and a shingle roof. The rafters are trimmed with a 1 × 10 fascia. A 2 × 8 subfascia ties the rafter tails (ends) together. The exterior wall is ¾″ plywood siding over ½″ OSB sheathing. The interior wall is finished with ½″ gypsum board.

Moldings for exterior and interior trim are commercially available in a wide range of shapes, sizes, and types of wood. Common moldings include casing, base, shoe, crown, mullion, apron, brick, cove, inside and outside corner, stop, and chair rail. Other moldings are also available. See Appendix.

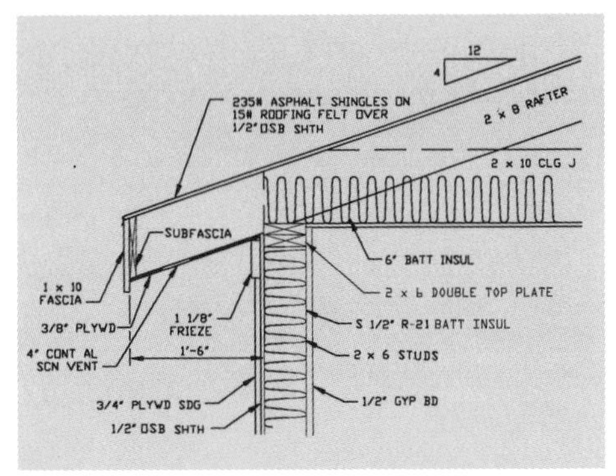

Figure 9-11. A cornice detail shows framing and finish members.

READING DETAILS

Plans for the Wayne Residence contain sheets with details. (Refer to Sheets 5 and 6 of the Wayne Residence.) The Kitchen Plan on Sheet 6 does not include additional details of the floor plan but does indicate cutting planes showing directions of sight for Elevations A-A, B-B, C-C, and D-D. The Stair Detail is shown on Sheet 7.

Panned Ceiling Detail

The Panned Ceiling Detail on Sheet 6 is a section view drawn to the scale of ¾″ = 1′-0″. The ceiling extends 2′-0″ from all walls and then slopes upward at a 45° angle to meet the center portion of the ceiling, which is 1′-0″ above the ceiling near the walls. The ceiling is framed with 2 × 4s and finished with ½″ gypsum board.

Deck Handrail Detail

The Deck Handrail Detail is an elevation view drawn at a ½″ = 1′-0″ scale. Dimensions for the deck are shown on the Floor Plan, Sheet 2. The detail provides dimensions for the handrail.

The deck rests on 2 × 8 joists attached to 4 × 4 pressure-treated posts. The deck floor is made of treated 2 × 6s. Handrail balusters are treated 2 × 2s, mitered at the top and bottom, spaced 6″ on center. Treated 2 × 4s and 2 × 6s form the handrail, which is 3′-2″ above the deck floor.

Fireplace Details

An elevation and a section of the brick fireplace are drawn as details at a ½″ = 1′-0″ scale. The elevation detail shows marble surrounding the firebox with a 2 × 10 × 6′-0″ mantel. The fireplace is 6′-0″ wide.

The section view of the fireplace shows a ceramic tile hearth extending 1′-6″ from the firebox opening. Brick symbols show firebrick around the firebox and face brick elsewhere.

Basement Wall Detail and Walkout Detail

The Basement Wall and Walkout Details, shown side by side on Sheet 5, are drawn at a ½″ = 1′-0″ scale. The Foundation/Basement Plan on Sheet 1 shows the stepped basement wall and the Plot Plan on Sheet 6 shows the lot slope.

The front and sides of the house are 4″ brick veneer over a 2 × 6 framed wall. The rear of the house is 12″ lap siding over a 2 × 6 framed wall for the portion around the walkout basement. The rear wall below grade is 8″ thick concrete. The 6 on 12 sloped roof is framed with 2 × 8 rafters spaced 16″ OC. Ceiling and floor joists are 2 × 10s spaced 16″ OC. Oriented strand board sheathing is covered with felt and cedar shakes.

The brick veneer walls are framed with 2 × 6s over which ½″ insulated sheathing is applied. A ½″ air space separates the face brick from the sheathing. The brick veneer walls are rated R-21. The rear wall is also framed with 2 × 6s that are faced with ½″ foam board insulation. The exterior finish for the rear wall of the house is 12″ lap siding. The rear wall is also rated R-21.

The footing for the 10″ concrete basement wall measures 8″ × 20″. Three #4 rebar are placed continuously in the foundation footing and #4 rebar are placed horizontally every 2′-0″ to reinforce the foundation. The concrete wall contains #4 rebar placed horizontally and vertically at 2′-0″ OC.

The footing for the rear wall of the house is 8″ × 16″ with #4 rebar. A 2′-0″ foundation wall supports the 4″ concrete slab that forms the basement floor. The slab is reinforced with 6 × 6—W1.4 × W1.4 welded wire reinforcement. A 6 mil plastic sheet provides a vapor barrier to resist moisture penetration.

The main floor of the house is supported by 2 × 10 floor joists spaced 16″ OC. Batt insulation is placed 30″ in from all exterior walls. A ¾″ tongue-and-groove plywood deck is nailed to the floor joists, covered with carpet, and trimmed with baseboard molding.

Trus Joist, A Weyerhaeuser Business

Engineered wood products such as wood I-beams and parallel strand lumber (PSL) beams may require specific installation details.

Details 9 Sketching

Name ____________________ Date ____________

Sketching 9-1

Sketch an elevation showing details of the vanity wall in the space provided. Include the following details:

1. 4′-0″, two-door vanity, 30″ high
2. 12″ × 12″ × 8′-6″ soffit above vanity
3. 6″ backsplash on vanity
4. 48″ × 48″ mirror above vanity
5. Floor-to-ceiling height of 8′-1″

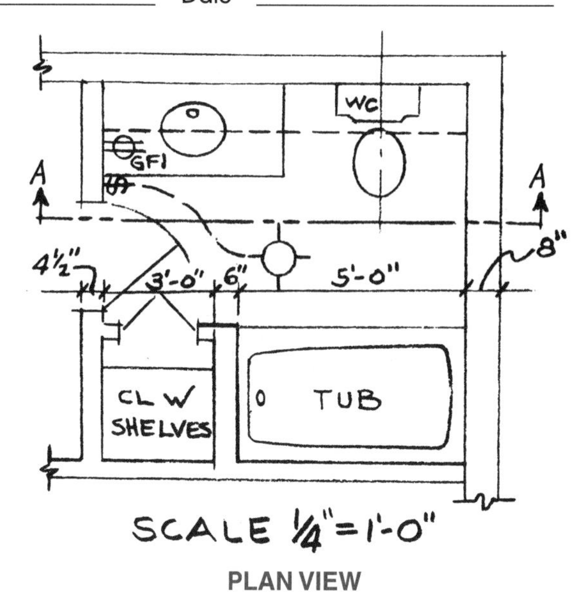

PLAN VIEW

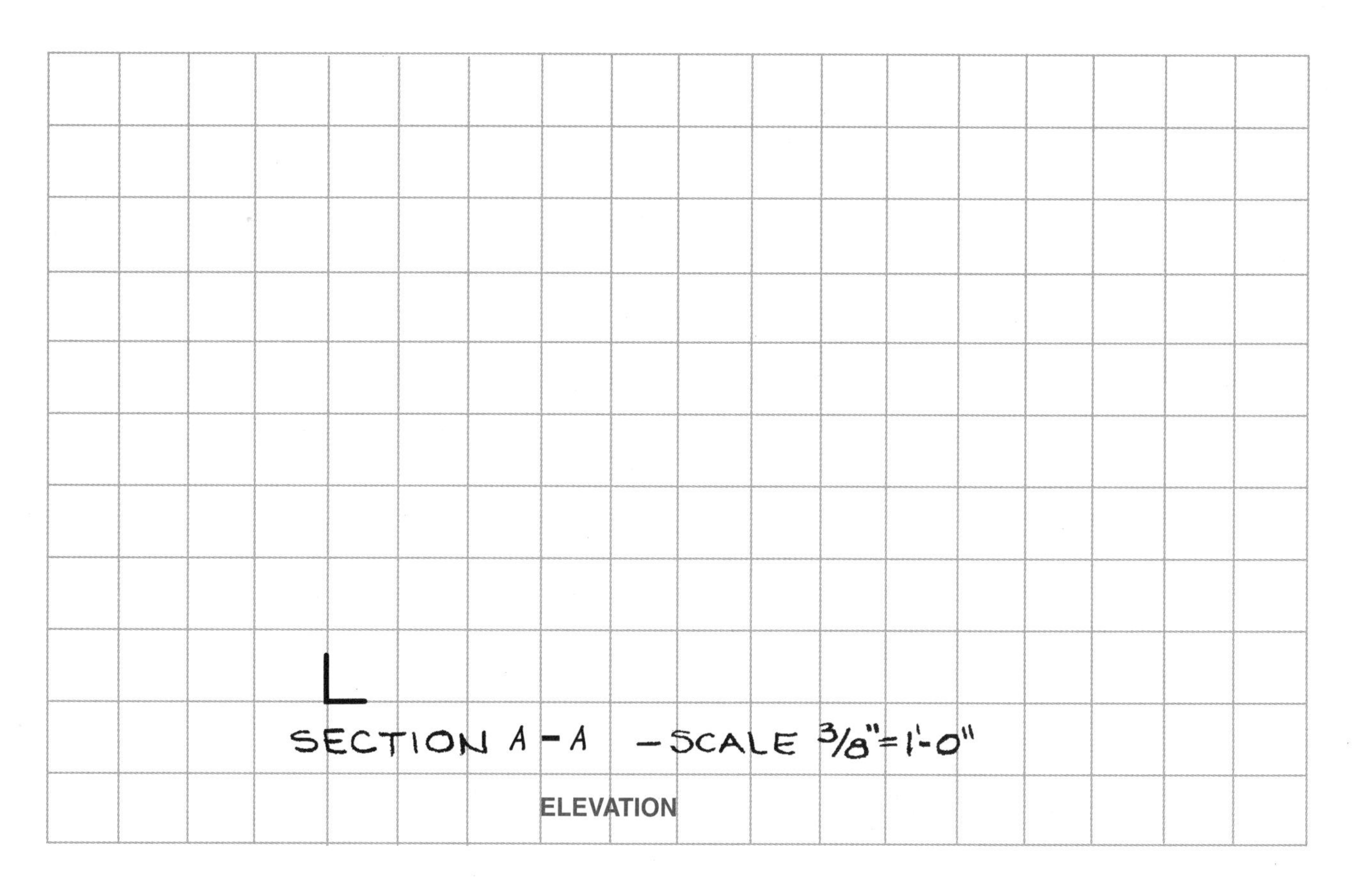

ELEVATION

Sketching 9-2

Sketch full-size details of the moldings in the space provided. Refer to Appendix.

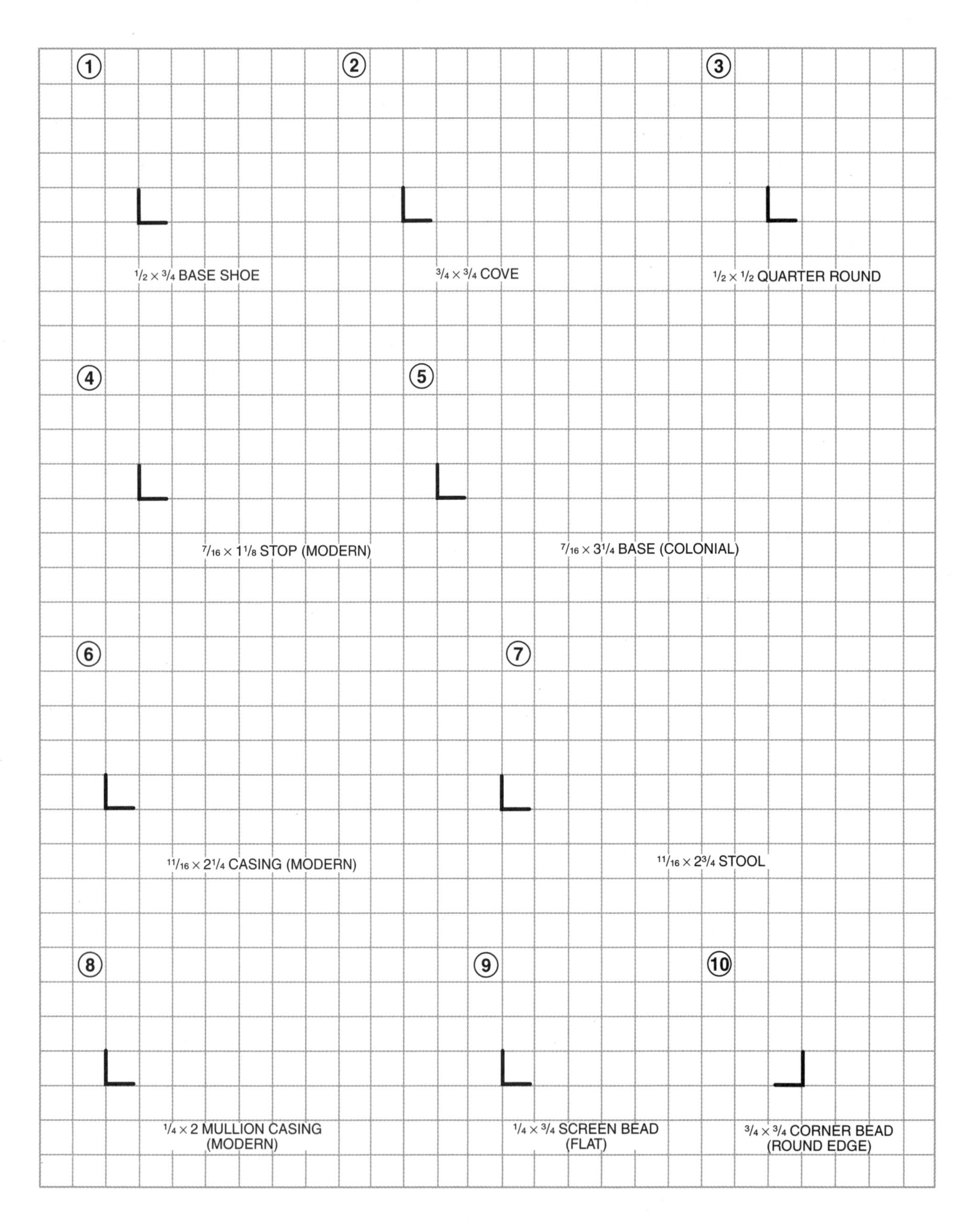

Details 9

Review Questions

Name ______________________________ Date ______________

Identification 9-1

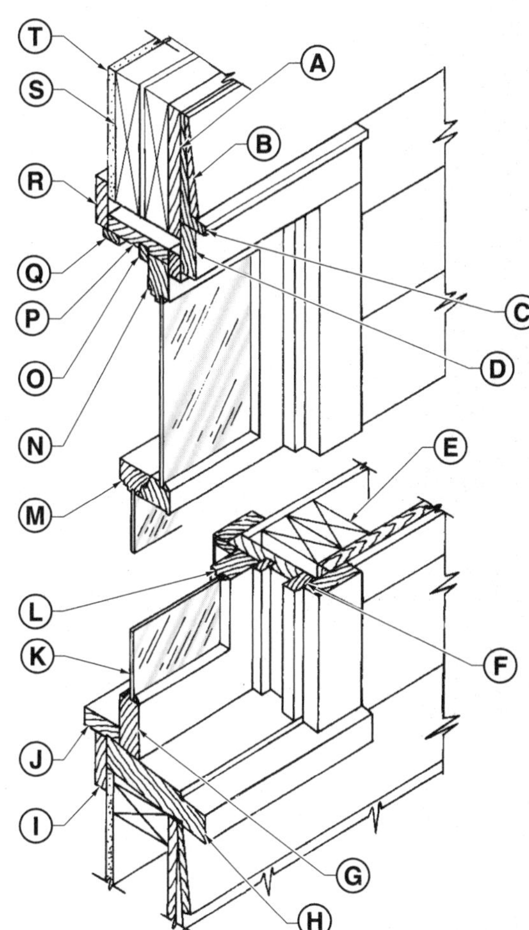

________________ **1.** Studs

________________ **2.** Siding

________________ **3.** Sash bottom rail

________________ **4.** Apron

________________ **5.** Glass light

________________ **6.** Jamb

________________ **7.** Inside casing

________________ **8.** Sheathing

________________ **9.** Sash top rail

________________ **10.** Sash stile

________________ **11.** Stop

________________ **12.** Outside casing

________________ **13.** Meeting rail

________________ **14.** Sill

________________ **15.** Gypsum board

________________ **16.** Drip cap

________________ **17.** Stool

________________ **18.** Parting strip

________________ **19.** Header

________________ **20.** Blind stop

Completion

Refer to Details on page 203.

_______________ **1.** The fireplace hearth is ___″ above the finished floor level.

_______________ **2.** The fireplace opening is ___ high.

_______________ **3.** The fireplace hearth is supported by a(n) ___ slab.

_______________ **4.** The fireplace chimney is faced with ___ brick.

_______________ **5.** The fireplace hearth is a piece of ___.

_______________ **6.** The fireplace mantel is made of ___.

_______________ **7.** The fireplace mantel is ___ above the finish floor level.

_______________ **8.** ___ panel doors are used on Kitchen cabinets.

_______________ **9.** The number of wall cabinet doors shown is ___.

_______________ **10.** The dishwasher is located to the ___ of the Kitchen sink.

_______________ **11.** Pass-through doors from the Kitchen to the breakfast bar are ___ above the finish floor level.

_______________ **12.** The Kitchen cabinets are made of ___.

_______________ **13.** Wall cabinets are ___″ tall.

_______________ **14.** Base cabinets are ___″ tall.

_______________ **15.** The number of base cabinet doors shown is ___.

_______________ **16.** The number of wall ovens shown is ___.

_______________ **17.** A(n) ___ cabinet is placed above the vanity.

_______________ **18.** A(n) ___ ceiling is shown over the bathtub.

_______________ **19.** ___ tile is 4′-0″ above the finished floor level behind the water closet and vanity.

_______________ **20.** The garage roof has a(n) ___ on 12 slope.

_______________ **21.** Plywood used to deck the Garage roof is ___″ thick.

_______________ **22.** The Garage soffit is covered with ___″ plywood.

_______________ **23.** ___ plates are used to tie chords and web members together.

_______________ **24.** Web members are made of ___.

_______________ **25.** Firecode gypsum board on the Garage ceiling is ___″ thick.

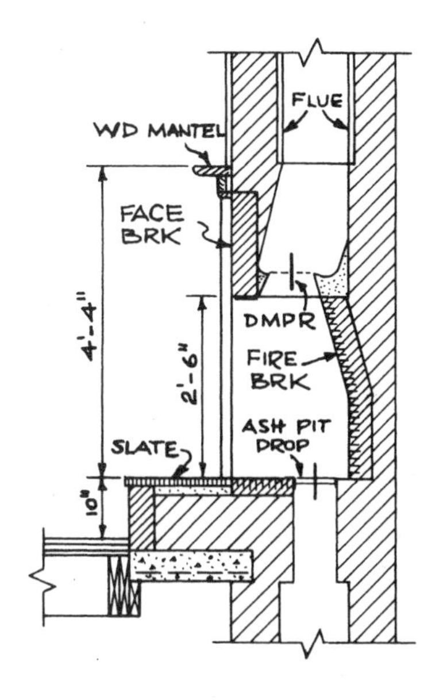

FIREPLACE ELEVATION

BATHROOM ELEVATION

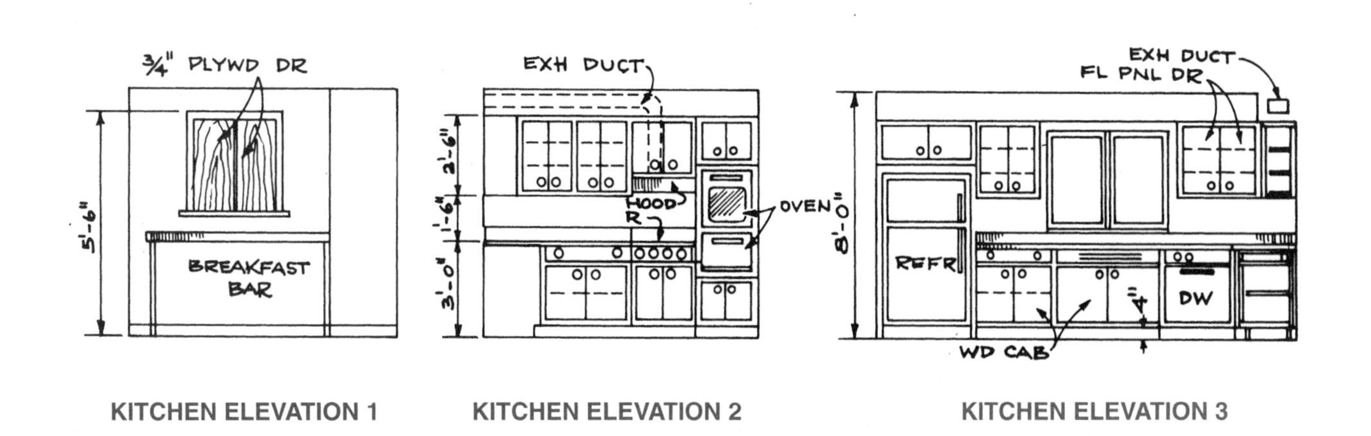

KITCHEN ELEVATION 1

KITCHEN ELEVATION 2

KITCHEN ELEVATION 3

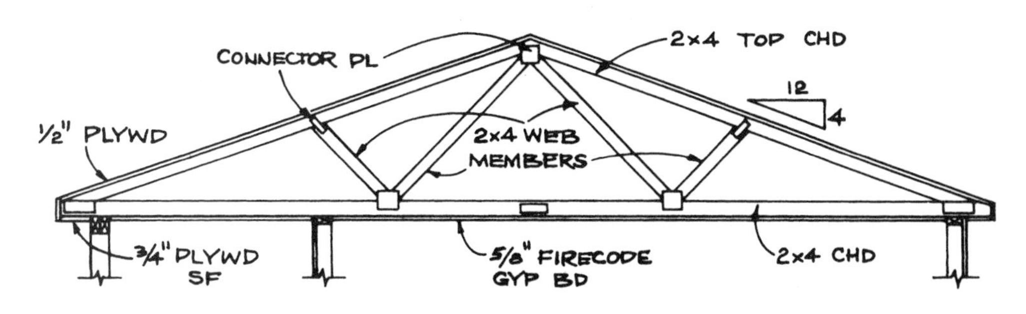

TRUSSED RAFTER DETAIL FOR GARAGE

True-False

T F **1.** Details are scaled plans, elevations, or sections drawn to a larger scale to show special features.

T F **2.** Details are usually drawn to a larger scale than the view from which they are taken.

T F **3.** A preferred scale for details on residential plans is ¼″ = 1′-0″.

T F **4.** Generally, dimensions are not repeated on plans unless necessary for clarity.

T F **5.** Size dimensions locate a particular feature in relation to another feature.

T F **6.** Arrows on a cutting plane line indicate the direction of sight for the section.

T F **7.** Specific references are required for typical details.

T F **8.** The plan view of a kitchen and its elevations are related.

T F **9.** Details of windows are drawn as sections through the head, sill, and jamb.

T F **10.** Interior details are commonly drawn of cabinets, millwork, and trim.

Identification 9-2

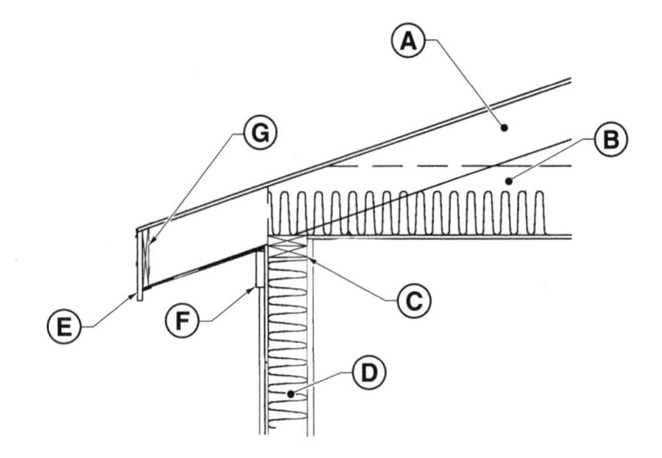

____________ **1.** Double top plate

____________ **2.** Rafter

____________ **3.** Batt insulation

____________ **4.** Ceiling joist

____________ **5.** Fascia

____________ **6.** Frieze

____________ **7.** Subfascia

Details 9

Trade Competency Test

Name ______________________________ Date ______________

Refer to Sheets 5 and 6 of Wayne Residence.

Multiple Choice

______________ **1.** Regarding the Basement Wall Detail, ___.

A. 4″ rigid foam insulation board is placed over the foundation wall
B. the foundation footing is 8″ thick
C. a 5″ drain tile carries water away from the foundation footing
D. floor-to-ceiling height in the basement is 7′-0″

______________ **2.** Regarding the Basement Wall Detail, ___.

A. insulation in the main floor wall is rated R-11
B. the main floor wall is framed with 2 × 4s at 16″ OC
C. 2 × 8 floor joists are spaced 16″ OC
D. none of the above

______________ **3.** Regarding the Basement Wall Detail, ___.

A. the finished main floor is carpet over ½″ plywood
B. a 2 × 4 truss chord is shown
C. a ½″ air space is maintained between the brick veneer and sheathing
D. 2 × 12 ceiling joists are specified

______________ **4.** Regarding the Basement Wall Detail, ___.

A. the foundation footing is 8″ × 20″
B. batt insulation is placed to 24″ in from exterior walls
C. #5 rebar is placed horizontally at 2′-0″ OC in the foundation footing
D. 235 lb asphalt shingles cover the roof

______________ **5.** Regarding the Walkout Detail, ___.

A. the roof overhang extends 16″ past the rear wall
B. the roof has a slope of 6 on 12
C. the subfloor is ¾″ OSB
D. all of the above

______________ **6.** Regarding the Walkout Detail, ___.

A. the minimum foundation wall height is 2′-0″
B. 1″ foam board insulation is used
C. WWR in the concrete slab is 4 × 4, W1.4 × W1.4
D. the scale is ¾″ = 1′-0″

______________ **7.** Regarding the Fireplace Details, ___.

A. a 1 × 10 is shown for the mantel
B. the slate hearth extends 1′-6″ in front of the fireplace
C. paneling is used to finish the wall above the fireplace
D. the firebox opening height is 29″

_______________ **8.** Regarding the Fireplace Details, ___.

A. face brick surrounds the firebox
B. the firebox is 36″ wide
C. a damper is installed above the smoke shelf
D. none of the above

_______________ **9.** Regarding the Panned Ceiling Detail, ___.

A. the scale is ¾″ = 1′-0″
B. the panned ceiling slopes up at a 30° angle
C. 2 × 6s are used to frame the panned ceiling
D. ⅝″ gypsum board is installed over the framing

_______________ **10.** Regarding the Deck Handrail Detail, ___.

A. the handrail is 3′-0″ above the deck floor
B. 2 × 10 joists support the deck floor
C. balusters are spaced 8″ apart
D. 2 × 2s are used as balusters

Completion

_______________ **1.** Fireplace Details are shown in elevations and ___.

_______________ **2.** The Deck Handrail Detail is drawn to a ___ scale.

_______________ **3.** Roof rafters are spaced ___″ OC.

_______________ **4.** ___ fill surrounds the drain tube at the front and sides of the house.

_______________ **5.** The Walkout Detail shows ___″ rigid foam insulation within 30″ of the exterior wall.

_______________ **6.** All windows are to be double-glazed high ___ glass.

_______________ **7.** Posts for the Deck are treated ___.

_______________ **8.** The hearth extends ___″ from the fireplace.

_______________ **9.** The foundation footing of the Walkout Detail measures ___.

_______________ **10.** The concrete slab in the basement is ___″ thick.

_______________ **11.** ___ grade plywood is used on the soffit.

_______________ **12.** Interior walls on the main floor are finished with ___″ gypsum board.

_______________ **13.** The Fireplace Details are drawn to a(n) ___ scale.

_______________ **14.** The total width of the fireplace is ___.

_______________ **15.** The concrete foundation wall of the basement is ___″ thick.

_______________ **16.** ___ molding and carpet finish the main floor.

_______________ **17.** The finished ceiling height of the main floor is ___.

_______________ **18.** Batt insulation on the rear wall provides an insulation value of ___.

Trade Information 10

Many trade-specific tasks are required in residential construction to complete a house in an orderly and efficient manner. Carpentry, masonry, electrical work, plumbing, and HVAC installation are all utilized in residential construction. A general contractor coordinates the work to ensure that construction is completed according to plans and specifications.

CARPENTRY

Carpentry is used in many aspects of residential construction, including erecting forms for walkways, driveways, basement and garage floors, and foundation footings and walls. Other carpentry work involves framing walls and roofs, installing floors, and applying exterior and interior trim. Information required for carpentry work is generally included in the specifications and schedules and on plot plans, floor plans, sections, details, and elevations.

Concrete Foundation Work

Concrete foundation work begins by establishing the corners of a building in relation to the point of beginning, which is shown on the plot plan. Stakes are set up to indicate the size and depth of the excavation. After rough excavation of a building lot, batterboards are erected and strings are stretched to indicate the exterior face of the foundation wall. Batterboards are set back from the excavation and are used for verifying dimensions and elevations at several stages during footing and foundation form construction. See Figure 10-1.

Concrete for footings is usually placed first. Footing formwork is positioned so that the foundation wall is centered on the footings and the tops of the footings are level and at the proper height. Rebar is placed in the footings when the concrete is placed to increase the tensile strength of the concrete. A *rebar* is a deformed steel bar used to reinforce concrete structural members. After the concrete for the footings is placed, a keyway is formed in the footings. A *keyway* is a groove formed in fresh concrete that interlocks with concrete from the foundation wall.

Anchor bolts are positioned in fresh concrete for fastening the sill plate.

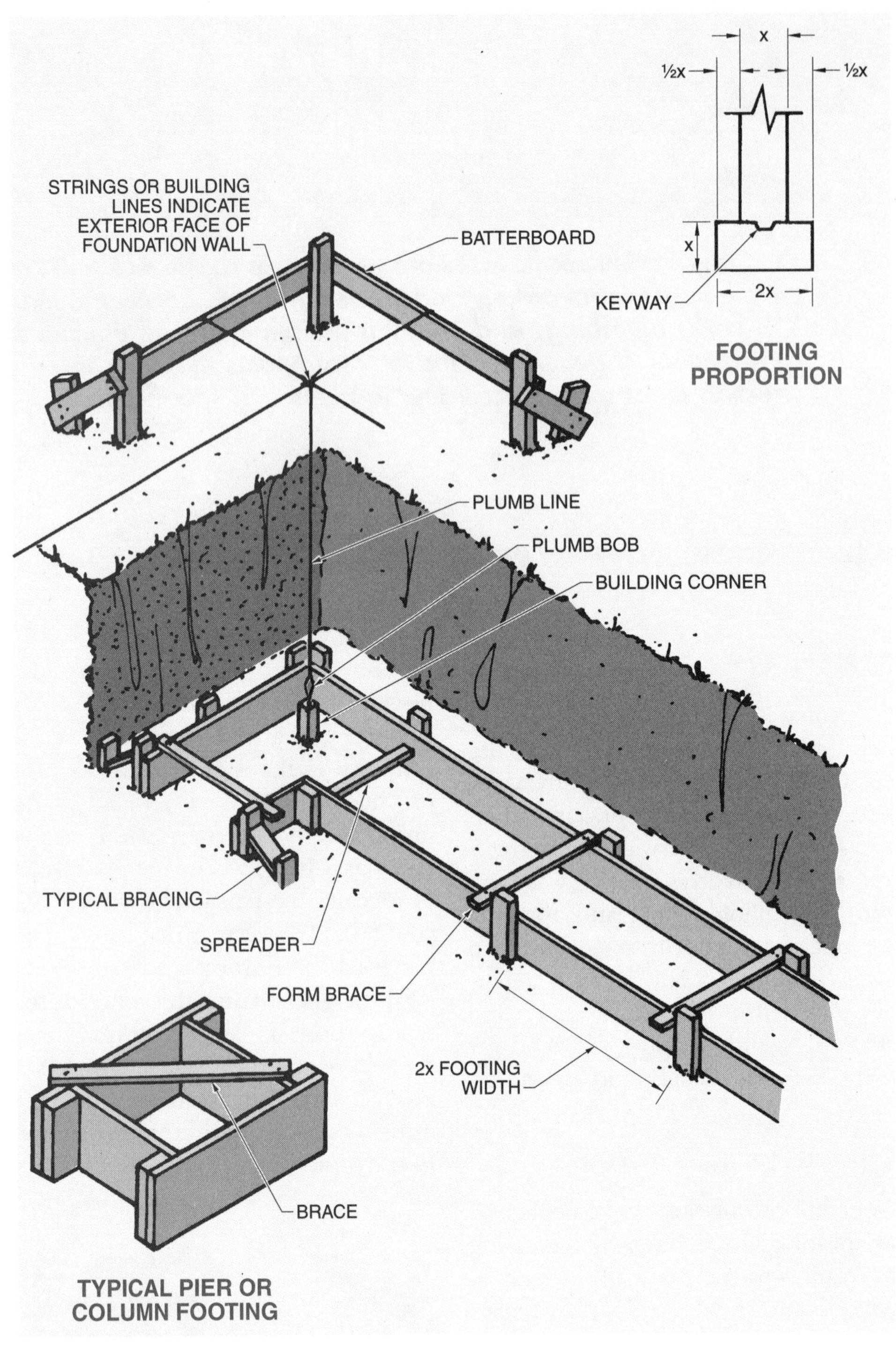

Figure 10-1. An intersection of building lines stretched between batterboards establishes the exterior face of a foundation wall.

High-density overlay (HDO) plywood is exterior grade plywood with a resin-impregnated fiber veneer. High-density overlay plywood is commonly used for concrete forming.

Foundation wall formwork is erected after the concrete for the footings has set and hardened. In general, foundation walls for a full-basement house are 8″ thick and high enough to provide adequate headroom in the basement.

Foundation walls are formed using job-built or panel forms. See Figure 10-2. A *job-built form* is a foundation wall form constructed piece by piece on top of a footing. Bottom plates are fastened to the footing and studs are erected and properly braced. Sheathing is attached to the studs and braces are installed to secure the forms. High-density overlay (HDO) plywood, Plyform®, or fiberglass-reinforced-plastic (FRP) plywood is commonly used as sheathing.

Foundation Wall Forms

Figure 10-2

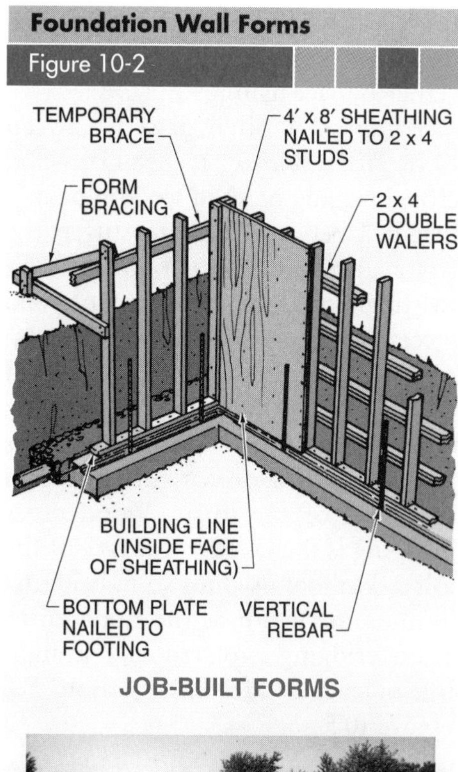

Figure 10-2. Job-built or panel forms are used to form foundation walls.

A *panel form* is a prefabricated concrete wall form consisting of a metal frame and metal or wood panel facing or wood studs and plates with wood panel facing. Panel forms are the most common form used in residential construction. Steel wedge form ties or snap ties with spreader cones hold the forms at a uniform distance and maintain proper form wall position when concrete is placed. The end studs are fastened together with duplex nails to form the corner of the foundation wall. A *duplex nail* is a double-headed nail designed to be pulled out easily to facilitate the stripping of forms.

Slab-at-grade foundations are commonly used in warm climates, where the shallow frost line does not present soil heaving and shifting problems. A *slab-at-grade foundation* is a ground-supported foundation system that combines short concrete foundation walls or a thickened concrete edge with a concrete floor slab. The slab rests on a bed of gravel and a vapor barrier placed over the ground. When slab-at-grade foundations are installed in cold climates, additional slab edge insulation is required. See Figure 10-3. Residential slabs are commonly 4″ to 8″ thick and may be thicker for heavier loads.

Vapor barrier joints should be lapped at least 4″. The vapor barrier should not be punctured when placed over the gravel base.

Floor, Wall, and Roof Construction

Floor construction begins after the concrete for the foundation footings and walls is set and hardened and the forms are stripped. Floor plans and section details through exterior walls are referred to in order to determine dimensions and materials to be used. Pressure-treated sill plates are attached to the top of the foundation wall. Floor joists are supported at the ends by the sill plates. Sizes and directions of the floor joists are indicated on the floor plan with notations. Floor joists typically run across the width of a house. Header joists are installed along the ends of the floor joists to maintain proper spacing and joist alignment.

A keyway in the footing interlocks with concrete from the foundation wall.

APA—The Engineered Wood Association

Depending on the span, joists may be supported with a steel beam or a solid, built-up, or engineered wood beam.

Longer joists require additional support along their span and may be supported with a steel beam or a solid, built-up, or engineered wood beam. Steel or wood posts may be required for intermediate support of beams depending on the length, size, and weight to be supported. See Figure 10-4. Floor joists should either butt together or be lapped over the intermediate beam. Butted joists are joined using wood or metal ties. Lapped joists are fastened together by face-nailing.

Bridging is often placed between floor joists. *Bridging* is metal or wood cross bridging or solid wood blocking installed between joists to stiffen the floor unit and prevent joists from twisting. Cross bridging is metal or wood pieces (commonly 1 × 3s or 1 × 4s) attached near the top of one joist and the bottom of the adjacent joist in a cross pattern. Metal cross bridging has nailing flanges driven into place in the faces of the joists. Wood cross bridging is toenailed to the face of the joists. Lumber used for solid bridging is the same width as the floor joists. Solid bridging is placed in a straight line or staggered for faster nailing. See Figure 10-5.

When installing cross bridging, the upper ends are secured to the joists prior to the subfloor being installed. The lower ends are fastened after all building loads are in place.

Slab-at-Grade Foundations

Figure 10-3

2 x 6 STUD
SHEATHING
2 x 6 PLATE
FLASHING
VAPOR BARRIER
FIBER CEMENT BOARD
2" RIGID WATERPROOF INSULATION
THICKENED EDGE
WARM CLIMATES

6" MINIMUM
GRADE
2 x 6 WITH ½"Ø x 12" ANCHOR BOLTS 48" OC (TYP)
RIGID INSULATION EXTENDS 2'-0" UNDER FLOOR
2 x 4 PLATE
SHORT FOUNDATION WALL
CARRY RIM WALL BELOW FROST LINE
FOOTING FOR BEARING WALL
ASPHALT FELT
VAPOR BARRIER
4" GRAVEL
COLD CLIMATES

Figure 10-3. Slab-at-grade foundations are common in warm climates and must be properly insulated in cold climates.

After the floor joists are placed, the subfloor is installed and walls are framed. Floor plans and wall details are referred to for proper dimensions. The length and height of walls, size and spacing of studs, and location and size of rough openings for doors and windows are determined. Framed walls are commonly laid out and constructed on the subfloor and raised into position. The walls are then plumbed, squared, and fastened to the framing members of the floor. See Figure 10-6. Room layout, particularly for bathrooms and kitchens, is critical, requiring additional reference to the floor plans.

Roofs are framed after the walls are in place and properly braced. Wall and roof framing sections and details indicate type and size of roofing material and roof overhang. Notations on floor plans indicate the size and direction of ceiling joists. Roof framing plans providing comprehensive roof framing information are included in the prints for houses with complex roof layouts. Elevations indicate the roof slope and finish materials to be applied.

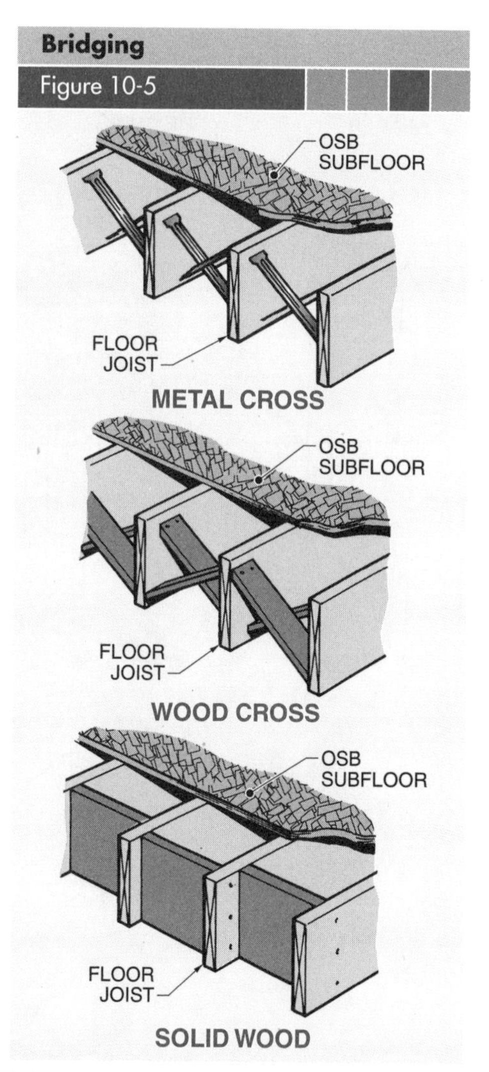

Figure 10-5. Bridging stiffens a floor unit and prevents joists from twisting.

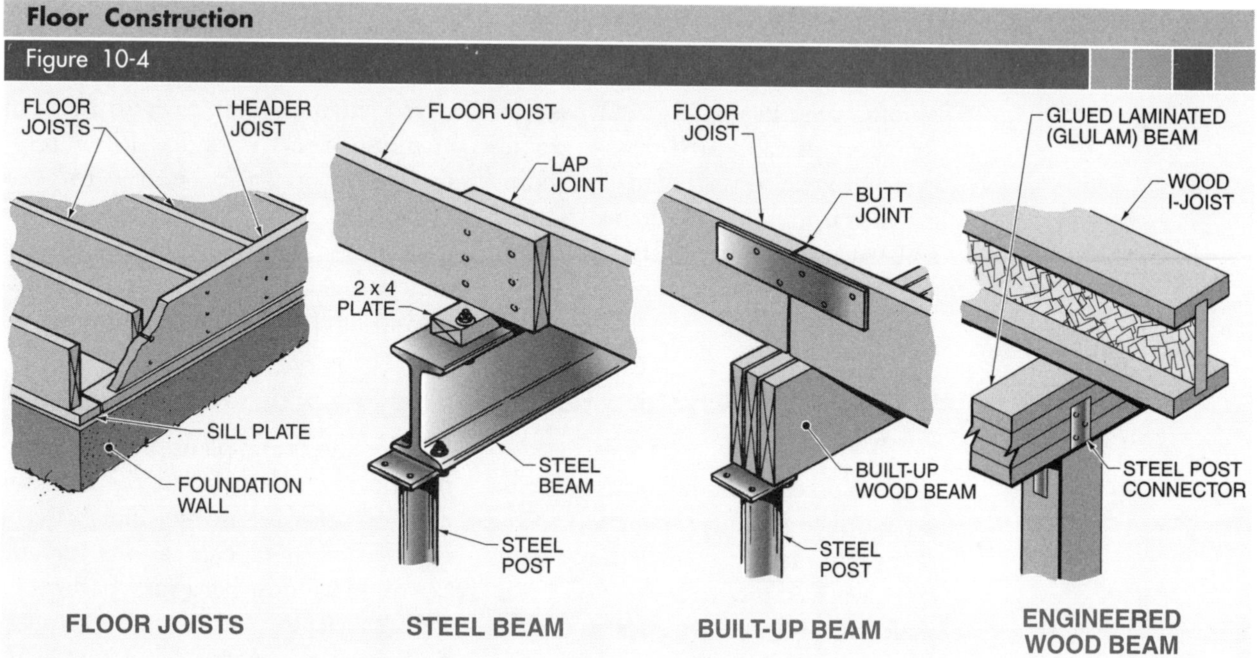

Figure 10-4. Notations on floor plans indicate size and direction of floor joists.

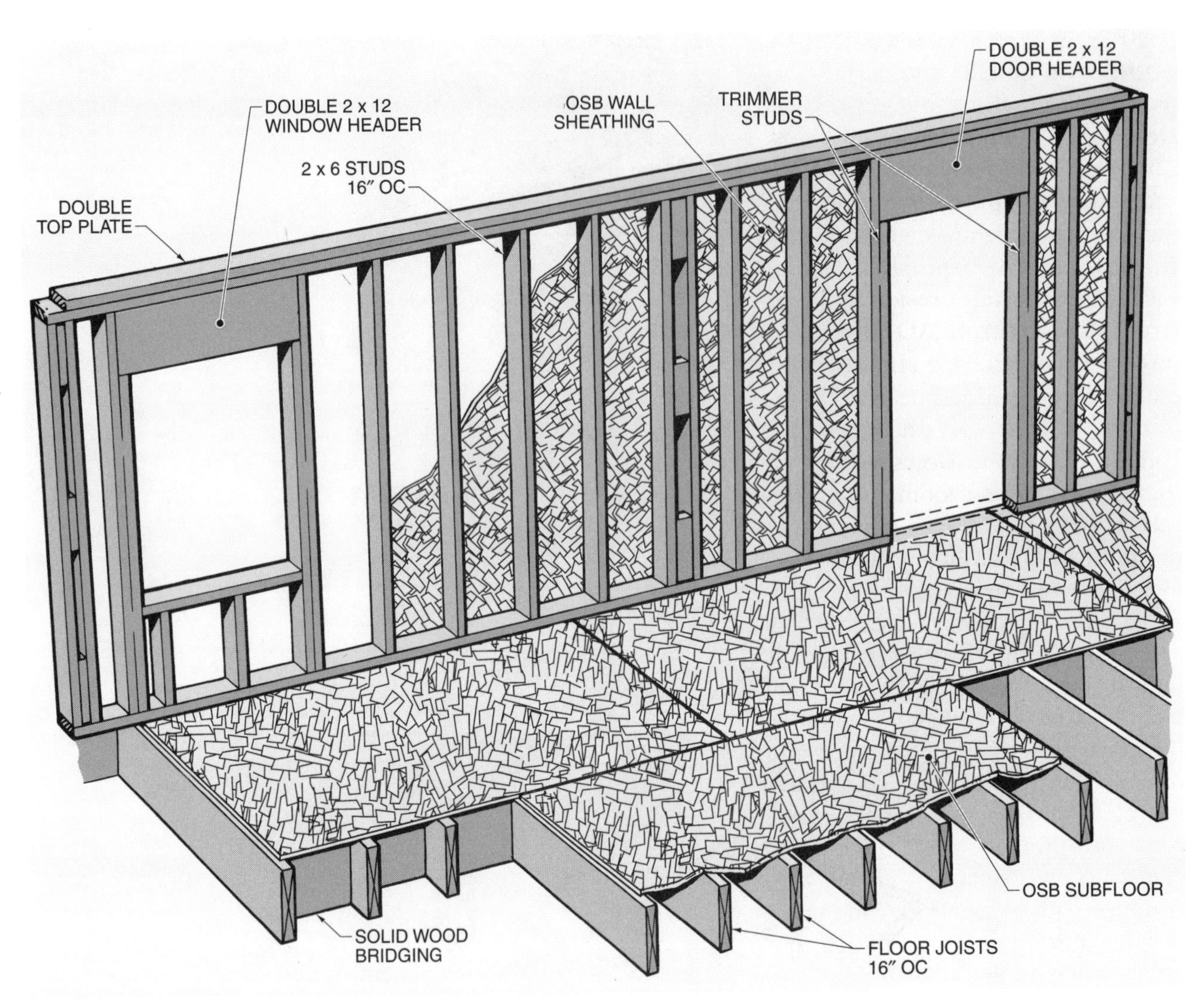

Figure 10-6. Floor plans and details provide wall framing dimensions.

Roofs are constructed with trusses or rafters. See Figure 10-7. In truss construction, prefabricated trusses are delivered to the job site and commonly installed using a mobile crane. In rafter construction, ceiling joists and rafters are cut and installed individually on the job site.

Walkway, Driveway, and Floor Formwork

Tradesworkers erect the forms for walkways, driveways, and garage floors. The forms are usually installed and concrete is placed toward the end of a construction project so the concrete does not become worn or damaged prior to occupation. Driveways and garage floors are often reinforced with welded wire reinforcement to help prevent cracking. See Appendix. Fiber reinforcement may also be used to reinforce slabs. After finishing, the surface is kept damp for several days as the concrete cures. Forms are typically removed after the second day.

Exterior and Interior Trim

Specifications and plans provide information needed to finish the exterior and interior of a house. Finish items detailed in the specifications include door and window hardware, bathroom fixtures, electrical fixtures, appliances, wall and floor materials and finishes, and similar items. Carpenters fit and apply all trim, such as window and door casings and exterior and interior molding.

In 1992, the Americans with Disabilities Act (ADA) was enacted to provide people with physical disabilities greater access to public and commercial buildings. While the ADA does not specifically apply to residential construction, a variety of ADA-compliant residential fixtures and appliances are available to accommodate people with physical disabilities. These fixtures and appliances provide users with additional maneuvering room and special features such as grab bars. The fixtures and appliances must be installed at the proper height and may require specialized fittings or trim to operate properly. Print dimensions should be carefully verified when constructing a house with ADA-compliant fixtures and appliances. For example, the standard height of a kitchen base cabinet with a sink is 36″ above the finished floor, while the height of an ADA-compliant kitchen base cabinet with a sink is 34″ maximum above the finished floor.

Ramps may be required to accommodate people with physical disabilities. The maximum slope of a ramp in new construction should be 1:12.

Trusses vs. Rafters

Figure 10-7

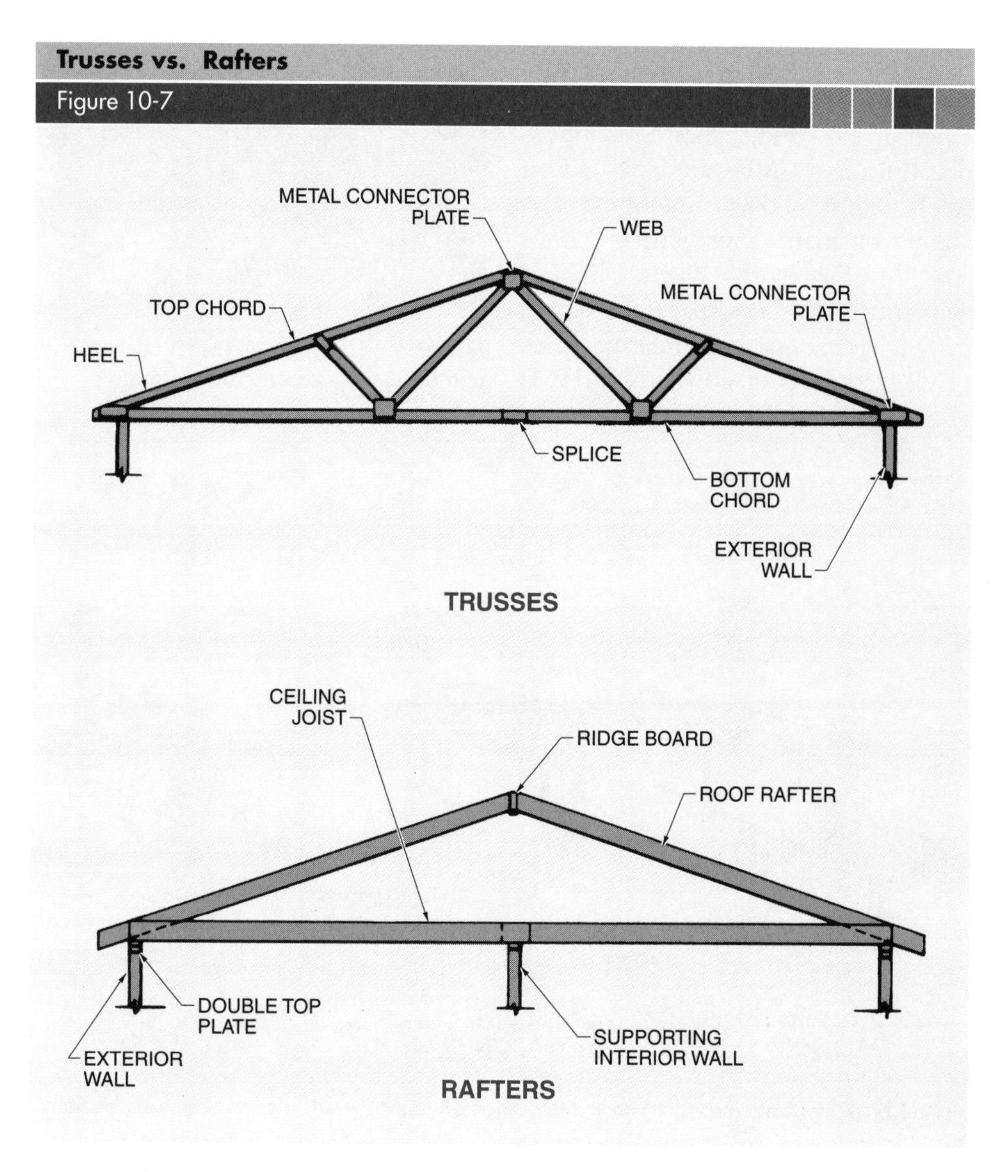

Figure 10-7. Roofs are supported by trusses or rafters.

MASONRY

Masonry work includes laying brick, concrete masonry units (CMUs), tile, and stone. Elevations identify the walls to be bricked and the types of brick to be used. Sections, often section details, through exterior walls show masonry information for the walls at windows and doors and the wall finish at the eaves. Floor plans provide dimensions showing the exact location of window and door openings. Masonry walls include brick, CMU, brick and CMU, and brick veneer over frame construction. See Figure 10-8.

Brick and Concrete Masonry Units

Brick and CMUs are available in a wide range of sizes and shapes. See Appendix. Brick and CMUs are indicated on prints with symbols. Common brick is shown on plan views with 45° lines spaced further apart, while face brick is shown with 45° lines spaced closer together. Horizontal lines indicate brick on elevations. A notation such as FACE BRICK is generally shown in open spaces between the horizontal lines.

Brick are either standard or modular in size. *Standard brick* is brick classified by its nominal size. For example, the nominal size of a face brick is 2″ × 4″ × 8″, but its actual size is approximately 2¼″ × 3¾″ × 8″. *Modular brick* is brick classified by its actual size and designed so every third horizontal joint falls on a multiple of 4″ (modular measure).

Mortar is placed between each brick course and between brick in each course.

Masonry Walls

Figure 10-8

HEADER COURSE
STRETCHER COURSE
INNER WYTHE
OUTER WYTHE
BRICK

STRETCHER
MORTAR JOINT
CONCRETE MASONRY UNITS

HEADER COURSE
CONCRETE MASONRY UNIT
BRICK WYTHE
BRICK AND CONCRETE MASONRY UNITS

BRICK
WALL TIE
AIR SPACE
SHEATHING
2 x 4 STUD
BRICK VENEER

Figure 10-8. Masonry walls include brick, concrete masonry unit (CMU), brick and CMU, and brick veneer over frame construction.

Brick is laid in various positions and bonds as noted on the elevations. *Brick bond* is the pattern formed by the exposed faces of the brick. See Figure 10-9.

Concrete masonry units are commonly used for foundations. Information on CMUs is shown in elevations with horizontal and vertical lines and a series of dots and small circles or triangles. Standard CMU size is 7⅝″ × 7⅝″ × 15⅝″. When a CMU is laid with ⅜″ mortar joints, it measures 8″ × 8″ × 16″ and conforms to modular measure.

Brick positions and bonds are noted on elevations.

ELECTRICAL

Supply conductors from a utility company to a premises are regulated by the *National Electrical Safety Code®* (*NESC®*) Service-entrance conductors and electrical work from the meter base to the service equipment and the dwelling are governed by the current edition of the *National Electrical Code®* (*NEC®*). The purpose of the NEC is the practical safeguarding of persons and property from hazards that may arise from using electricity. The NEC is published by the National Fire Protection Association (NFPA). The NEC is advisory; it is adopted by states, counties, and municipalities. The NEC is updated every three years. In addition to the NEC, local building code ordinances are enforced.

Service

Electrical service is grounded at the service equipment to provide a common grounding terminal for the grounded neutral, equipment grounding conductors, and the grounding electrode conductor. A main bonding jumper grounds the terminal bar to the metal enclosure connected to the grounding electrode conductor. Feeders and branch circuits are routed from the service equipment to subpanels and loads.

Electrical service to a dwelling unit (house) is commonly 120/240 V, 1φ (phase) and provided by an overhead service drop or an underground service lateral. Clearances for overhead

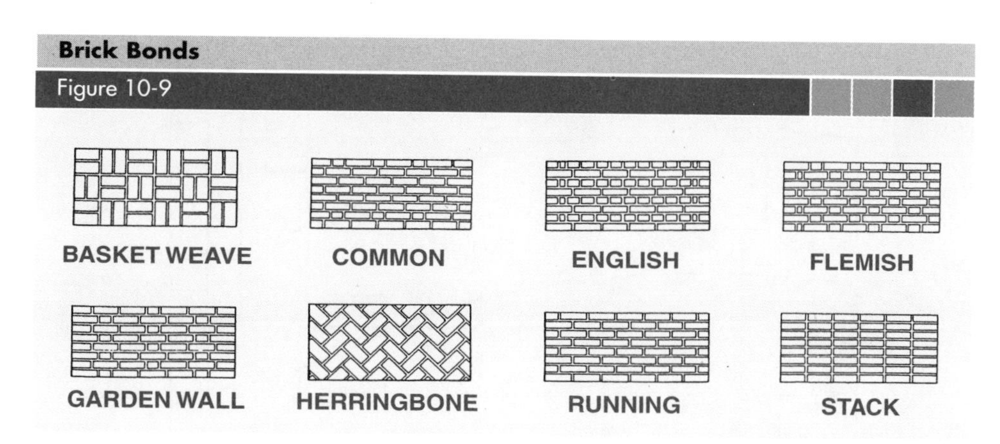

Figure 10-9. Brick bonds are noted on elevations.

Since 1959, the National Electrical Code® has been updated every three years.

drops are regulated by subsection 230.24(B) of the NEC and Table 232.1 of the NESC. Burial depths for service laterals are regulated by Table 300.5 of the NEC. Local building code ordinances may amend these clearances and depths. A power line is shown on a plot plan using a line consisting of a series of intermittent long and short dashes. See Figure 10-10.

Conductors

Copper (Cu) and/or aluminum (Al) conductors are used to wire dwellings. Copper conductors are more expensive than aluminum conductors but have less resistance (for the same size conductor) to current flow. Aluminum conductors in larger sizes are commonly used for service-entrance conductors and feeders to supply power to subpanels. Table 310.16 of the NEC identifies conductor sizes and ampacities for copper and aluminum or copper-clad aluminum conductors according to temperature ratings.

Service-Entrance Conductors. Service-entrance conductors shall have at least a 100 A (ampere) rating where the calculated load is 10 kVA (kilovolt amps) or greater. A 100 A service shall be provided if there are six or more two-wire branch circuits. The minimum size for service-entrance conductors for any dwelling is #6.

Service-entrance conductors are routed in a raceway or service-entrance cable. Insulation is TW, THW, THWN, XHHW, RHW, or other types listed for

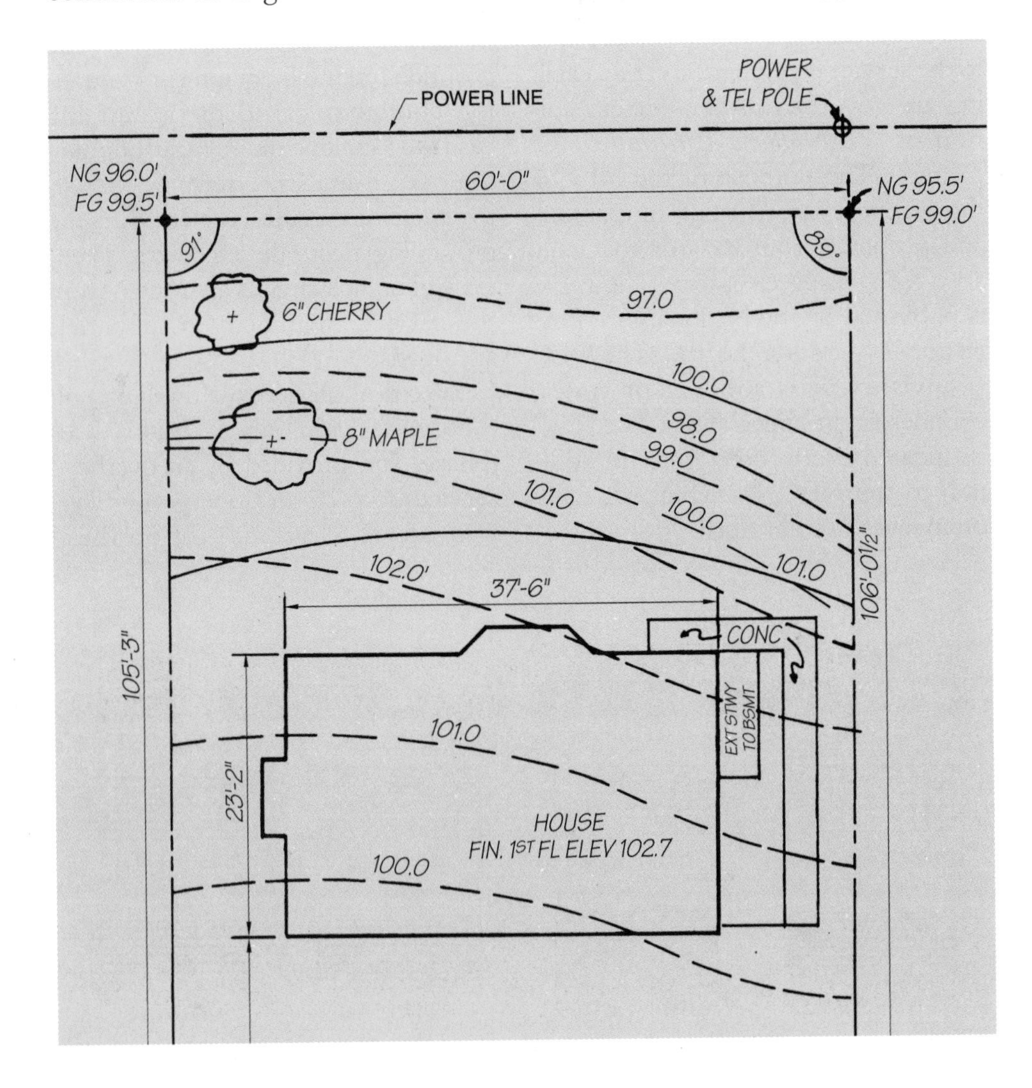

Figure 10-10. The location of power lines is shown on a plot plan.

Electrical systems are grounded to limit and stabilize voltages to ground.

wet locations. (See Table 310.13 of the NEC for conditions of use for each type of insulation.)

Service-entrance conductors are sized from the total VA (volt-amp) rating of the dwelling unit. The formula $I = VA/V$ is used to calculate amps, where I = amps, VA = volt-amps, and V = volts. Volt-amps are divided by volts to obtain ampacity ($I = VA/V$). For example, a calculated load of 36,000 VA on a 240 V, 1ϕ service has an ampacity of 150 A (36,000 VA ÷ 240 V = 150 A). Per Table 310.16 of the NEC, #1/0 THWN Cu conductors are required.

Feeder and Branch Circuit Conductors. Feeder circuits include all conductors between the service equipment and subpanels. Branch circuits include all conductors from the last overcurrent protection device to the load.

Feeder circuits can be any size required to supply a subpanel or taps to individual loads. The minimum size feeder circuit conductor is #10 per section 215.2 of the NEC. Wiring methods for feeder circuits include conductors in raceways, service-entrance (SE) cable, or metal-clad (MC) cable. Feeder circuit conductors are selected from the VA rating of the load served. For example, a subpanel on a 240 V, 1ϕ supply circuit with a connected load of 20,400 VA requires #4 THWN Cu conductors. The formula, $I = VA/V$, is used to find the ampacity of the conductors (20,400 VA ÷ 240 V = 85 A). The conductors are then selected from Table 310.16 of the NEC based on ampacity. An ampacity of 85 A requires #4 THWN Cu conductors at 75°C.

The minimum size branch circuit for a dwelling unit is 15 A. Branch circuits can be any size required to supply the load and can supply one load or a number of loads. Typical branch circuits in dwellings supply combination or single loads. General lighting and receptacle outlets are combination loads. An air conditioner, heating unit, food waste disposer, compactor, or similar appliance is a single load.

A 15 A branch circuit commonly has 10 outlets to supply power to lighting fixtures and general-purpose receptacles. A 20 A branch circuit commonly has 13 outlets to supply power to lighting fixtures and general-purpose receptacles. The minimum VA per outlet is 180 VA per subsection 220.12 of the NEC. Total outlets permitted is calculated by dividing 180 VA by 120 V to obtain 1.5 A, then dividing 1.5 A into the rating of the overcurrent protection device (OCPD). See Figure 10-11.

All 125 V, 15 A, and 20 A, 1ϕ receptacles installed in a kitchen where the receptacles serve countertop surfaces shall have GFCI protection.

Raceway Systems

A *raceway* is an enclosed channel designed to protect electrical conductors. Raceways are made of metal or insulating material. The types of raceway systems used for specific dwelling units are indicated in the print specifications and governed by electrical codes adopted for local use. Many cities permit nonmetallic-sheathed cable (commonly called Romex®) to be used for wiring dwelling units. Other cities may require that all conductors in residential construction be routed in metal conduit.

The most common raceway systems used in residential construction are electrical metallic tubing (EMT), electrical nonmetallic tubing (ENT), rigid nonmetallic conduit (PVC), and multiple-conductor cable assemblies such as AC armored cable and nonmetallic-sheathed cable. In addition, service-entrance cable (SE, ASE, and USE) is used in applications requiring protection from severe mechanical abuse. See Figure 10-12.

Electrical Metallic Tubing (EMT). Electrical metallic tubing (EMT), also known as thin-wall, is similar to rigid

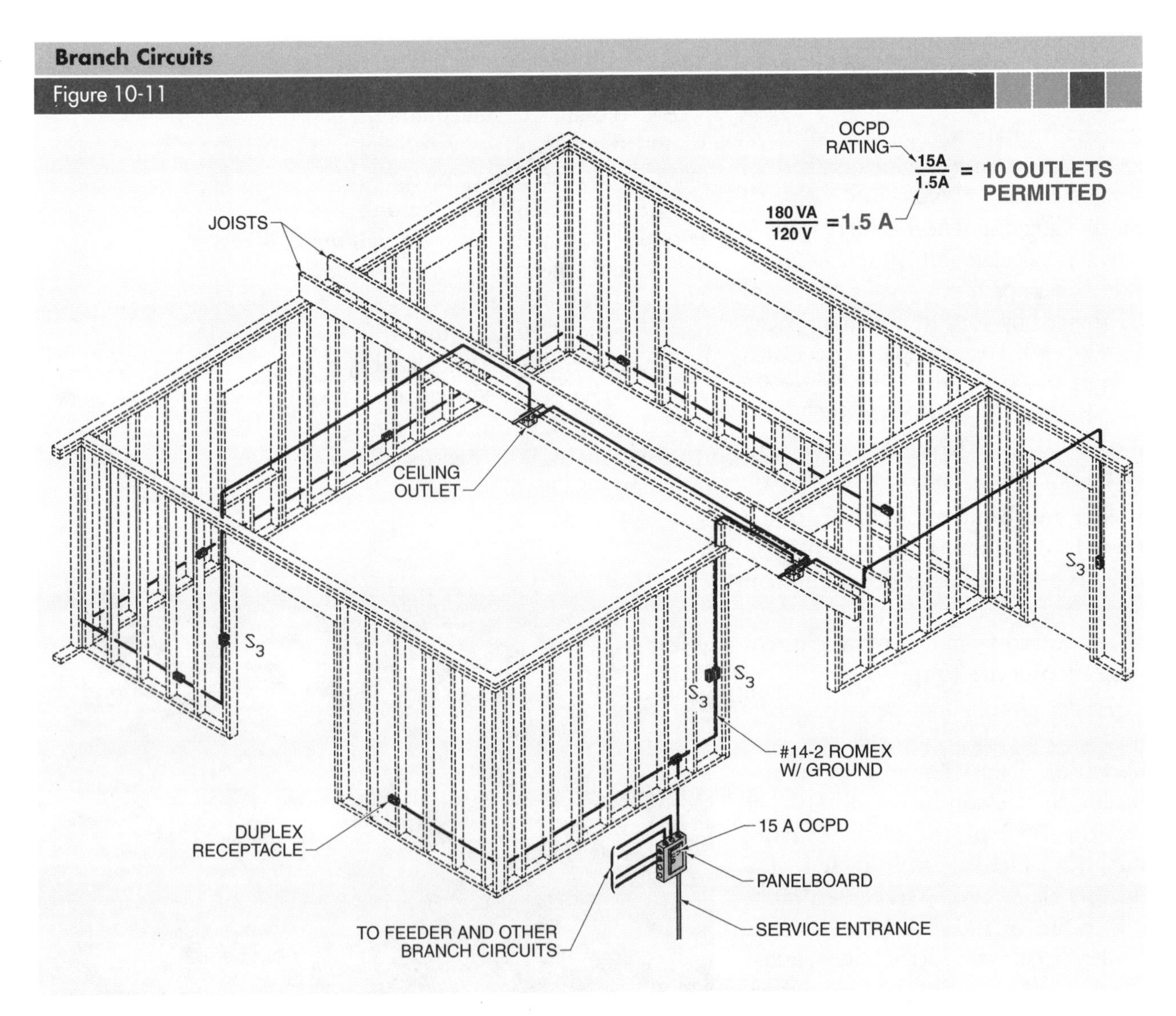

Figure 10-11. The number of outlets permitted on a branch circuit is determined by dividing the rating of the overcurrent protection device (OCPD) by 1.5 A. The NEC® does not limit the number of outlets.

metal conduit but is approximately 40% lighter. Thin-wall is used in exposed and concealed work. Thin wall may be used in most locations, but because of its light weight, thin wall, and compression or set screw couplings and connectors, it cannot be used where it could be subjected to severe physical damage. Thin-wall is available in 10′ lengths from ½″ to 4″ in diameter with ½″ and ¾″ sizes most commonly used in residential construction.

Thin-wall must be supported every 10′ and within 3′ of each outlet box, junction box, cabinet, or fitting. The total number of degrees in all bends in any run shall not exceed 360°. Studs and joists may be notched or drilled to accommodate EMT. Notches, which are preferred, should be as narrow as possible and no deeper than necessary to minimize weakening of the structural members.

Electrical Nonmetallic Tubing (ENT). Electrical nonmetallic tubing (ENT) is a flexible corrugated plastic raceway. Sizes of ½″ to 1″ diameter are available for installation in walls, floors, ceilings, and above suspended ceilings with a 15-minute fire rating. ENT may be bent by hand to a radius of not less than 4″ for a ½″ diameter and not less than 5¾″ for a 1″ diameter. The total number of degrees in all bends in any run shall not exceed 360°.

Electrical nonmetallic tubing must be supported every 3′ and within 3′ of each outlet box, junction box, cabinet, or fitting. All cut ends must be trimmed and bushings or adapters must be used to prevent abrasion.

Rigid Nonmetallic Conduit (RNMC). Rigid nonmetallic conduit (RNMC), also known as polyvinyl chloride (PVC), is waterproof, rustproof, and decay resistant. Types of rigid nonmetallic conduit are Schedule 40 and

Conductors are sized using the American Wire Gauge (AWG) numbering system. The smaller the AWG number, the larger the diameter of the conductor.

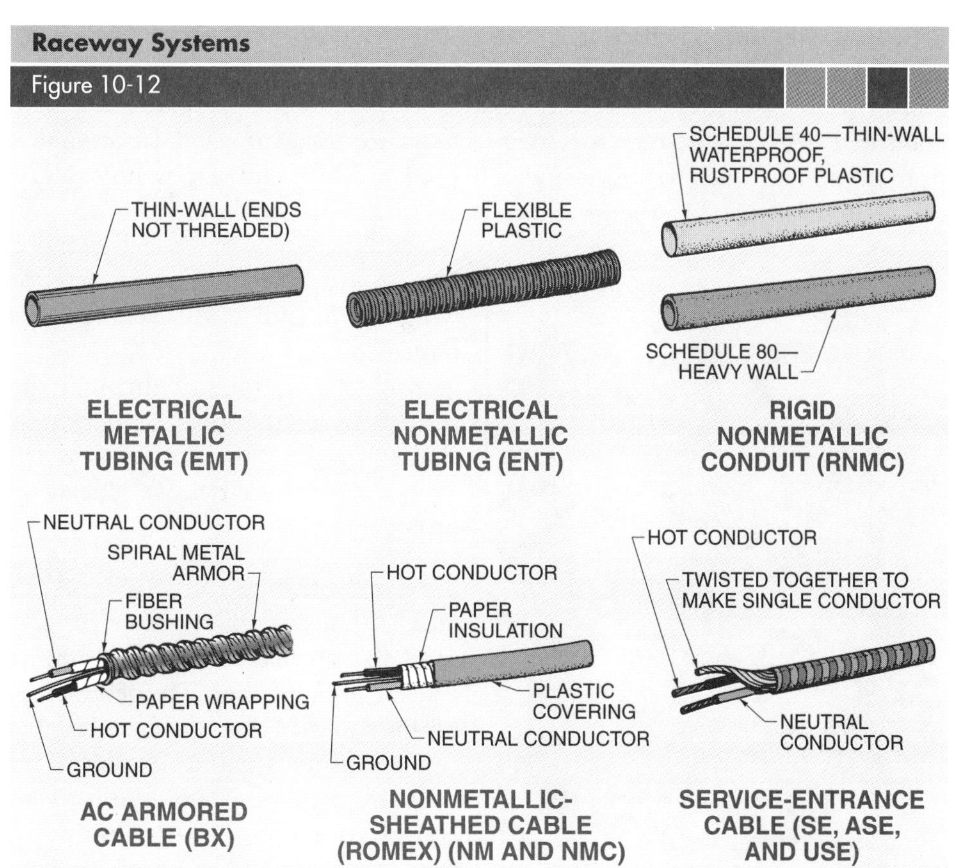

Figure 10-12. Raceway systems for one-family dwelling units include electrical metallic tubing, electrical nonmetallic tubing, rigid nonmetallic conduit, AC armored cable, nonmetallic-sheathed cable, and service-entrance cable.

Schedule 80. Schedule 40 (thin-wall) does not have the strength of metal conduit and should not be used where it may be subjected to severe physical damage. Schedule 80 (heavy wall) may be installed in most locations where rigid metal conduit may be used.

Polyvinyl chloride conduit can be cut with a handsaw and glued together with couplings and connectors. The total number of degrees in all bends in any run shall not exceed 360°. All field bends must be made with bending equipment. Supports shall be provided within 3′ of each box, cabinet, or connection. The maximum spacing between supports is based upon the diameter of the PVC. For example, the 3′ maximum spacing between supports is permitted in runs of ½″ to 1″ diameter PVC.

AC Armored Cable (BX). AC armored cable, also known as BX, is a fabricated assembly of insulated conductors in a flexible metallic enclosure. BX may be used in dry locations where it is not exposed to physical damage. An integral grounding conductor ensures a good path for fault current to return from the point of fault to the grounded conductor connected to the grounded busbar in the main service panel.

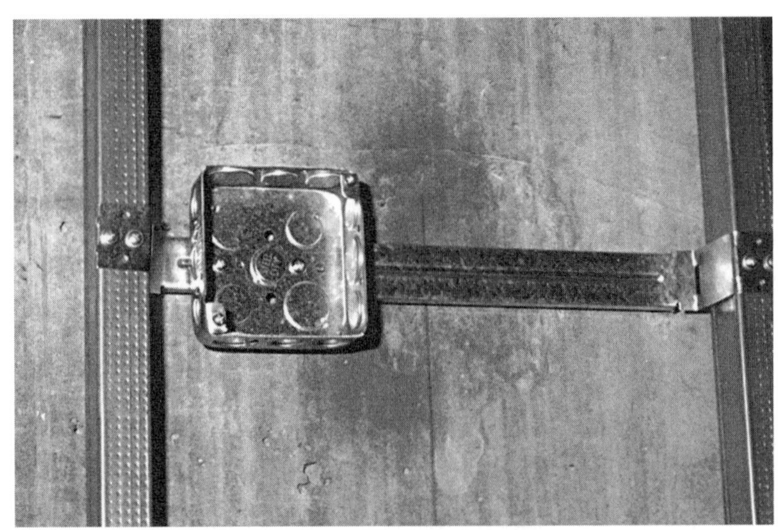

Boxes shall be securely supported.

BX must be supported every 4′-6″ and within 12″ of outlets and fittings and may be run through holes or notches in the studs. Holes must be at least 1¼″ from the edge of the stud so the cable is not penetrated when fastening gypsum board or other items to the wall. Notches must be covered with 1⁄16″ thick steel protector plates. Bends are permitted but shall be no greater than five times the diameter of the cable when measured along the inside edge. For example, the maximum bending radius of ½″cable shall not exceed 2½″.

Nonmetallic-Sheathed Cable (NM and NMC). Nonmetallic-sheathed cable, also known as Romex®, has two, three, or four conductors with a green insulated or bare grounding conductor. Romex is used to economically wire dwelling units and small commercial buildings. Romex is installed with staples or connectors and must be supported every 4′-6″ and within 12″ of outlets and fittings. Metal or nonmetallic boxes may be used with Romex.

NM cable has a flame-retardant and moisture-resistant outer jacket and is used only for interior wiring. NMC cable also has a flame-retardant and moisture-resistant outer jacket, but it can be used for interior or exterior wiring. NM and NMC shall not be buried in concrete.

Romex may be run through the centers of studs, joists, and rafters. Romex may be installed in notched studs when covered by a 1⁄16″ thick steel protector plate.

Service-Entrance Cable (SE, ASE, and USE). Service-entrance cable is used for service-entrance wiring or general interior wiring. The neutral may be insulated or bare when used for service-entrance wiring. An insulated neutral is required when used for general interior wiring. SE cable has a flame-retardant and moisture-resistant outer covering. SE cable is unarmored and will not withstand severe physical abuse. ASE cable has an armored coating for additional protection. USE cable is moisture-resistant with an unarmored coating and is designed for underground installation.

Service-entrance cables may be routed through studs, joists, and rafters. Holes drilled in wood members must not be closer than 1¼″ to the edge. Notched members shall be covered with a 1/16″ thick steel protector plate. Bends in unarmored cable shall not have a radius less than five times the diameter of the cable. The cable shall be supported at intervals not exceeding 4′-6″ and within 12″ of outlets and fittings.

Trus Joist, A Weyerhaeuser Business

When wood I-joists are installed, conductors are routed through prefabricated knockouts.

Plans and Schedules

Specifications describe the material and fixtures to be used when wiring a house. Trade names and catalog numbers are often provided as well. Floor plans show where lighting outlets, receptacles, and switches are located. Branch-circuit schedules provide information for lighting outlets, receptacles, and switches. Lighting fixture schedules provide information regarding the type and location of lighting fixtures required.

Dimensions are not indicated on floor plans to locate receptacles and outlets. Long dashed lines on a floor plan representing circuits connect receptacles, outlets, and switches to show outlet controls, not location. An electrician determines the exact location of the outlets, receptacles, and switches and installs them. No point along the floor line in any wall 2′-0″ or more in length shall be more than 6′-0″ from a receptacle in that wall space. Corners unbroken along the floor line by doorways, fireplaces, and similar openings are included. Receptacles, in general, should be spaced equal distances apart.

In kitchens, receptacles are installed in each countertop space of 12″ or more. Receptacles for fastened-in-place appliances, such as disposals, or receptacles for appliances occupying dedicated space, such as refrigerators, are not included. In bathrooms, at least one wall receptacle is installed adjacent to the basin. In addition, one receptacle each is required for the laundry area, basement, garage, and exterior of the house. All 15 A and 20 A, 125 V receptacles in bathrooms, in garages, and outdoors (with grade-level access) and receptacles serving countertops of kitchens shall have ground-fault circuit interrupter (GFCI) protection per subsection 210.8(A) of the NEC. In addition, receptacles in unfinished basements and within 6′-0″ of wet bar sinks shall have GFCI protection.

Provisions must be made for the proper placement and installation of electrical receptacles, switches, and similar items for people with disabilities. ANSI/ICC A117.1, *Accessible and Usable Buildings and Facilities,* contains specifications to make buildings accessible and usable for people with disabilities. For example, ANSI/ICC A117.1 specifies the side reach height range of a wheelchair-bound

Unless otherwise indicated, all wall receptacles should be installed an equal distance from the floor.

Ground-fault circuit interrupter (GFCI) receptacles that are installed in covers must be attached to the cover with more than a single screw.

individual as 9″ to 48″. Outlets and switches should not be installed below 9″ or above 48″ from the finished floor.

Branch circuit schedules and lighting fixture schedules supplement information shown on the floor plan. See Figure 10-13. The service drop is located on the northwest corner of the house. Service-entrance conductors run in rigid galvanized conduit to the meter and then to the main power panel. All branch circuits are protected by fuses or circuit breakers. Circuits 1

Electrical Plans

Figure 10-13

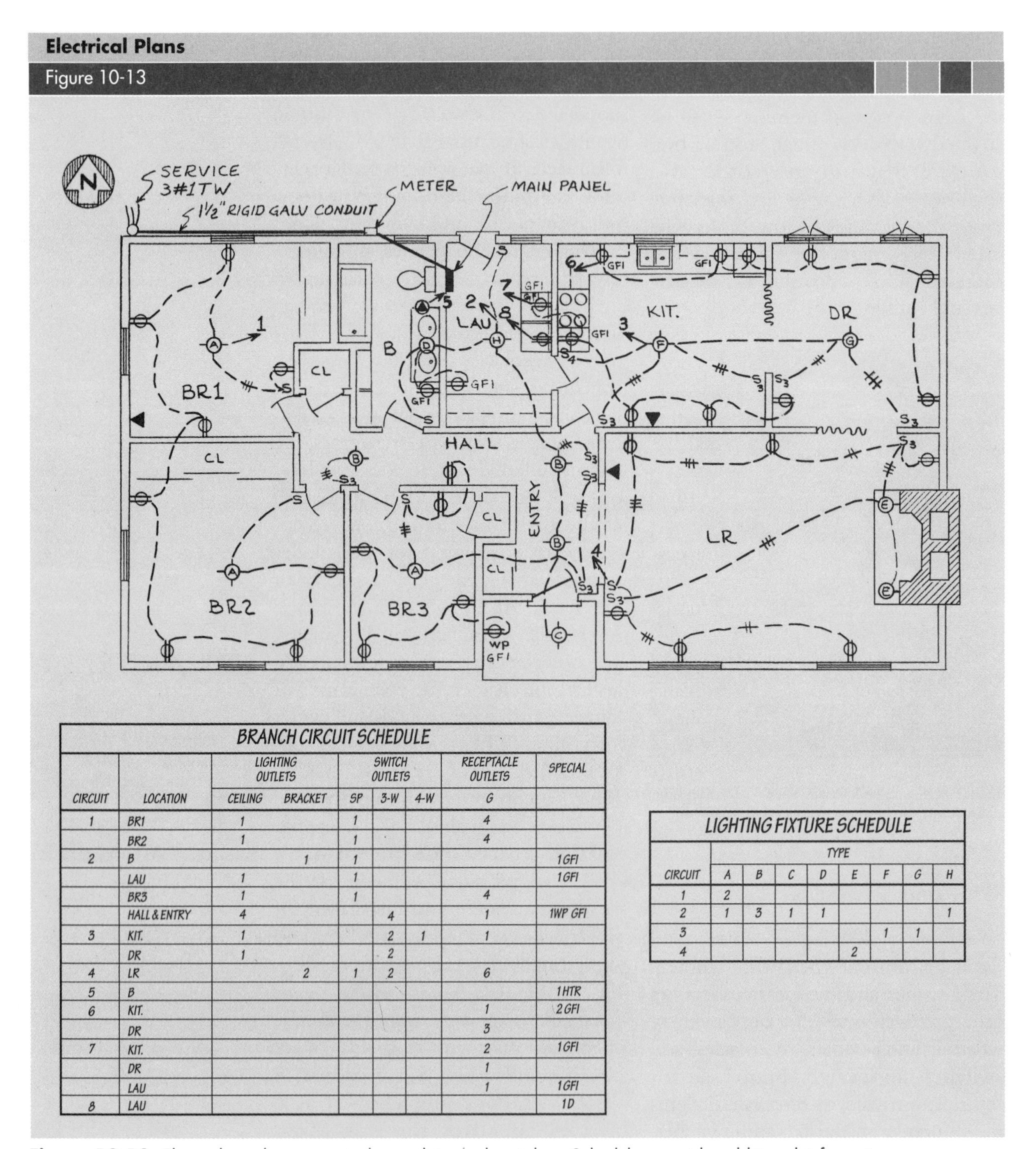

BRANCH CIRCUIT SCHEDULE

CIRCUIT	LOCATION	LIGHTING OUTLETS		SWITCH OUTLETS			RECEPTACLE OUTLETS	SPECIAL
		CEILING	BRACKET	SP	3-W	4-W	G	
1	BR1	1		1			4	
	BR2	1		1			4	
2	B		1	1				1GFI
	LAU	1		1				1GFI
	BR3	1		1			4	
	HALL & ENTRY	4			4		1	1WP GFI
3	KIT.	1			2	1	1	
	DR	1			2			
4	LR		2	1	2		6	
5	B							1HTR
6	KIT.						1	2GFI
	DR						3	
7	KIT.						2	1GFI
	DR						1	
	LAU						1	1GFI
8	LAU							1D

LIGHTING FIXTURE SCHEDULE

CIRCUIT	TYPE A	B	C	D	E	F	G	H
1	2							
2	1	3	1	1				1
3						1	1	
4					2			

Figure 10-13. Floor plans show receptacles, outlets, and switches. Schedules provide additional information.

through 8 are shown. Circuit 1 includes all receptacles and lighting outlets for BR 1 and BR 2. The arrow with the number 1 indicates a home run to circuit 1 on the main power panel. In BR 1, one type A ceiling outlet is controlled by a single-pole wall switch. Four wall receptacles are installed per section 210.52 of the NEC. In BR 2, one type A ceiling outlet is controlled by a single-pole wall switch and four wall receptacles are shown.

Voice/data/video systems are becoming increasingly common in residential construction. A *voice/data/video (VDV) system* is a low-voltage electrical system designed for use with various types of information technology equipment, including telephone systems, computer systems and networks, television video systems for cable and/or satellite dish systems, and security and fire alarm systems. These systems are typically detailed and installed by specialty contractors or suppliers familiar with these systems.

Voice/data/video drawings show connections to information sources such as incoming telephone lines and video feeds. Locations for connections to the systems in various rooms and locations throughout the house are shown with symbols on a floor plan. Security and alarm system suppliers not only install and maintain the wiring and connections at the residence, but also provide monitoring services to alert police and fire departments in case of a breach of the systems.

PLUMBING

Plumbing is installed according to information provided on plan views, specifications, and piping drawings. This information is used in conjunction with applicable building code ordinances to determine installation requirements and procedures. House design and structural configuration are considered when installing plumbing lines. For example, the direction of joists and the location of plumbing fixtures have a bearing on pipe locations.

Installation of plumbing occurs during many stages of the construction process. Connections to the water supply, sanitary sewer, and storm water drainage systems are completed following initial excavation. Basement drains are placed before the concrete for the basement slab is placed. Slab-at-grade foundations require water supply and waste water piping to be installed before concrete is placed.

Rough-in plumbing work is completed after rough framing is completed. The insides of walls can be easily accessed for installation of supply, wastewater, and vent piping. Plumbing fixtures for the finish plumbing work are installed after the walls are covered.

Piping

The major components of a plumbing system are water supply piping, sanitary drainage and vent piping, and storm water drainage piping. See Figure 10-14. In some regions, natural gas piping may also be installed to the house from the main supply.

Herbert C. Hoover developed the first plumbing code, known as the "Hoover Code," in the 1920s.

Sioux Chief Manufacturing Company, Inc.

Steel protector plates are commonly installed on studs and top and bottom plates to prevent fasteners from puncturing piping.

Plumbing Systems

Figure 10-14

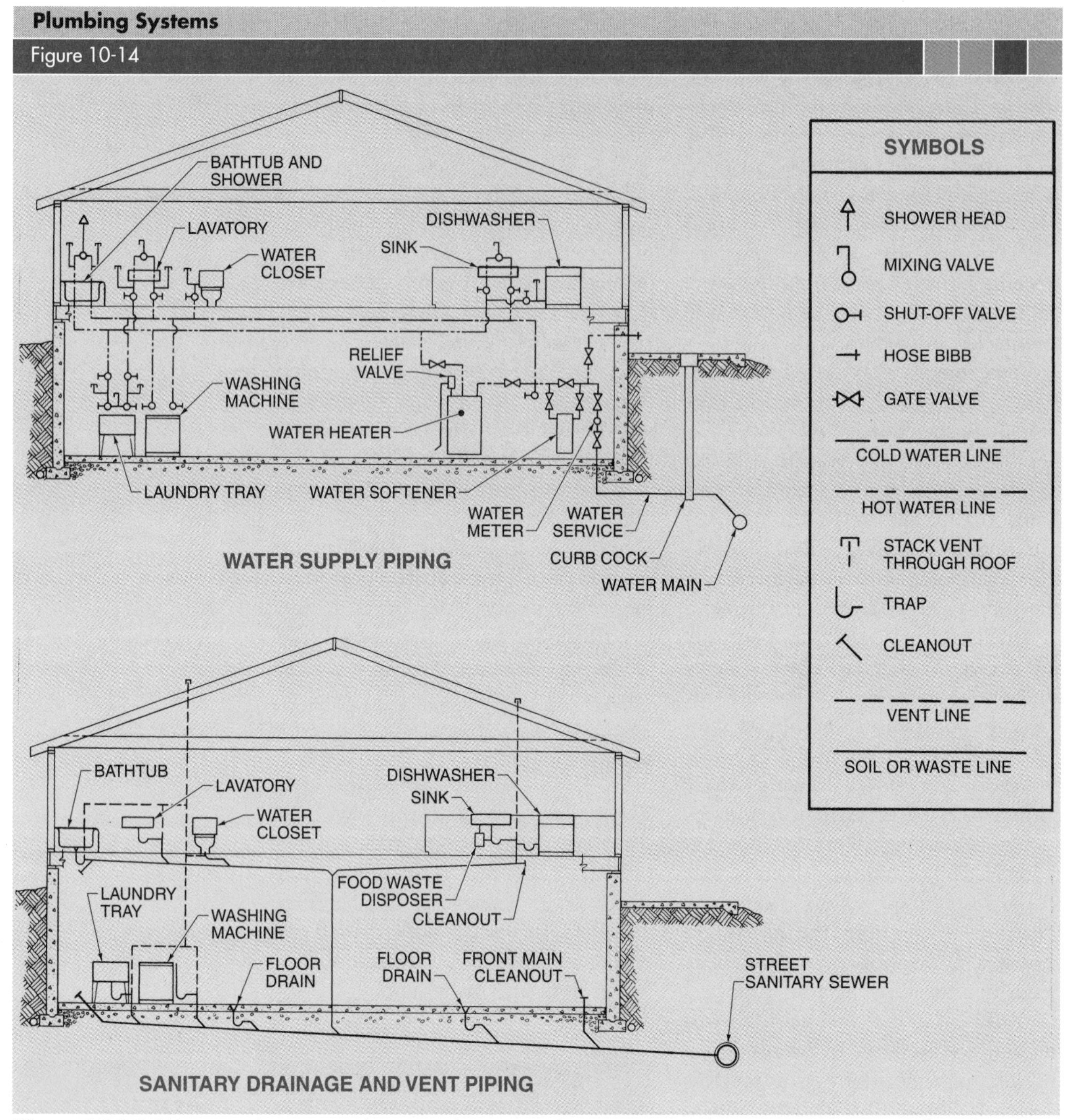

Figure 10-14. Water supply piping and sanitary drainage and vent piping are the primary components of residential plumbing systems.

Water Supply Piping. Water supply piping conveys potable water to the points of use in the house. *Potable water* is water free from impurities in amounts that could cause disease or harmful effects.

The water supply is under pressure as it enters the house from a city water main or a well pump. The water flows through a water meter, which measures the amount of water used. The cold water supply is piped throughout the house to points of use. One branch of the cold water supply is diverted to a water heater, where it is heated. Hot water is piped from the outlet side of the water heater to points of use throughout the house, often parallel to the cold water line. Fixtures at each hot and cold water outlet control the flow of water.

Sanitary Drainage and Vent Piping. The sanitary drainage system conveys wastewater and waterborne waste from plumbing fixtures to the sanitary sewer or septic system. Sanitary sewers receive wastewater that must be treated before being released into the fresh water supply.

A water seal is provided on waste piping at each plumbing fixture using a plumbing trap to prevent sewer gas from entering a house. See Figure 10-15. Local building codes may require pressurized testing of the waste piping before inspection approval and issuance of an occupancy permit.

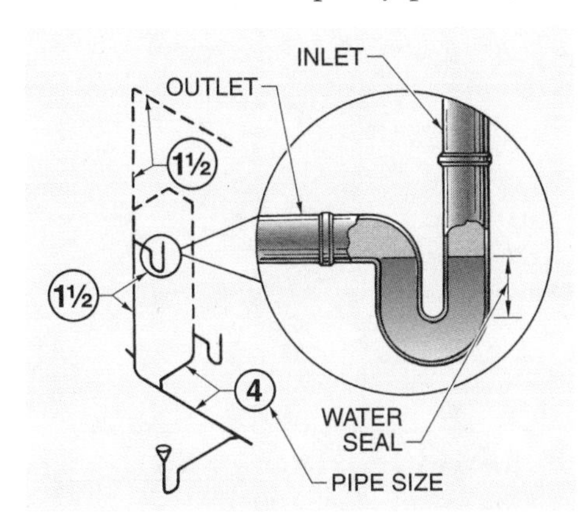

Figure 10-15. Traps provide a water seal on waste piping to prevent sewer gas from entering a house.

In a city or municipality sanitary sewer system, waste piping is routed directly from the house to the municipal sewer system. In rural areas, a municipal sewer system may not be available. Wastewater and waterborne waste is routed to a septic tank and an absorption (leach) field. Solid waste collects in the septic tank and is broken down by chemical action. Wastewater flows from the septic tank and is distributed into the soil of the absorption field.

Vent piping provides air circulation within the sanitary drainage system to equalize pressure in the system and permit the proper flow of wastewater and waterborne waste. In addition, vent piping allows gases and odors to flow from the plumbing system through roof vents.

Storm Water Drainage Piping. Storm water drainage piping carries rainwater and other precipitation to a storm sewer or other place of disposal. Proper storm water drainage systems prevent the accumulation of storm water around a house and leakage into the house. Storm water is collected and routed away from the house using drain tile placed around the foundation. Storm water must be distributed and absorbed in the soil according to local ordinances. Many local ordinances prohibit storm water from being directed to storm sewers.

In some areas, water is collected and routed to a sump well installed in the basement floor of a house. A sump pump mounted in the sump well ejects the storm water from the well through a pipe extending through the foundation wall to the outside, away from the building.

Roughed-in Piping. Rough-in is the installation of pipes and plumbing fittings that will be concealed after construction is complete. Knowing the exact location of roughed-in piping is critical to ensuring that fixtures are installed in the proper final location.

Leaks, spills, and waste materials must be cleaned up, properly disposed of, or recycled to prevent pollution of the storm water drainage system.

Topcon Laser Systems, Inc.

Drainage piping must be properly graded so water and waterborne waste flows to the point of disposal.

Specifications provide the manufacturer name and model numbers of fixtures to be installed. Dimensions for water supply and sanitary drainage and vent piping required for proper installation of plumbing fixtures are provided on rough-in sheets from the manufacturer. See Figure 10-16. Specifications state that a Kohler K-3310-4 stainless steel kitchen sink is to be installed. The rough-in sheet shows that sink drains are $11\frac{7}{8}''$ from the back edge of the sink top and the sink is $7\frac{1}{2}''$deep. A $1\frac{1}{2}''$ outlet pipe extends through the wall $12\frac{13}{16}''$ below the top edge of the sink.

Figure 10-16. Manufacturer rough-in sheets provide dimensions for the proper installation of plumbing fixtures.

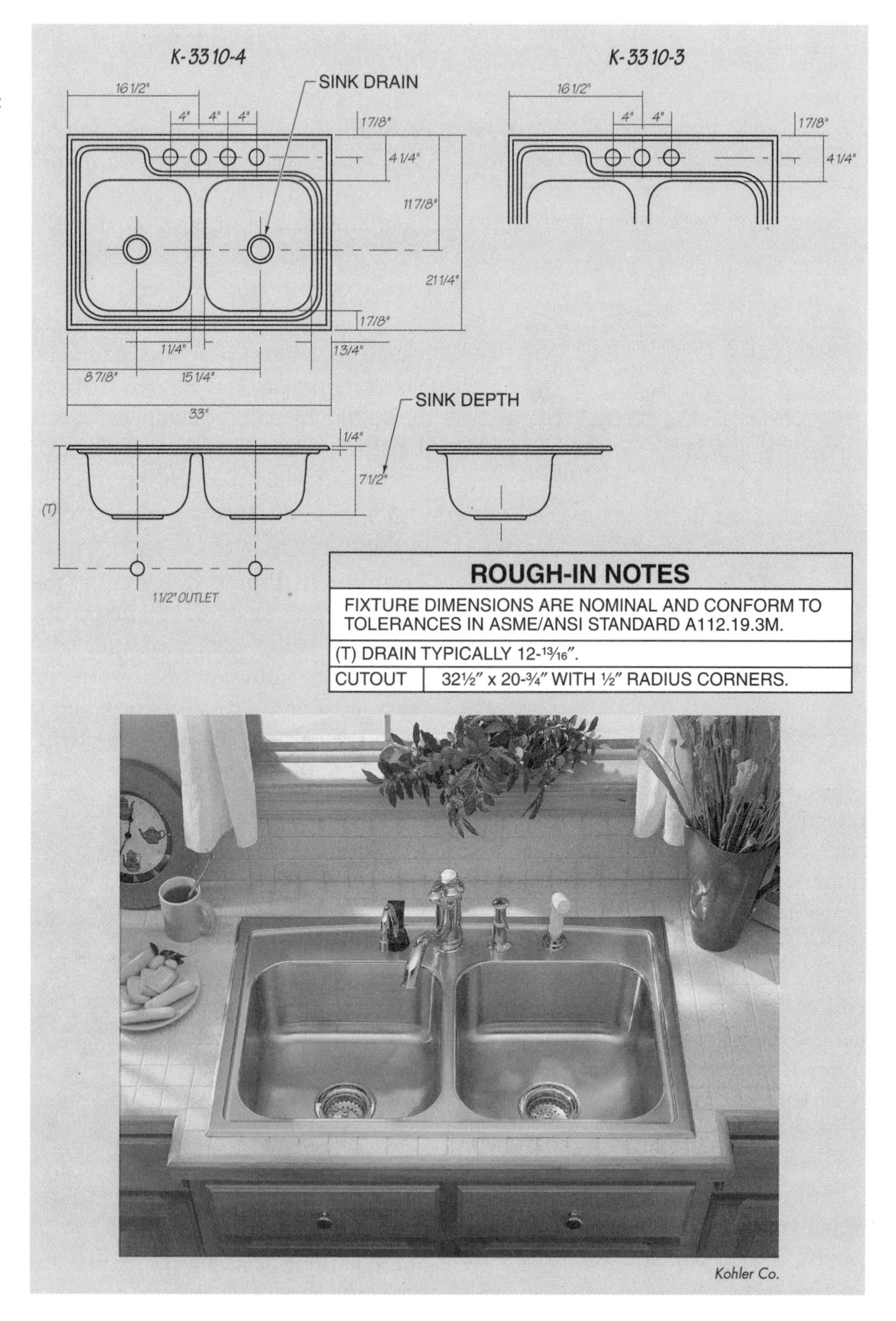

Kohler Co.

In some cases, a rough-in sheet will not include water supply and sanitary drainage rough-in dimensions. The variety of faucet selections or job conditions prohibits listing the exact rough-in dimensions. Standard dimensions accepted in the trade are used in these instances. For example, standard dimensions for roughing in a double-bowl sink specify the waste opening at approximately 14″ above the finished floor and the hot and cold supply water through the wall at approximately 16″ above the finished floor.

Plans and Specifications

Plan views provide information regarding the location of plumbing fixtures, waste piping, sanitary drains, and storm water drains. See Figure 10-17. Plumbing fixtures such as lavatories, water closets, and bathtubs are shown using symbols. Hose bibbs are located on exterior walls to provide connections for garden hoses.

Basement and foundation plans provide the location of floor drains and drain tile. Plot plans provide information regarding drainage fields and connections to the supply water and sanitary sewer lines. Sections and interior elevation details supplement information provided on plan views.

Specifications provide information such as fixture manufacturers, model numbers, and number of plumbing fittings required. Some specifications are included on the prints as notations. Other specifications may be sheets bound in a separate document or forms supplied by the contractor or a government agency. If there is a discrepancy between the specifications and the prints, the specifications take precedence.

Standard specification forms are commonly used for residential construction. The appropriate information is written in spaces provided on the form.

To accommodate people with physical disabilities, ADA-compliant plumbing fixtures may be required. Water closets must be positioned to provide adequate clearance for a wheelchair-bound individual to maneuver within the bathroom. The top edges

Plan Views

Figure 10-17

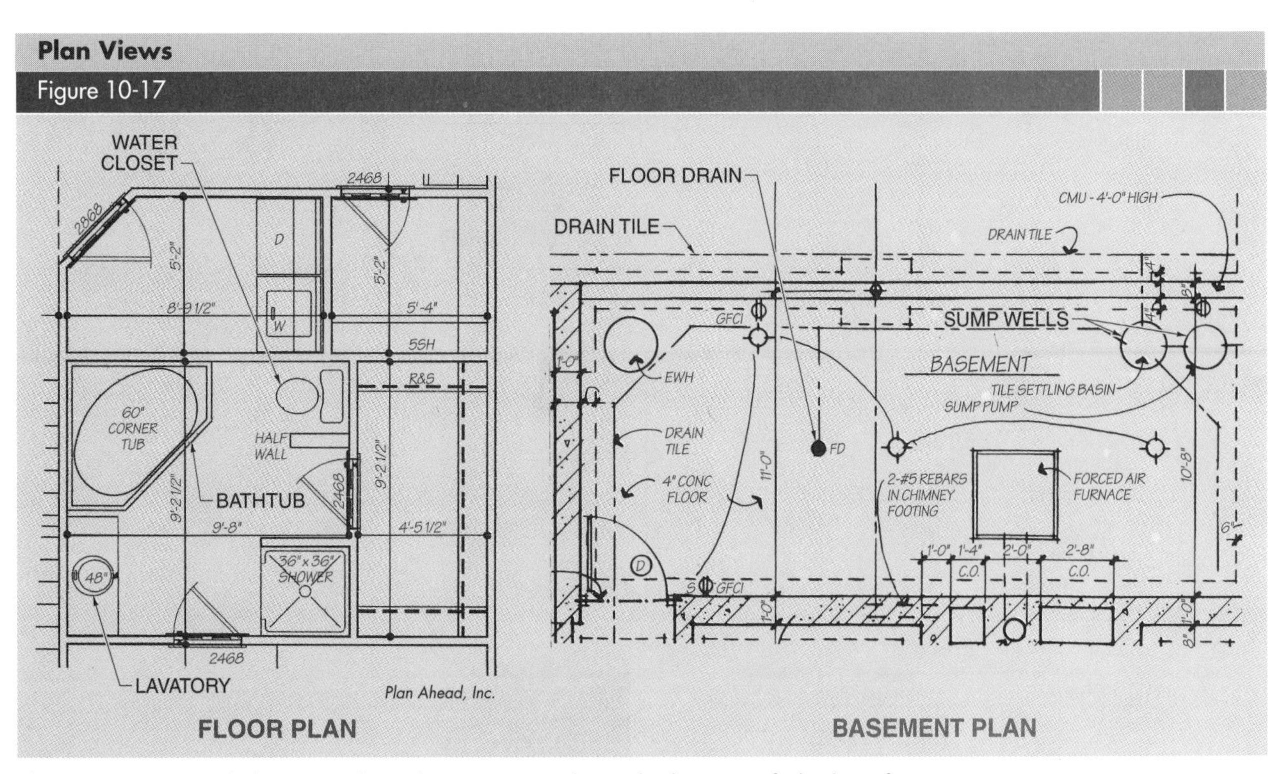

Figure 10-17. Symbols are used on plan views to indicate the location of plumbing fixtures.

of lavatories and kitchen sinks must be 34″ maximum from the finished floor and provide adequate knee and toe clearance. Grab bars may need to be installed along the bathtub or water closet.

Piping Drawings. Piping drawings show the layout of a plumbing system. Piping drawings are more common for larger or custom-built houses and may not be provided in a set of prints for smaller dwellings. An architect, contractor, subcontractor, or supplier may provide piping drawings, depending on the size of the job and local code requirements.

Piping drawings are not drawn to scale, but the general location of plumbing components is shown on the drawings. Different linetypes are used to indicate the difference between waste and vent piping and between hot and cold water supplies. The lines are also used with symbols to show the installation of required plumbing fittings. A piping drawing can be drawn as an elevation or as an isometric. See Figure 10-18.

An elevation piping drawing is drawn in section with interior walls removed. Plumbing fixtures and piping are shown in cross section in the installed position. Isometric piping drawings are three-dimensional drawings showing piping and plumbing fixtures drawn with vertical and 30° lines. Vertical lines represent vertical pipes and 30° lines represent horizontal pipes. A comprehensive isometric piping drawing includes all piping and fittings for a plumbing system. Piping drawings are useful when ordering the pipe and fittings required for a job.

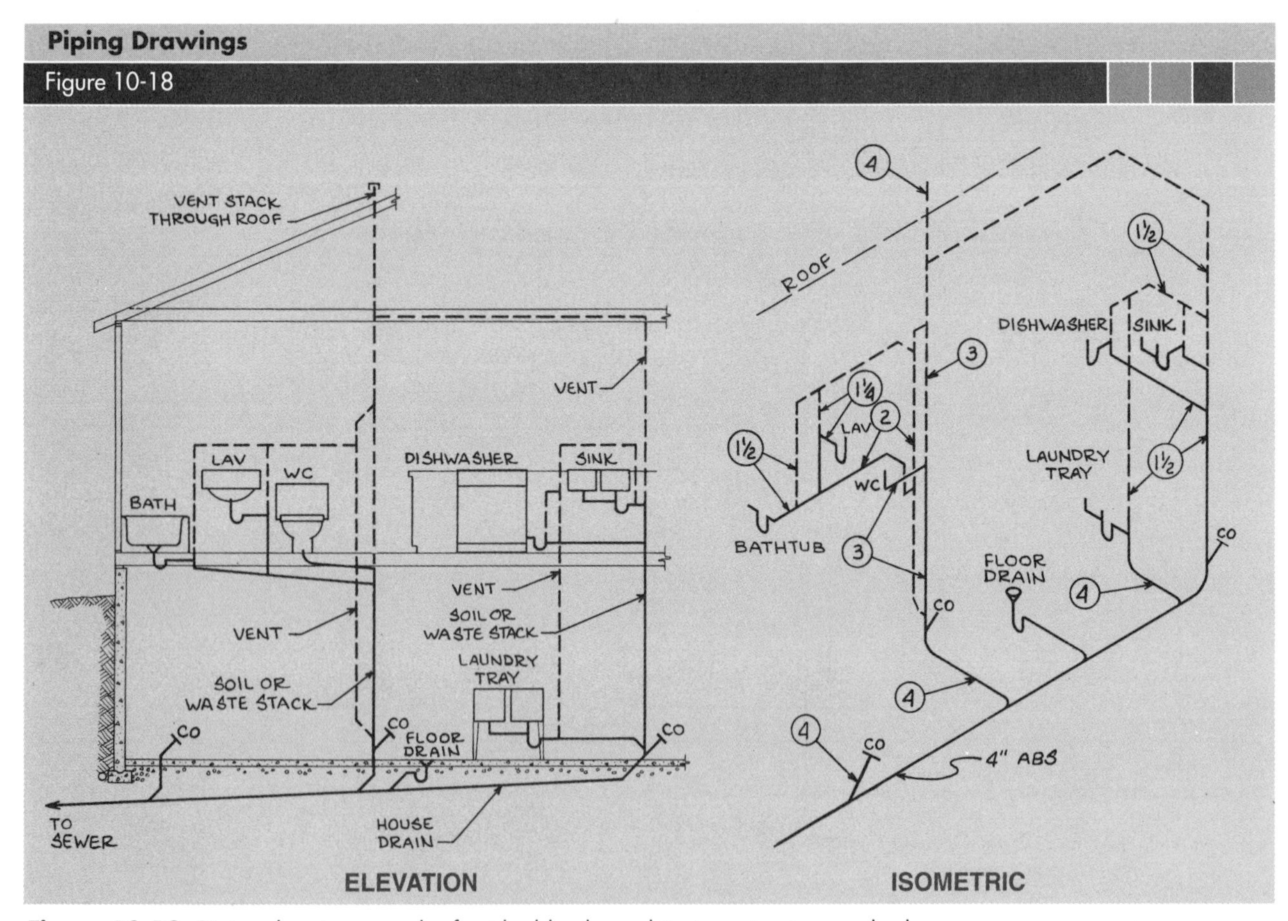

Figure 10-18. Piping drawings may be furnished by the architect, contractor, or plumber.

HEATING, VENTILATING, AND AIR CONDITIONING

Heating, ventilating, and air conditioning (HVAC) systems supply the heat, ventilation, and cooling necessary to provide comfort to occupants of a house. An HVAC system also controls humidity, introduces fresh air, and cleans and circulates air throughout the house to provide maximum comfort.

Installation occurs after rough framing but before wall finish materials are applied. Installation of an HVAC system involves installation of ductwork, piping for hot water heating systems and gas lines, and electrical wiring for the HVAC system.

Two common types of residential heating systems are forced warm air and radiant heating systems. Radiant heating systems are classified as hot water (hydronic) and electric systems. Alternate heating systems, such as solar or geothermal systems, may also be used to efficiently heat homes.

Forced Warm Air Heating

A *forced warm air heating system* is a residential heating system that uses a blower to draw air (return air) from rooms through return air grilles and ductwork. The return air passes through a furnace where it is heated. The heated air (supply air) passes through the plenum and is distributed to the desired rooms through supply ductwork and registers.

Supply air ducts are commonly installed between floor joists or under slab-at-grade floors. Supply air ducts branch from a trunk supply line centrally located in the house. Supply air registers are generally located close to the exterior walls of the house with return air grilles and ductwork located near the center of the house. This arrangement of supply registers and return air grilles allows the proper circulation of return air from rooms to the furnace and of supply air back to the rooms. See Figure 10-19.

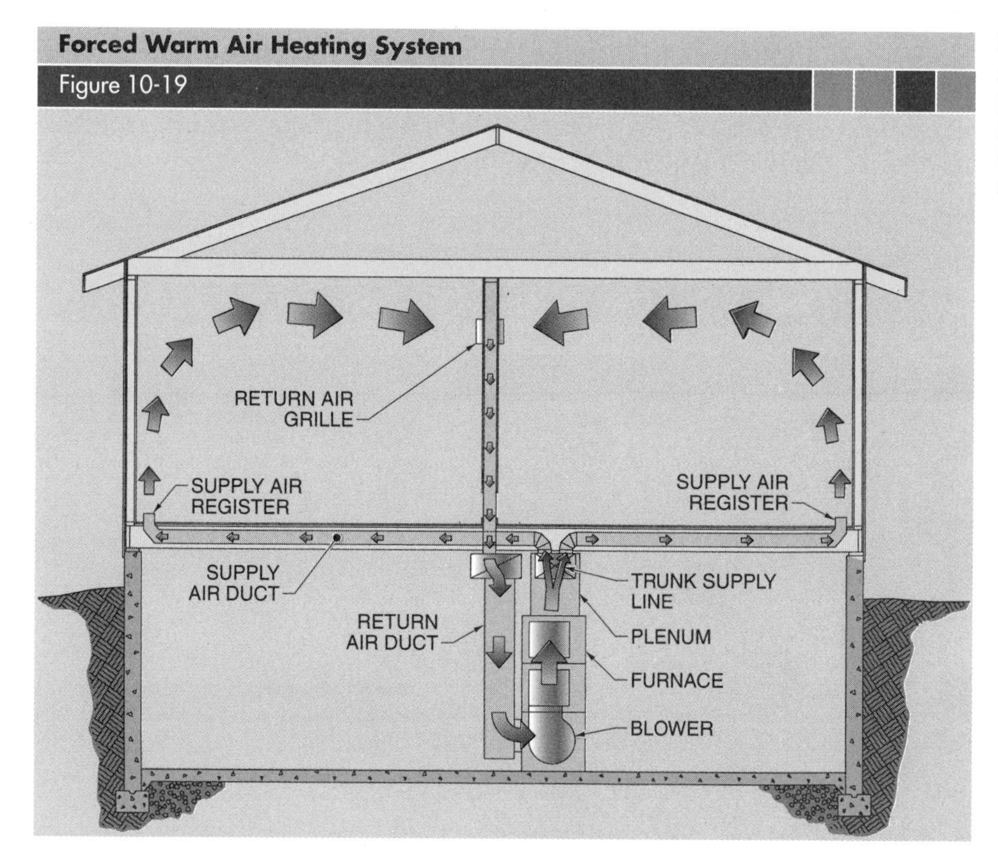

Figure 10-19. Supply air registers and return air grilles are located to provide maximum comfort in a forced warm air heating system.

The major components of a forced warm air heating system are the furnace, blower, ductwork, and unit controls. A *furnace* is a piece of equipment that heats the air circulating in a forced warm air heating system. Furnaces differ in design and operation based on the method used to generate heat. Electricity or fuels such as natural gas or fuel oil are commonly used to generate heat in a furnace. Furnaces using fuel have a burner, combustion chamber, heat exchanger, and unit controls. The burner is the heat-producing element of the furnace. The combustion process occurs in the combustion chamber. Heat is transferred to the heat exchanger where the air is heated as it passes through. Electric furnaces have resistance heating elements in place of the burner, combustion chamber, and heat exchanger.

A *blower* is a piece of equipment that draws air from rooms and moves it through the furnace where it is heated before being conveyed back into the rooms. The most common blower is the centrifugal blower, which has a blower wheel that rotates rapidly within a sheet metal enclosure. As the blower wheel rotates, air is drawn through the center and discharged out the side to the plenum. Ductwork directs supply air from the furnace into rooms to be heated, and return air from building spaces to the furnace. Supply air registers and return air grilles are located at the ends of the ductwork runs.

Duct size is expressed in inches of diameter for round ducts and in inches of width and height for rectangular ducts. The width is always given first for rectangular ducts.

Supply air ductwork must be sized to carry the proper amount of air to the supply air registers. The design of the ductwork system is selected for economy and efficiency. Furnaces, blowers, and ductwork are all sized by the cubic feet of air per minute (cfm) required to properly condition the air in the living space. See Figure 10-20.

A perimeter radial ductwork system is the simplest and most cost-effective ductwork system for residential construction. An extended plenum ductwork system is useful in long house designs such as ranch-style houses. A perimeter loop ductwork system is commonly used in cold climates.

Ductwork is made from galvanized sheet metal, fiberglass, or reinforced aluminum-faced materials. All ductwork must comply with applicable building codes.

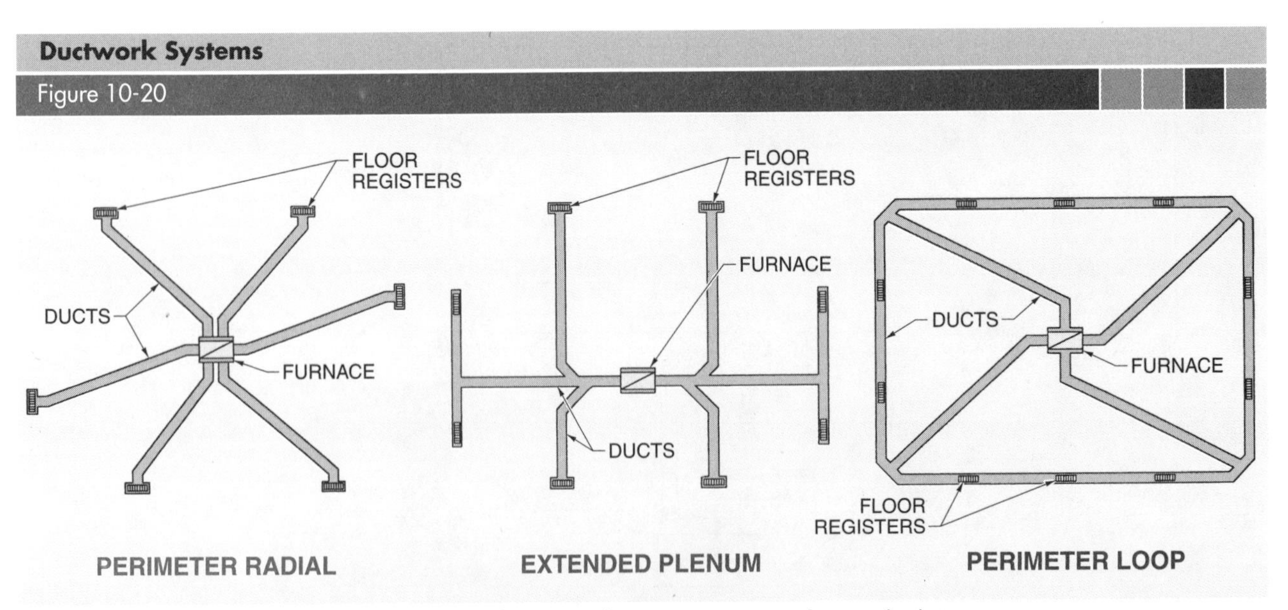

Figure 10-20. Ductwork systems must be sized to carry the proper amount of air to the living space.

Supply air registers are installed on supply ducts. Supply air registers are sized to supply and direct the proper amount of heated air to a desired room. See Figure 10-21. Return air grilles in forced warm air heating systems are located and sized to return as much air from building spaces as supplied. Fewer return air grilles are generally installed than supply air registers. However, return air grilles are usually larger and more centrally located than supply air registers.

Unit controls are a set of controls installed on a furnace by the manufacturer or installer to maintain safe and efficient operation of the furnace. Unit controls are classified as power controls, operating controls, and safety controls. *Power controls* are controls located in the electrical conductors leading to the furnace and blower. Power controls include the conductors, overcurrent protection devices, overload protection devices, and disconnecting means to the heating unit. Power controls must be installed by a licensed electrician according to provisions of the NEC®.

Operating controls are controls that cycle equipment ON and OFF as required. Operating controls include a step-down transformer (in a low-voltage control system), a thermostat, a blower control, and operating relays or contactors that turn components ON or OFF. A step-down transformer reduces the voltage available for use in the control system. A thermostat is a temperature-sensing electrical switch that turns the furnace ON and OFF. The blower control operates the blower used to circulate air through the furnace and to the desired rooms.

Safety controls are controls that prevent injury to personnel or damage to equipment in the event of equipment malfunction. Safety controls include

Supply air registers are designed to deliver air in a specific pattern.

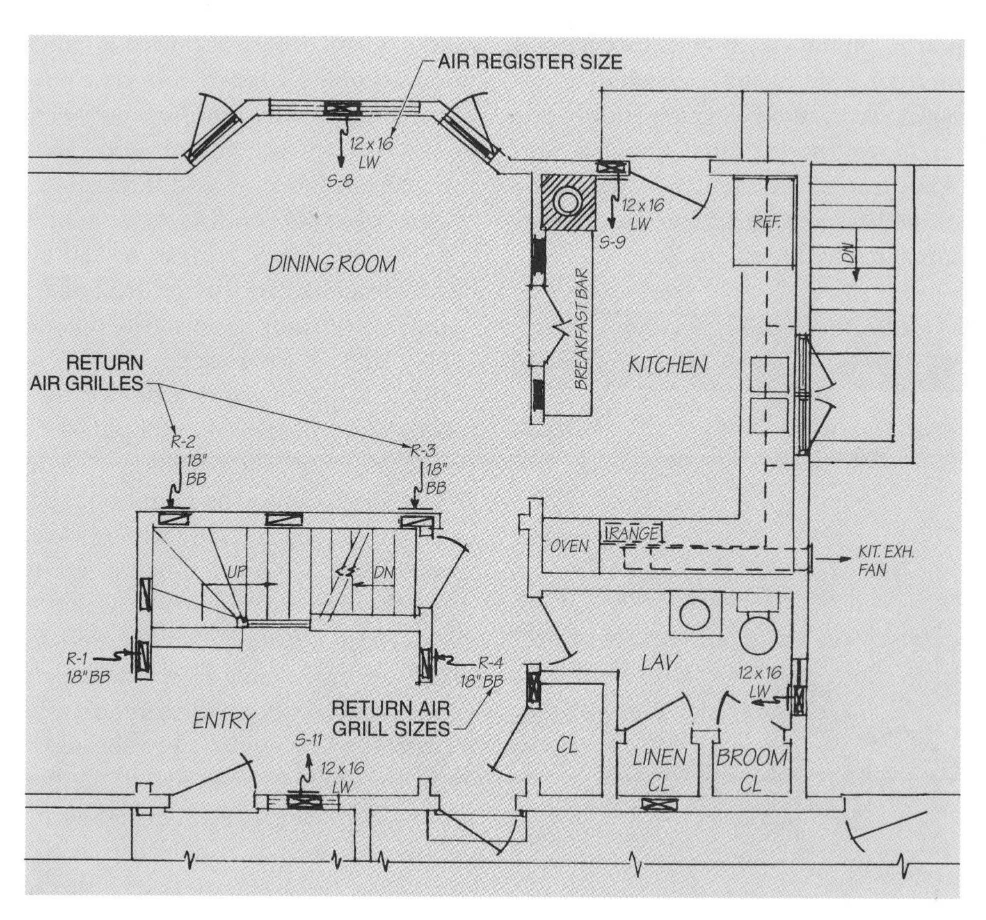

Figure 10-21. Sizes of supply air registers and return air grilles are specified on floor plans.

limit and combustion safety controls, such as thermocouples. Motor overloads may also be required in some furnaces.

A thermocouple is a device used to measure temperature differences.

Radiant Heating Systems

A *radiant heating system* is a heating system that transfers heat to the living space via hot water tubing or electrical cables embedded in the floor or ceiling. Radiant heating systems are hot water (hydronic) and electric systems.

Hot Water Radiant Heating. A *hot water radiant heating (hydronic) system* is a radiant heating system in which water is heated in a boiler at a central location and then distributed to the desired rooms through tubing. Hot water heating systems consist of a boiler, compression tank, circulating pump, piping, terminal devices, and unit controls. Water is heated in the boiler, distributed through supply water piping to terminal devices, and flows through the return water piping to the boiler to be reheated and redistributed. See Figure 10-22.

Figure 10-22. A circulating pump is located in return water piping and provides the pressure required to pump heated water through a system.

Hot Water Heating System

Figure 10-22

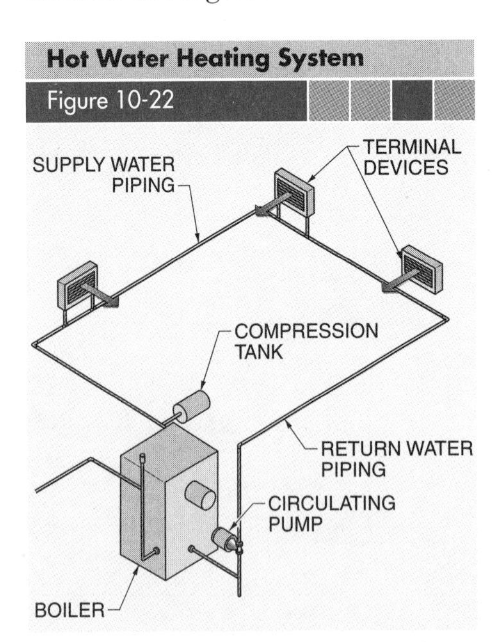

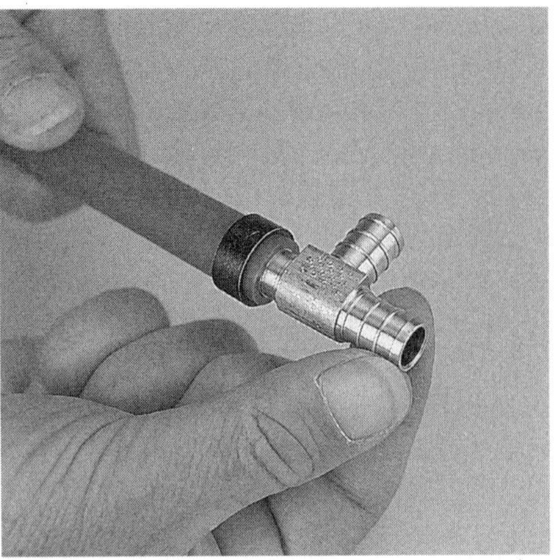

Vanguard Piping Systems, Inc.

Cross-linked polyethylene (PEX) tubing and fittings may be used to route the water in a hot water radiant heating system.

A *boiler* is a closed tank connected to an energy source that heats water to a high temperature. Boilers are classified according to the method of heat generation. Fuels commonly used include natural gas and fuel oil. Electric heating elements are also used if other fuels are unavailable or less cost efficient. Water heated by the sun (solar) or hot water from the ground (geothermal) can also be used if feasible.

A *compression tank* is a tank that absorbs and relieves pressure caused by water expansion when heated. A compression tank is installed on the supply side of the boiler.

A *circulating pump* is a device that moves water through the supply water piping in a hot water heating system. A circulating pump is commonly located in the return water piping close to the boiler. Circulating pump size is determined by the volume of water in the system. *Piping* is tubing used to distribute hot water from the boiler to terminal devices and to return water back to the boiler to be reheated.

In the rooms to be heated, the hot water passes through terminal devices. A *terminal device* is a device that extracts heat from hot water to heat the air in the desired rooms. Terminal devices include radiators, convectors, blower coils, and radiant panels. A fan or blower may be used to circulate the heated air efficiently. The water returns to the boiler through return water piping. Unit controls control the boiler, circulating pump, and terminal devices and are similar to unit controls for a forced warm air heating system.

Piping for hot water radiant floor heating systems is installed in frame buildings or in slab-at-grade floors. In a frame building, flexible plastic tubing is placed between sleepers below the subfloor. Care is taken to ensure that the tubing is not punctured when the subfloor is installed. For slab-at-grade floors, tubing is set in place before the concrete is placed. See Figure 10-23. Heat from the hot water is radiated through the concrete and into the living area.

Figure 10-23. Tubing for a hot water heating system is installed before concrete is placed for a slab-at-grade foundation.

Hot water heating systems in residential construction are commonly one-pipe systems. In a one-pipe system, hot water is distributed to the desired rooms using branches and risers. The amount of hot water flowing through each terminal device is controlled by a valve setting or a thermostat. Two-pipe systems use separate pipes for supply and return water. Two-pipe systems provide a more consistent water temperature to all terminal devices as cooled water is returned in a separate pipe to the boiler. The two-pipe system requires additional piping and often is not cost effective for residential construction.

Electric Radiant Heating. An *electric radiant heating system* is a radiant heating system in which heat is generated when electricity meets resistance as it flows through embedded heating cables or baseboard units. Electric radiant heating is used in panel systems and baseboard units. Panel systems have thin wires embedded in prefabricated floor, wall, or ceiling panels. The amount of electricity flowing through the wires in the panels is controlled by a thermostat. See Figure 10-24. Baseboard units are installed after the walls are finished and are mounted to the wall or floor. Air is heated as it contacts the heated electric coils. Some baseboard units contain water or another heat-retaining fluid to extend the time that heat is retained and transferred to the air.

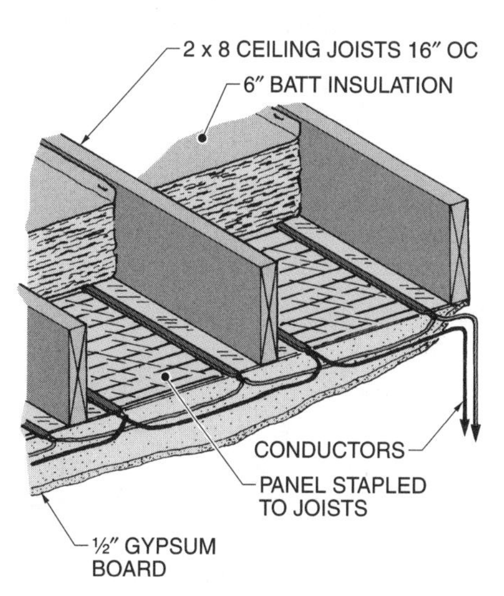

Figure 10-24. Radiant electric heating panels have thin wires embedded within each panel that radiate heat.

Alternate Heating

Alternate heating uses natural sources that do not require a combustion process or electricity to produce heat. Solar energy and geothermal heat are two alternate heating sources. *Solar energy* is the energy available from the sun in the form of sunlight. Solar energy can be changed to thermal energy and used to heat air and water. Solar energy systems collect and store solar energy. Well-designed solar energy collection and storage systems maximize the amount of thermal energy received. See Figure 10-25. Solar energy can also be converted to electrical energy for use in the house.

Geothermal heat is heat that is derived from heat contained within the earth. Heat contained within the earth is maintained at a constant temperature (50°F to 70°F) in the soil and rocks below the frost line. With geothermal heating systems, ground-source heat pumps pull heat energy from the earth and convey it through a house. A typical ground-source heat pump uses flexible tubing filled with water, refrigerant, or an antifreeze solution. As the fluid flows through the tubing in the ground, it absorbs heat energy from the earth. The warmed fluid is pumped into and provides heat for the living space. As the energy is transferred to the living space, the fluid flows through return tubing into the earth to repeat the heating process.

Solar Heating System

Figure 10-25

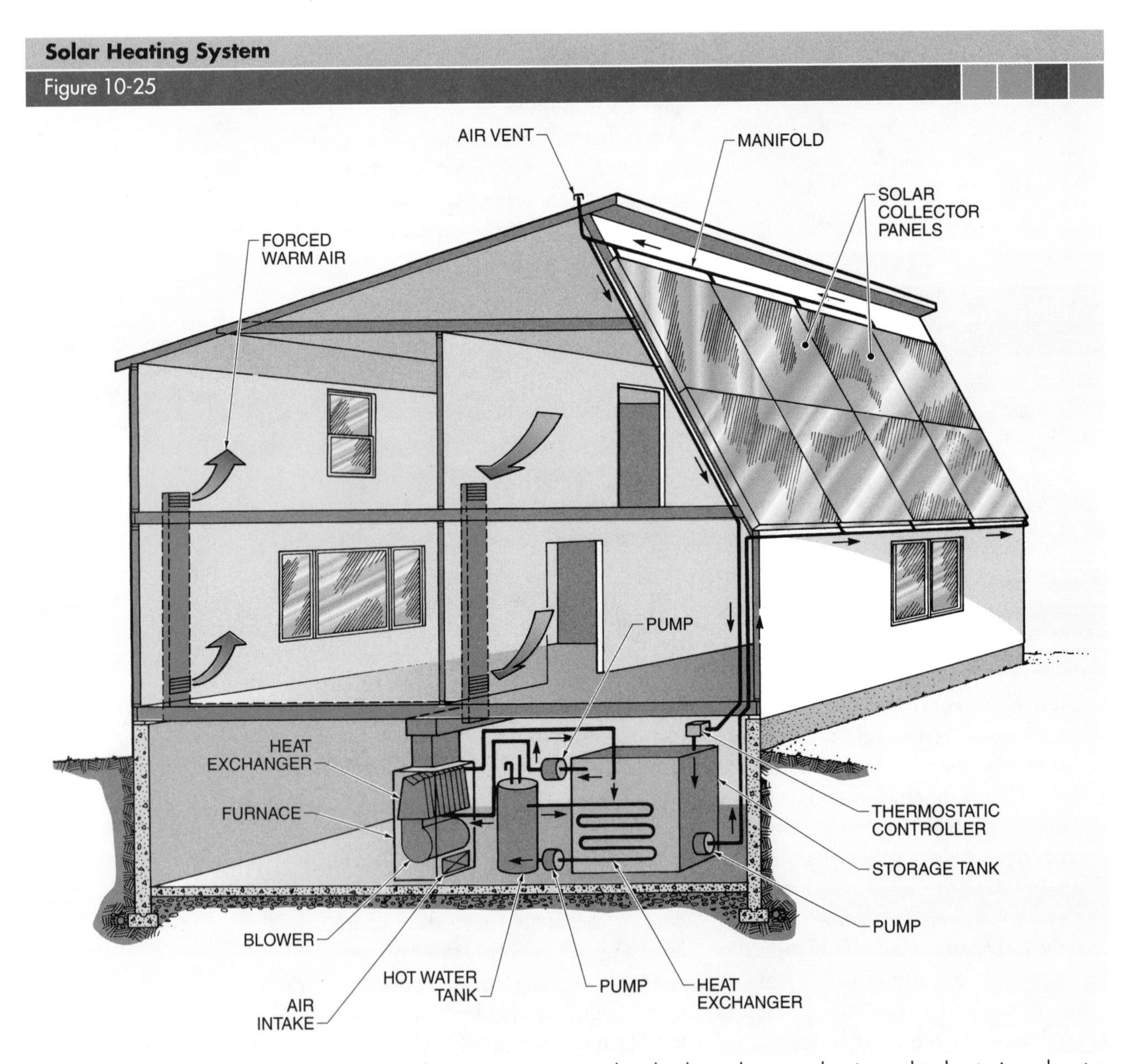

Figure 10-25. In a solar heating system, solar energy is converted to the thermal energy that is used to heat air and water.

Air Conditioning

Air conditioning is the process used to control humidity, temperature, and/or cleanliness of air within a house. An *air conditioner* is a piece of equipment used to control humidity and temperature and remove airborne impurities in a house. An air conditioner consists of a cooling coil or an evaporator, an electric compressor, and a condenser that work together to cool the air. Cooled air is circulated through the desired rooms where it absorbs heat and is returned to the air conditioner. Two types of air conditioning systems used in residential construction are split systems and package systems. See Figure 10-26.

In a split system, also known as central air conditioning, the condenser and evaporator are in separate locations. The condenser is typically located on a concrete slab outside the house. The evaporator is installed on the discharge side of the furnace blower. Heat is absorbed from the air as it passes through the fins of the evaporator. A condensate drain line directs water that has condensed to the drain. The ductwork of the forced air system conveys the cooled air throughout the house. When designing and sizing a split HVAC system, additional flow resistance caused by the evaporator must be considered.

Package systems have the condenser and evaporator installed in the same unit. Supply air and return air are ducted to the unit. The unit can be mounted in the attic, in the garage, or through foundation walls. Some package air conditioning systems have integral heating system components.

Plans and Specifications

The plot plan, foundation plan, and floor plan are commonly used for HVAC installation. The plot plan shows the location of the air conditioner and the natural gas supply if required. The foundation plan gives information regarding ventilation fan locations, furnace, boiler, air conditioning equipment locations, and ductwork. The floor plan indicates the sizes and locations of supply and return ducts. If required, furnace, boiler, and air conditioning information is also provided on the floor plan. The

Air Conditioning Systems

Figure 10-26

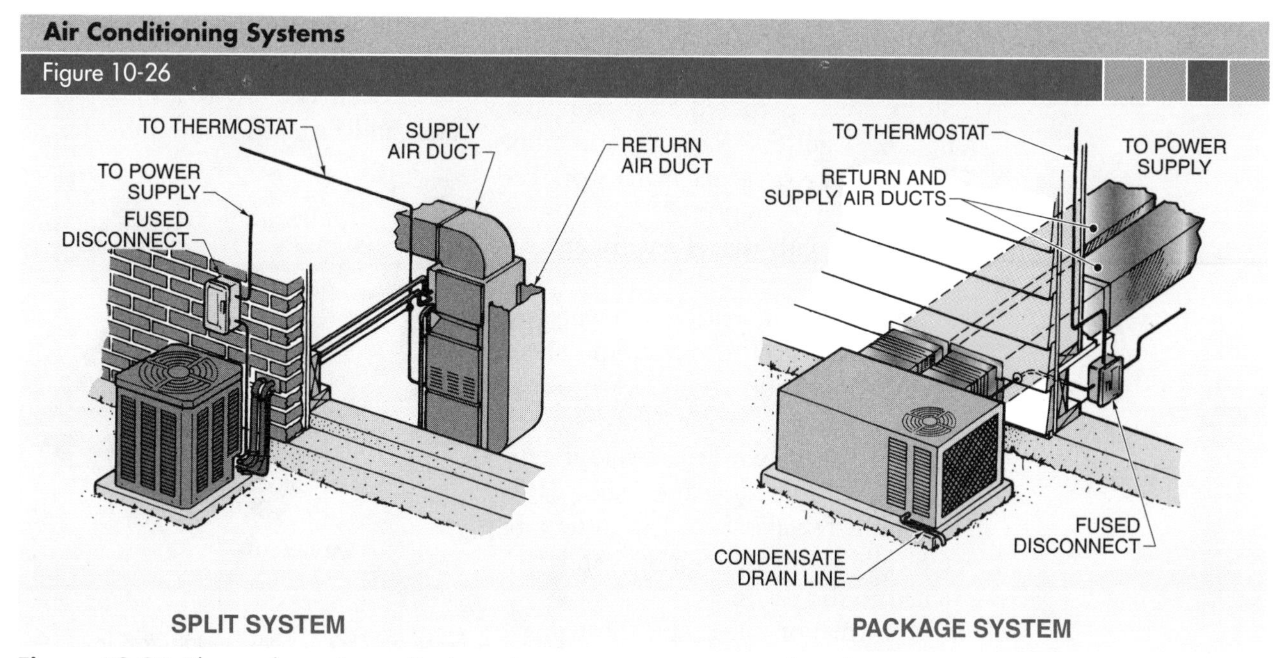

Figure 10-26. The condenser in a split air conditioning system is separate from the evaporator. Components of a package system are in the same unit.

direction of floor joists and studs in partitions and walls must be determined before piping or ductwork can be run. Information regarding the HVAC system is shown using dimensions and symbols.

Additional information may be provided on mechanical schedules, ductwork layouts provided by the architect or HVAC contractor, or specifications. An engineer may be contracted by the architect or contractor to determine the specific equipment, ductwork, and/or piping requirements of the system.

Specifications for an HVAC system include information that is not included in other parts of the prints. Heating, ventilating, and air conditioning equipment is described in the specifications by the manufacturer name, model number, and type of system. The output capacity and materials used for piping and supports are also included.

Self-tapping screws, designed especially for sheet metal work, tap their own mating threads as they are driven into the material.

SHEET METAL WORK

Sheet metal products and other sheet materials are applied on both the exterior and interior of a house. Sheet metal products used on the exterior of the house provide protection against weather and moisture infiltration. Storm water is routed away from the house by gutters, flashing, and trim. The majority of sheet metal work required on the interior of the house is associated with the HVAC ductwork system.

Many sheet metal components are prefabricated. However, some sheet metal applications require a product to be fabricated on site. This requires knowledge of pattern development, cutting, and bending the materials to the needed shapes. Seamless gutters are typically fabricated on site and fastened to the eaves.

Information regarding sheet metal work is included on foundation plans, exterior elevations, floor plans, and details. Additional information may also be provided in the specifications or by the architect or contractor. Termite shields, shown on a wall section, may be installed along the top of the foundation wall in regions with large insect populations. Termite shields are becoming unnecessary in most regions as pressure-treated wood and perimeter soil treatments provide adequate termite protection.

Sheet metal work is required when installing roofing to ensure moisture protection and proper flow of storm water. Flashing and crickets prevent storm water from seeping under roofing materials. See Figure 10-27. Flashing is required at any change in direction of a roof, including along adjacent walls and roof projections. Step flashing is overlapped on the sides of a chimney or adjacent wall. Flashing is also used on built-up roofing as a transition between adjacent vertical and horizontal surfaces. A sheet metal cricket (saddle) is fabricated and placed along the upper edge of a chimney to prevent ice and debris from accumulating. The dimensions of a cricket are determined by the location of the chimney and the pitch of the roof. Custom sheet metal gutters or special ornamental vents may also be fabricated and installed at the job site.

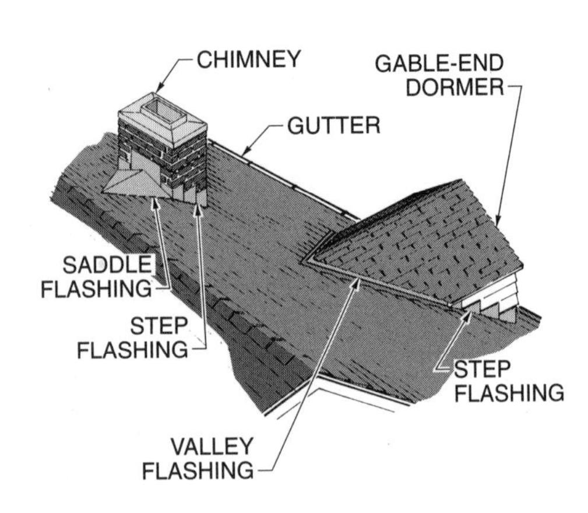

Figure 10-27. Sheet metal provides moisture protection and proper flow of storm water.

Trade Information 10

Review Questions

Name ______________________ Date ______________

Multiple Choice

______________ **1.** Regarding concrete foundation work, ___.

A. concrete for footings is usually placed separately from the foundation wall
B. rebar is only used in foundation walls
C. foundation wall forms require few braces
D. all of the above

______________ **2.** Regarding forms for concrete foundation walls, for a full-basement house ___.

A. carpenters erect all forms before concrete for footings is placed
B. the walls are generally 8″ thick
C. walers are vertical pieces that strengthen the form walls
D. job-built forms are the most common in residential construction

______________ **3.** Regarding floor joists, ___.

A. long joist spans may be supported by steel beams or solid, built-up, or engineered wood beam
B. lapped joists should be face-nailed together
C. joist size and direction are given on floor plans
D. all of the above

______________ **4.** Regarding masonry walls, ___.

A. elevations indicate the type of brick to be used
B. floor plans show the brick bond
C. brick and CMUs are shown on floor plans with the same symbol
D. standard brick is classified by actual size

______________ **5.** Regarding electrical conductors, ___.

A. copper conductors have more resistance than aluminum conductors of the same size
B. service-entrance conductors are sized from the total VA rating of the dwelling unit
C. branch-circuit conductors run from the service equipment to the subpanel
D. Table 310.16 of the NEC gives conductor sizes and ampacities based on conductor length

______________ **6.** Regarding branch circuits, a ___ A branch circuit can have up to ___ outlets.

A. 15; 12
B. 15; 18
C. 20; 16
D. none of the above

______ **7.** Regarding the plumbing system, ___.
A. water seals are provided on waste piping
B. supply water is under pressure
C. storm water drainage piping conveys rainwater
D. all of the above

______ **8.** Regarding plumbing, ___.
A. floor plans provide information regarding connections with the city sanitary sewer
B. piping drawings may be drawn as elevations or isometrics
C. vent piping is not required in residential construction
D. potable water must be treated before use

______ **9.** Regarding furnaces, ___.
A. supply air registers of a forced warm air heating system are generally located near the center of the house
B. they are sized by the cubic feet per hour
C. operating controls cycle the furnace ON or OFF
D. it is common to have more return air grilles than supply air registers

______ **10.** Regarding alternate heating systems, ___.
A. no combustion process is required to produce heat
B. solar energy can be converted to thermal energy
C. geothermal systems derive heat from heat contained within the earth
D. all of the above

True-False

T F **1.** The point of beginning is shown on the plot plan.

T F **2.** Slab-at-grade foundations are commonly used in warm climates where the frost line presents no heaving and shifting problems.

T F **3.** Slabs for residential construction are commonly 2″ to 6″ thick.

T F **4.** Floor joists commonly run along the full length of a house.

T F **5.** Straight-line solid bridging is preferred for faster nailing.

T F **6.** Roofs are framed after walls are in place and properly braced.

T F **7.** Elevations indicate the roof slope and finish materials.

T F **8.** Standard brick is classified by its nominal size.

T F **9.** Brick bond is the pattern formed by the exposed faces of brick.

T F **10.** A lateral electrical service is run from a pole to a dwelling.

T F **11.** The minimum size service-entrance conductor for a dwelling is #6.

T F **12.** The minimum size branch circuit for a dwelling is 20 A.

T F **13.** Raceways for electrical conductors are always made of metal.

T F **14.** EMT must be supported every 10′ and within 3′ of each outlet box, junction box, cabinet, or fitting.

T F **15.** Romex® may have a green insulated or bare grounding conductor.

T F **16.** Potable water is free of impurities in amounts that could cause harm.

T F **17.** Piping drawings are commonly drawn to the same scale as the floor plans.

T F **18.** Hot water heating systems are also known as radiant warm air systems.

T F **19.** Water expands when heated.

T F **20.** Solar energy can be converted to electrical energy.

PRINTREADING 11

Plans for residential construction consist of plot plans, floor plans, elevations, details, and sections. Plot plans provide information about the building lot. Floor plans show the overall shape and size of the house and provide location and size dimensions. Elevations show the exterior of the house. Details and sections provide specific structural information. Various plans within a set of prints relate to each other to provide comprehensive information.

PRINTREADING

Plans for residential construction are drawn following accepted drafting practices using conventional methods or CAD. Prints are produced from the working drawings. The prints and specifications are used by contractors and subcontractors to determine bids. Tradesworkers refer to the prints throughout construction to determine materials, location dimensions, and size dimensions.

Symbols and abbreviations are used to standardize plans and conserve space. Plot plans show the size and shape of the building lot and the location of the house on the lot. Utilities and streets are also shown. Plot plans are drawn to smaller scales. For example, a common scale for residential plot plans is 1″ = 20′-0″. Floor plans are generally the first drawings to be completed. The most commonly used scale for floor plans is ¼″ = 1′-0″. Elevations show the exterior of the house. The most commonly used scale for elevations is also ¼″ = 1′-0″. Cutting planes on plan views refer to sections and details. Sections provide additional information that cannot be easily shown on plan views. Details are drawn to larger scales than plan views to clearly show the type and size of construction materials required to complete the house. Window and door schedules identify the specific type and size of windows and doors required. Windows and doors are referenced in schedules.

Tasks performed by tradesworkers are coordinated to facilitate efficient construction. Many tasks require the completion of rough-in work and return follow-up finish work to complete the job. For example, electrical work requires a rough-in of all circuits after stud walls are erected and installation of receptacles, switches, and fixtures after finished walls are in place.

A general contractor secures a building permit, and local building officials inspect the work of the various trades during construction. The house must pass a final inspection before occupancy is permitted.

The Final Review for *Printreading for Residential Construction—Part 1* should be completed before taking the exams based on the Stewart Residence. The Final Review is based on the contents of the text.

Review prints in the order in which a building is constructed– site plan, foundation plan, floor plan(s), and exterior elevations.

STEWART RESIDENCE

The Stewart Residence is a contemporary house on a sloping lot located on Leawood Drive in Columbia, Missouri. The house was designed by Hulen & Hulen Designs and the plans were drawn by Pam Hulen. The complete set of plans contains eight sheets. The Stewart Residence includes the following plans:

Sheet 1

Basement Level Floor Plan
Plot Plan

Sheet 2

Floor Plan
Door Schedule
Window Schedule
Electrical Symbols Legend

Sheet 3

South Elevation
East Elevation
Roof Plan

Sheet 4

North Elevation
West Elevation
Typical Wall Detail

Sheet 5

Kitchen Plan Detail
Elevation Detail 1/5
Elevation Detail 2/5
Elevation Detail 3/5
Elevation Detail 4/5
Elevation Detail 5/5
Section 6/5

Sheet 6

Section 1/6
Section 2/6
Elevation Detail 3/6

Sheet 7

Basement Level Mechanical Plan
Section A-A

Sheet 8

Main Level Mechanical Plan
Cold and Hot Water Piping Diagram
Sanitary and Vent Piping Diagram

Exams for the Stewart Residence include the following:

- Plot Plans
- Floor Plans
- Elevations
- Details
- Sections
- Mechanical Plans and Plumbing Diagrams
- Final Exam

Printreading 11

Final Review

Name ______________________________ Date ______________

______________ **1.** A township contains ___.
A. 36 sq mi
B. 36 quarter sections
C. 6 furlongs
D. all of the above

______________ **2.** The denominator of a fraction is the number ___ the fraction bar.
A. to the left of or above
B. to the right of or below
C. above
D. none of the above

T F **3.** The size and direction of floor joists is generally indicated on elevations.

T F **4.** Rebar increases the strength of concrete.

T F **5.** Elevations are pictorial drawings showing interior views of a building.

______________ **6.** A dimension showing the distance from a building corner to the center of a rough opening for a window is a(n) ___ dimension.

______________ **7.** The abbreviation 7 R DN refers to the number of ___ descending to a lower floor level.

T F **8.** A true blueprint has white lines on a blue background.

______________ **9.** Studs in ___ framing run full length from the sill plate to the double top plate.

______________ **10.** The North Elevation is the ___.
A. elevation facing north
B. direction a person faces to see the north side of the house
C. elevation facing south
D. none of the above

______________ **11.** Solid bridging is placed in a straight line or ___ to facilitate nailing.

T F **12.** Cutting planes for floor plans are taken 6′-0″ above the finished floor.

T F **13.** True North is designated on a plot plan to show house orientation on the lot.

______________ **14.** The window size is indicated ___.
A. on elevations
B. in a window schedule
C. with the width stated first
D. all of the above

T F **15.** Roof slope is shown on exterior elevations.

______ **16.** Details are ___ drawn to a larger scale to show additional information.
A. sections
B. elevations
C. plan views
D. all of the above

______ **17.** The ___ size of a piece of wood is its size before planing.

T F **18.** Quadrilaterals are three-sided plane figures.

______ **19.** The dry diazo process utilizes ___ to produce prints.
A. chlorine
B. ammonia
C. bromine
D. water

T F **20.** The radius of a circle is twice its diameter.

______ **21.** The ___ is the lowest member in a platform-framed structure.

______ **22.** Regarding CAD, ___.
A. drawings cannot be shared on the Internet
B. layering facilitates the drawing of plans for specific trades
C. a mouse is an output system
D. none of the above

T F **23.** Floor plans show the shapes, sizes, and relationships of rooms.

______ **24.** Detail 4/7 is found on Sheet ___.

T F **25.** A riser is the vertical portion of a stair step.

T F **26.** A two-story house with finished basement requires two floor plans.

______ **27.** Regarding angles, a(n) ___.
A. right angle contains 90°
B. straight line contains 180°
C. obtuse angle contains more than 90°
D. all of the above

______ **28.** Regarding the Americans with Disabilities Act (ADA), the ADA ___.
A. does not specifically apply to residential construction
B. was enacted in 1992
C. provides greater accessibility to public and commercial buildings for people with physical disabilities
D. all of the above

T F **29.** An isosceles triangle contains 3 different angles.

T F **30.** Exterior doors must be at least 1¾″ thick.

______ **31.** C size paper is ___.
A. 8½″ × 11″
B. 11″ × 17″
C. 17″ × 22″
D. 18″ × 24″

__________ **32.** In an oblique ___ drawing, receding lines are drawn at one-half scale.

__________ **33.** The locations of streets, easements, and utilities are shown on ___.

A. sections
B. details
C. elevations
D. none of the above

T F **34.** GFCI receptacles must be placed adjacent to kitchen sinks.

T F **35.** A gable roof has a double slope in two directions.

__________ **36.** A(n) ___ scale is used to draw plot plans.

A. architect's
B. civil engineer's
C. mechanical engineer's
D. none of the above

T F **37.** Air conditioning units remove moisture from air in the house.

T F **38.** In regard to roofs, unit rise is the horizontal increase in height per foot of run.

__________ **39.** Natural grade is ___.

A. the slope of the lot before rough grading
B. the slope of the lot after rough grading
C. shown on plot plans with solid lines
D. none of the above

__________ **40.** A(n) ___ section is made by passing a cutting plane through the short dimension of a house.

__________ **41.** Dimension lines may be terminated with ___.

A. arrowheads
B. slashes
C. dots
D. all of the above

T F **42.** The preferred method of dimensioning framed walls is to dimension to the outside face of stud corner posts.

T F **43.** Roof slope is the relationship of unit rise to unit run.

T F **44.** All triangles contain 180°.

__________ **45.** The architect's scale ___.

A. is graduated in decimal units
B. contains a ruler and nine scales
C. may be triangular in shape
D. none of the above

T F **46.** A lateral electrical service is run underground.

T F **47.** The minimum size branch circuit for a dwelling is 15 A.

__________ **48.** Size dimensions for rooms are most clearly shown on ___.

A. plot plans
B. floor plans
C. sections
D. details

______ **49.** ___ framing is the most common type of framing for residential construction.

T F **50.** A light is a pane of glass.

T F **51.** Floor joists may be lapped.

T F **52.** Concrete masonry unit partitions are generally dimensioned to the face of the finish material.

______ **53.** A line 2¼″ long on a plan drawn at the scale of ¾″ = 1′-0″ represents a dimension of ___ in the finished house.

T F **54.** The apex of a triangle drawn on a window on an elevation points to the hinged side.

T F **55.** Termite shields or treated wood serves as a barrier against termites.

T F **56.** A 15 A, 120 V circuit contains 8 VA.

T F **57.** Supply air registers of a forced warm air heating system are generally located near the center of a house.

______ **58.** The scale most commonly used on floor plans is ___.
- A. ⅛″ = 1′-0″
- B. ¼″ = 1′-0″
- C. ⅜″ = 1′-0″
- D. ½″ = 1′-0″

______ **59.** Compass directions are commonly used to ___.
- A. name exterior elevations
- B. locate dimensions for rough openings of doors and windows
- C. denote roof slope
- D. determine swale on building lots

______ **60.** Horizontal lines of isometric drawings are drawn ___° above the horizon.
- A. 15
- B. 30
- C. 45
- D. 90

______ **61.** A flat roof must slope at least ___″ per foot for proper water runoff.
- A. 1⁄16
- B. ⅛
- C. ½
- D. ¾

T F **62.** A dimension line indicates size or location.

T F **63.** The abbreviation for bedroom is BDR.

T F **64.** A wood stud partition with ⅜″ drywall on each side of 2 × 4 studs is 4½″ thick.

______ **65.** A(n) ___ is a strip of privately owned land set aside for placement of public utilities.
- A. set aside
- B. planter strip
- C. right-of-way
- D. easement

T F **66.** Standard brick is classified by its nominal size.

T F **67.** Copper conductors are more expensive than aluminum conductors.

T F **68.** A cornice detail provides framing and finish information.

Identification 11-1

______ **1.** Concrete (elevation)

______ **2.** Glass (elevation)

______ **3.** Rough framing member

______ **4.** Sliding doors (plan view)

______ **5.** Pedestal lavatory

______ **6.** Common brick (plan view)

______ **7.** Ceiling light

______ **8.** Hose bibb

______ **9.** Brick veneer (plan view)

______ **10.** Switch

Ⓐ Ⓑ Ⓒ Ⓓ Ⓔ Ⓕ Ⓖ

HB

Ⓗ Ⓘ

S

Ⓙ

Identification 11-2

______ **1.** Isosceles triangle

______ **2.** Hexagon

______ **3.** Acute angle

______ **4.** Square

______ **5.** Equilateral triangle

______ **6.** Eccentric circles

______ **7.** Concentric circles

______ **8.** Rectangle

______ **9.** Rhombus

______ **10.** Octagon

Ⓐ Ⓑ

LESS THAN 90°

Ⓒ

(two equal sides, two equal angles)

Ⓓ

(three equal sides, three equal angles)

Ⓔ

(eight sides)

Ⓕ

(six sides)

Ⓖ

(four equal sides, opposite angles equal)

Ⓗ

(opposite sides equal, four 90° angles)

Ⓘ

(four equal sides, four 90° angles)

Ⓙ

Identification 11-3

______ 1. Baseboard
______ 2. Vertical rebar
______ 3. Horizontal rebar
______ 4. Sill plate
______ 5. Floor truss
______ 6. Brick
______ 7. Foundation footing
______ 8. Brick tie
______ 9. Foundation wall
______ 10. Earth

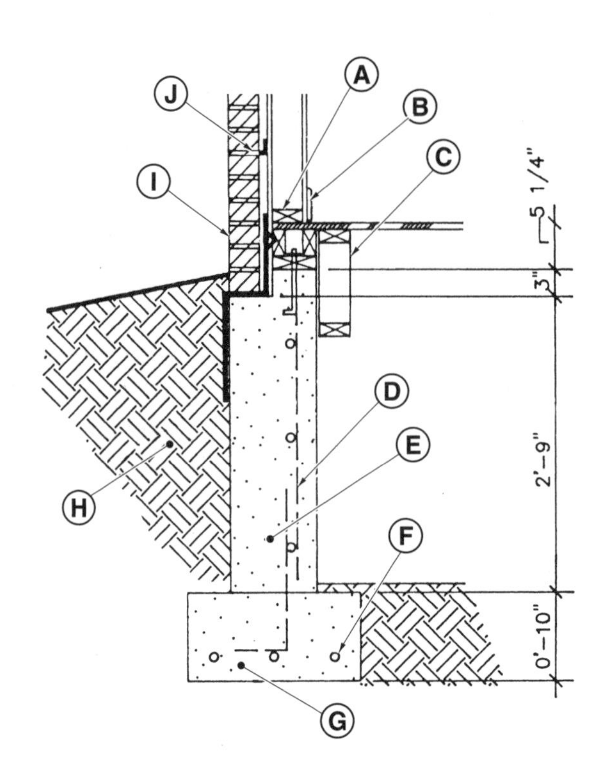

Completion

______ 1. ¼″ = 1′-0″
______ 2. ¼″ = 1′-0″
______ 3. ¼″ = 1′-0″
______ 4. ¼″ = 1′-0″
______ 5. ⅜″ = 1′-0″

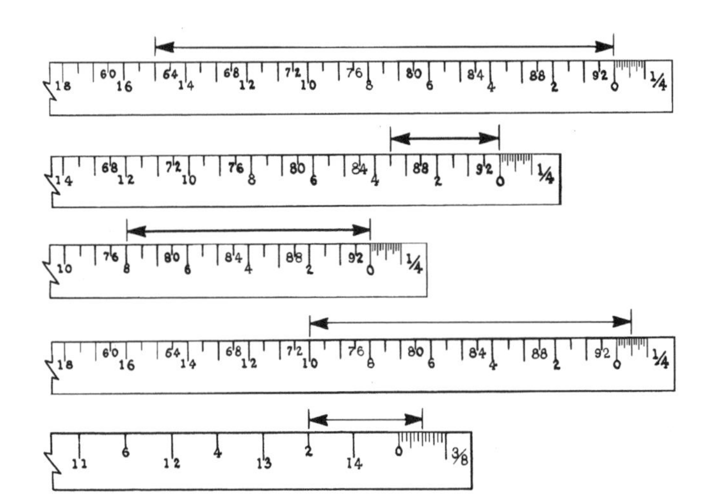

Identification 11-4

______ 1. Interior door
______ 2. Exterior door
______ 3. Casement window
______ 4. Double-hung window
______ 5. Earth
______ 6. Steel

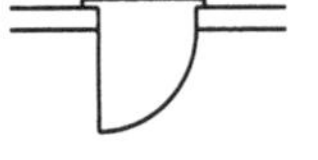

Printreading 11

Exam—Plot Plans

Name ______________________ Date ______________

Refer to Stewart Residence.

T F **1.** The Plot Plan is drawn to the scale of ¼″ = 1′-0″.

______ **2.** The lot for the Stewart Residence generally slopes down from ___to ___.
A. southeast; northwest
B. northeast; southwest
C. northwest; southeast
D. none of the above

T F **3.** PH drew the Plot Plan for the Stewart Residence.

______ **4.** The lot for the Stewart Residence contains ___ sq ft.
A. 514
B. 11,025
C. 15,960
D. 23,104

______ **5.** The front of the Stewart Residence faces ___.

______ **6.** The finish floor elevation of the main level is ___ above the finish floor elevation of the garage.

T F **7.** A 24′ existing pine tree in the front yard has a 20′-0″ diameter drip.

______ **8.** The southeast corner of the garage is ___ from the point of beginning and ___ from the east property line.
A. 15′-0″; 40′-0″
B. 40′-0″; 15′-0″
C. 40′-0″; 96′-0″
D. 105′-0″; 152′-0″

______ **9.** The general shape of the lot for the Stewart Residence is a ___.
A. rhombus
B. rhomboid
C. trapezoid
D. trapezium

T F **10.** The Den is located on the south side of the Stewart Residence.

T F **11.** Leawood Drive is parallel to the east property line of the Stewart Residence.

______ **12.** The northeast corner of the lot is approximately ___ above the point of beginning.
A. 6′-0″
B. 8′-0″
C. 10′-0″
D. 12′-0″

______ **13.** The lot for the Stewart Residence is ___ wide and ___ deep measured along the property lines.
A. 84′-0″; 96′-0″
B. 96′-0″; 105′-0″
C. 105′-0″; 152′-0″
D. none of the above

______ **14.** The finish floor elevation of the Den is ___.

______ **15.** A(n) ___ on the finish grade diverts surface water away from the northwest corner of the Stewart Residence.

______ **16.** A Deck is located on the ___ side of the Stewart Residence.

______ **17.** The driveway slopes approximately ___ from the Garage entrance to Leawood Drive.
A. 2′-0″
B. 4′-0″
C. 6′-0″
D. 8′-0″

______ **18.** An easement for utilities is located along the ___ property line.

T F **19.** Three existing trees are shown on the lot for the Stewart Residence.

______ **20.** The elevation for the point of beginning is ___.

______ **21.** The lot for the Stewart Residence is steepest near the front ___.

______ **22.** The finish floor elevation of the garage is ___.

______ **23.** The west property line is ___ long.

______ **24.** The south and east property lines meet at an angle of ___.

______ **25.** The Stewart Residence is set back ___ from the front property line.

Identification

______ **1.** Tree
______ **2.** Property line
______ **3.** Finish grade
______ **4.** Utility easement
______ **5.** Point of beginning

Ⓐ Ⓑ Ⓒ

14′

• ELEV 0′-0″

Ⓓ Ⓔ

Printreading 11

Exam—Floor Plans

Name ______________________ Date ______________

Refer to Stewart Residence.

______________ **1.** The overall dimensions of the Stewart Residence, not including the Deck, are 42′-5″ × ___.

______________ **2.** Floor plans for the Stewart Residence are drawn to the scale of ___.

______________ **3.** The main electrical panel is located in the ___.

______________ **4.** The Future Bath is stubbed in for a ___, ___, and ___.
- A. lavatory; water closet; shower
- B. lavatory; water closet; bathtub
- C. double-bowl vanity; water closet; bathtub
- D. none of the above

______________ **5.** Notations on the Basement Level Floor Plan indicate that floor joists for the main level are ___.

T F **6.** The incandescent fixture in the dining room is controlled by three-way switches.

______________ **7.** Regarding the bedrooms, ___.
- A. Bedroom 3 has a walk-in closet
- B. Bedroom 2 is larger than Bedroom 3
- C. fluorescent fixtures are located in each bedroom
- D. the Master Bedroom has four recessed can fixtures

______________ **8.** Regarding the Den, ___.
- A. bypass doors separate the Den from the Entry
- B. bay windows are located in the west wall
- C. overall room dimensions are 13′-0″ × 15′-8½″
- D. six duplex outlets are located 12″ above the finish floor

______________ **9.** Regarding the Kitchen, ___.
- A. a 3′ square skylight is located in the panned ceiling
- B. the refrigerator is located on the north wall
- C. casement windows provide a view of the front yard
- D. the cooktop is mounted in the island cabinet

______________ **10.** Regarding the Living Room, ___.
- A. built-in cabinets and shelves flank the fireplace
- B. all duplex outlets are located 12″ above the finish floor
- C. ceiling lighting is controlled by three-way switches
- D. casement windows with a circle top are located in the east wall

______________ **11.** Regarding the Dining Room, ___.
- A. sliding glass doors lead to the Deck
- B. a ceiling fan with light fixture is controlled by a wall switch
- C. all electrical receptacles are GFCI protected
- D. none of the above

_______ **12.** Regarding the Entry, ___.

A. a coat closet is to the right of the front door
B. a stairway with 7 R UP leads to the main level
C. four skylights provide natural light
D. a 12″ sidelight is located on the hinged side of the door

_______ **13.** Regarding the bathrooms, ___.

A. both bathrooms on the main level are the same size
B. the Master Bath contains a linen closet
C. the Master Bath has a pull chain fixture
D. none of the above

_______ **14.** Regarding the Future Family Room, ___.

A. a full glass door leads to the Deck
B. a telephone outlet is adjacent to the fireplace
C. the overall size is 14′-9½″ × 17′-4½″
D. no windows are shown in the west wall

_______ **15.** Regarding the future Workshop, ___.

A. casement windows provide natural light
B. floor joists above run east to west
C. all duplex outlets are 42″ above the finish floor
D. none of the above

_______ **16.** Regarding the Garage, ___.

A. a 7′-0″ × 16′-0″ overhead door is shown
B. the lower level of the house is three steps up
C. an overhead steel beam is supported by 3″ diameter steel columns
D. overall dimensions are 13′-1½″ × 23′-2½″

T F **17.** A low storage area is located beneath the Den.

T F **18.** Two cutting planes on the Floor Plan refer to sections shown on Sheet 6.

_______ **19.** A total of ___ telephone outlets is installed in the Stewart Residence.

_______ **20.** The hallway on the main level is ___ wide.

_______ **21.** ___ lighting fixtures in the kitchen are controlled by three-way switches.

T F **22.** The walk-in closet in the Master Bedroom has two shelves with rods.

T F **23.** The lighting fixture in the storage closet of the Future Family Room is controlled by a pull chain.

T F **24.** The concrete slab beneath the stairway is thickened to support the additional weight.

_______ **25.** The finish floor level of the low storage area is ___″ above the finish floor level of the garage.

_______ **26.** A(n) ___ fixture near the fireplace in the Future Family Room is controlled by a wall switch.

_______ **27.** The attic is entered through a scuttle measuring ___.

_______ **28.** The Deck measures 10′-0″ × ___.

_______ **29.** Closets in Bedrooms 2 and 3 are ___ deep.

_______________ **30.** The hall linen closet is ___ wide.

_______________ **31.** A panned ceiling is shown in the ___.

_______________ **32.** The ceiling in the ___ Room is vaulted.

_______________ **33.** The concrete foundation wall beneath the den is ___″ thick.

T F **34.** The water heater is located in the Laundry Room.

_______________ **35.** The north wall of the Kitchen projects ___ beyond the rest of the north wall of the Stewart Residence.

T F **36.** The Hall Bath contains two GFI outlets.

T F **37.** Bay windows are located in the Living Room.

T F **38.** The front stairs contain 12 risers.

T F **39.** The linen closet in the Master Bath has no lighting fixture.

T F **40.** Two Garage door opener outlets are shown on the Garage ceiling.

_______________ **41.** A total of ___ duplex outlets are located 12″ above the finish floor in the walls of the Future Family Room.

_______________ **42.** H windows in the Future Family Room are centered ___ from the northwest corner of the house.

_______________ **43.** E windows in the Workshop are centered ___ from the northeast corner of the house.

T F **44.** A floor drain is shown in the center of the Garage floor.

T F **45.** Two H windows provide natural light in the Laundry Room.

T F **46.** The hallway on the main floor is 21′-1¼″ long.

T F **47.** Ceiling fixtures on the Kitchen ceiling are fluorescent.

_______________ **48.** The service-entrance conductors enter the house on the ___ side.

_______________ **49.** Bedroom 3 contains approximately ___ sq ft.

A. 93
B. 103
C. 133
D. 153

_______________ **50.** The Master Bedroom contains approximately ___ sq ft.

A. 117
B. 137
C. 157
D. 177

_______________ **51.** The storage closet beneath the stairway measures 3′-5¾″ × ___.

_______________ **52.** D windows in the Den are centered ___ from the framed southwest corner.

T F **53.** A cased opening with arch leads from the Kitchen to the Dining Room.

_______________ **54.** The main entry is ___ wide.

_______________ **55.** The bay for A windows in the Den is ___ wide.

_______________ **56.** Four recessed can lighting fixtures are shown in the ___.
A. Den
B. Living Room
C. Master Bath
D. all of the above

T F **57.** Wrought iron handrails are shown for the entry steps.

T F **58.** Refrigerator space in the Kitchen is provided on the east wall.

T F **59.** All bedroom closets have bypass doors.

T F **60.** The wall receptacle on the north wall of the Living Room is split-wired.

Identification

_______________ **1.** Ceiling fixture

_______________ **2.** Floor drain

_______________ **3.** Ceiling fan with light

_______________ **4.** Wall sconce fixture

_______________ **5.** Casement window

_______________ **6.** Telephone

_______________ **7.** Single-pole switch

_______________ **8.** 120 V receptacle

_______________ **9.** Brick

_______________ **10.** Concrete

FD
Ⓐ Ⓑ
Ⓒ Ⓓ
Ⓔ Ⓕ
S
Ⓖ Ⓗ
Ⓘ Ⓙ

Printreading 11

Exam—Elevations

Name ______________________________ Date ______________

Refer to Stewart Residence.

______________ **1.** All elevations are drawn to the scale of ___.

______________ **2.** The ___ Elevation shows the front of the Stewart Residence.

______________ **3.** Regarding the South Elevation, ___.
- A. the fireplace chimney is centered on the bay windows
- B. sidelights flank the entry door
- C. all exterior walls are brick veneer
- D. one skylight is centered on the entry door

______________ **4.** Regarding the East Elevation, ___.
- A. the roof slopes 4 in 12
- B. an ACA-treated handrail leads to the entry door
- C. exterior walls are brick veneer and 4″ lap siding
- D. a 3′-0″ diameter glass block window and one-half glass door provide natural light in the Garage

______________ **5.** Regarding the North Elevation, ___.
- A. the Deck extends completely across the back side of the Stewart Residence
- B. 4 × 4 treated posts set in 12″ diameter concrete piers support the Deck
- C. exterior walls are brick veneer and 8″ lap siding
- D. the roof is finished with cedar shakes

______________ **6.** Regarding the West Elevation, the ___.
- A. Den floor and Basement floor are at the same level
- B. chimney is brick veneer with a 4″ limestone cap
- C. roof slope is 5 in 12
- D. none of the above

T F **7.** The Deck floor level is level with the main floor level.

T F **8.** Overhead Garage doors are vinyl-clad metal.

______________ **9.** A total of ___ C windows are shown on the East Elevation.

______________ **10.** Exterior walls are either brick veneer or ___″ vinyl siding.

______________ **11.** The O window has a(n) ___ top.

______________ **12.** The roof has a total of ___ skylights.

T F **13.** Front entry steps contain a landing.

T F **14.** Deck steps are attached to the rear wall.

_______ **15.** The front entry has a(n) ___-panel, insulated metal door.

T F **16.** All concrete foundation footings are at the same level.

_______ **17.** Regarding the deck, ___.
A. four 12″ diameter concrete piers are shown
B. stairs are straight run with two handrails
C. stairs are located on the west end of the Deck
D. none of the above

T F **18.** A combination of gable and shed roofs is shown for the Stewart Residence.

T F **19.** Glass blocks are shown on the North and East Elevations.

_______ **20.** The Basement has a clear ceiling height of ___.

_______ **21.** B windows are 4′-9″ × ___.

_______ **22.** The front entry steps are made of ___.

_______ **23.** Exterior wall finish above the Garage roof is ___.

_______ **24.** Each Garage door is 9′-0″ wide × ___ high.

_______ **25.** The K skylight is visible on all elevations except the ___ Elevation.

Matching

_______ **1.** Pitch symbol

_______ **2.** Finished wood

_______ **3.** Brick veneer

_______ **4.** Batt insulation

_______ **5.** Centerline

_______ **6.** Diameter

_______ **7.** Finish grade

_______ **8.** Gravel fill

_______ **9.** Dimension line

_______ **10.** Rigid foam insulation

(A) (B) 12 5 (C)

(D) (E) (F)

Ø

(G) (H) (I) (J)

Printreading 11

Exam—Details

Name ______________________ Date ____________

Refer to Stewart Residence.

______ **1.** Elevation Detail 2/5 shows the ___ wall of the Kitchen.

______ **2.** Kitchen base cabinets contain a total of ___ drawers.

______ **3.** The refrigerator cabinet has an opening ___″ wide.

T F **4.** Wall countertops have a 4″ backsplash.

T F **5.** Soffits over the wall cabinets are 14″ deep.

T F **6.** A full glass, insulated metal door leads to the Deck.

______ **7.** Regarding the wall cabinets, ___.
- A. two 21″ × 30″ cabinets are required
- B. three 33″ × 30″ cabinets are required
- C. all wall cabinets are 12″ deep
- D. none of the above

______ **8.** Regarding the base cabinets, ___.
- A. the drawer base is 14″ wide
- B. all base cabinets are 3′-0″ high
- C. an 8″ backsplash is attached to the countertop
- D. none of the above

T F **9.** Elevation 3/5 shows details of the oven cabinet.

______ **10.** Kitchen cabinet elevation details are drawn to the scale of ___.

______ **11.** The Kitchen Plan is drawn to the scale of ___.

______ **12.** Regarding the Kitchen Plan, ___.
- A. the panned ceiling is shown with dashed lines
- B. Detail 3/5 shows the sink
- C. the refrigerator is centered between the oven and broom pantry
- D. none of the above

______ **13.** Regarding the section detail of the skylight, ___.
- A. the skylight is 3′-0″ in diameter
- B. 6″ batt insulation surrounds the skylight shaft
- C. the scale is ⅜″ = 1′-0″
- D. galvanized sheet metal is used as flashing

T F **14.** The broom pantry is placed between the lazy Susan pantry and the cooktop.

______ **15.** The breakfast bar is ___″ lower than the island counter.

_______ **16.** All wall cabinets, except for the hood cabinet, are ___ high.

_______ **17.** The soffit over the broom pantry is ___ deep.

T F **18.** The microwave is placed in the oven cabinet above the oven.

_______ **19.** Vertical clearance between the base cabinet countertops and wall cabinets is ___.

T F **20.** The sink base cabinet is 36″ wide.

_______ **21.** Soffits over the wall cabinets are ___ high.

T F **22.** The lazy Susan pantry is shown on Elevation Detail 1/5.

_______ **23.** Regarding the section detail of the skylight, ___.

A. panned ceiling framing is 2 × 4s
B. the cutting plane for the section is shown on the Kitchen Plan
C. 6″ batt insulation surrounds the skylight framing
D. all of the above

_______ **24.** The opening for the oven is ___″ wide.

T F **25.** The vertical clearance between the countertops and wall cabinets is 1′-6″.

Matching

_______ **1.** Cabinet door hinged on left

_______ **2.** Cabinet door hinged on right

_______ **3.** Rough wood

_______ **4.** Cutting plane line

_______ **5.** Batt insulation

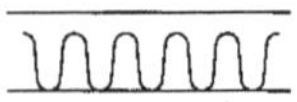

Ⓐ Ⓑ Ⓒ Ⓓ Ⓔ

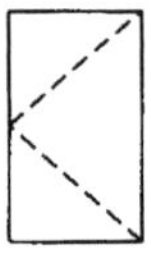

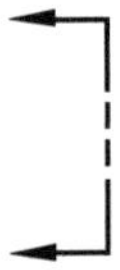

Exam—Sections

Name ______________________ Date ____________

Refer to Stewart Residence.

______ **1.** The cutting plane for Section 1/6 is shown on Sheet(s) ___.

______ **2.** Sections are drawn to the scale of ___.

T F **3.** Sections are shown on Sheet 6 of 7.

T F **4.** Roof slope for the Stewart Residence is 5 in 12.

______ **5.** The finish floor level of the Den is ___ below the finish floor level of the Living Room.

______ **6.** Floor-to-ceiling height of the Dining Room is ___.

T F **7.** The Workshop is directly below the Master Bedroom.

______ **8.** Section 1/6 is a(n) ___ section.

______ **9.** Section 2/6 is a(n) ___ section.

______ **10.** Regarding the chimney, ___.

A. face brick is used in the Future Family Room and the Living Room
B. native stone is used in the Living Room
C. a 12″ × 48″ mantle is installed over the fireplace
D. none of the above

______ **11.** Regarding the Workshop, ___.

A. two hollow-core doors are shown
B. six-panel exterior doors lead to the backyard
C. one steel-clad door is shown
D. all of the above

T F **12.** Door 8 provides an entrance to the Master Bedroom.

______ **13.** The finish floor level of the low storage area is ___ below the finish floor level of the Future Family Room.

T F **14.** North and south walls of the living room are 10′-0″ high.

T F **15.** The Living Room contains a vaulted ceiling.

T F **16.** Sections of the Stewart Residence were drawn by PH.

______ **17.** The stairway from the Garage to the lower level contain ___ risers.

______ **18.** A(n) ___ beam in the Garage supports floor joists above.

T F **19.** The floor-to-ceiling height of the Future Family Room is 7′-9½″.

T F **20.** Two casement windows are shown on the section of the Dining Room.

__________ **21.** Regarding the fireplaces, one is located on the ___.

A. south wall of the Living Room
B. east wall of the Future Family Room
C. east wall of the Living Room
D. north wall of the Future Family Room

__________ **22.** Closet doors for Bedroom 3 are ___ doors.

T F **23.** The size for Dining Room windows is 4′-9″ × 5′-0⅜″.

T F **24.** The circletop window in the living room contains six pie-shaped lights.

T F **25.** The floor-to-ceiling height in all bedrooms is 8′-1⅛″.

Matching

__________ **1.** Concrete

__________ **2.** Native stone

__________ **3.** Brick veneer

__________ **4.** Steel beam

__________ **5.** Pitch symbol

Ⓐ

Ⓑ Ⓒ

12
7

Ⓓ Ⓔ

Printreading 11 Exam—Mechanical Plans and Plumbing Diagrams

Name ______________________ Date ______________

Refer to Stewart Residence.

T F **1.** The Mechanical Room does not contain a floor drain.

______ **2.** The air conditioner condenser coil is ___.
- A. shown on Sheet 8
- B. attached to the furnace with a flex connection
- C. indicated by a circle on the north wall
- D. located directly above the furnace

T F **3.** The flue is 30″ × 8″.

______ **4.** The diameter of the waste vent from the Master Bath to the main vent is ___″.
- A. 1½
- B. 2
- C. 3
- D. 4

______ **5.** The flue in the outside wall of the Master Bath is connected to the ___.
- A. furnace
- B. hot water tank
- C. air conditioner
- D. vent piping for the Master Bath double sink

______ **6.** There are ___ air outlets from the HVAC system in the Den.

______ **7.** The main water service pipe is ___″ diameter.

______ **8.** The water supply line to the dishwasher is ___″ diameter.

______ **9.** The size of the return air duct at the point where it drops to enter the furnace is ___.
- A. 15 × 15
- B. 23 × 12
- C. 18 × 8
- D. 22 × 19

______ **10.** The abbreviation CO on the sanitary and vent piping diagram designates ___.

______ **11.** The duct at the north wall of the Future Family Room is ___.
- A. 12″ high × 8″ wide
- B. 12″ high × 6″ wide
- C. 8″ high × 12″ wide
- D. 6″ high × 12″ wide

T F **12.** One 10 × 6 duct in the Future Family Room is capped for future use.

T F **13.** No ducts extend to the upper floors from the wall between the Future Family Room and the low storage area.

T F **14.** The gas line is attached to the furnace on the west side of the furnace.

______ **15.** The liquid and suctions lines shown on the north wall are related to the ___.
A. hot water tank
B. floor drain
C. air conditioner
D. furnace

T F **16.** All air supply ductwork shown is 8″ high.

T F **17.** Ductwork must be lowered between the Workshop and Garage to extend under the supporting beam.

______ **18.** The 8 × 4 duct at the east end of the Garage supplies air to the ___.
A. Garage
B. Master Bedroom
C. Bedroom 2
D. Bedroom 3

______ **19.** The floor condensate drain from the air conditioning coil is ___″ in diameter.

______ **20.** The gas meter is on the ___ wall.
A. north
B. south
C. east
D. west

______ **21.** The capacity of the furnace is ___ cfm.

T F **22.** There are three hose bibbs shown on the water piping diagram.

T F **23.** Both bathtub waste drain pipes are 2″ in diameter.

______ **24.** In the note 5 TONS CLG CAP, the abbreviation CLG is ___.

T F **25.** A ½″ water supply line is provided for the refrigerator.

T F **26.** A funnel drain is located in the basement floor.

______ **27.** The cold water supply connection to the hot water tank is ___″.
A. ½
B. ¾
C. 1
D. 1½

______ **28.** The abbreviation WC on the cold and hot water piping diagram designates ___.

______ **29.** The abbreviation VTR on the sanitary and vent piping diagram designates ___.

Printreading 11

Final Exam

Name ______________________________ Date ______________

Refer to Stewart Residence.

T F **1.** Plans for the Stewart Residence contain six sheets.

T F **2.** PH of Hulen & Hulen Designs drew all plans for the Stewart Residence.

T F **3.** The scale for each sheet of the Stewart Residence plans is given in a title block on each sheet.

______________ **4.** The Stewart Residence is a ___.
A. one-story house with full basement
B. two-story house with full basement
C. one-and-one-half story house with living space in the attic
D. none of the above

______________ **5.** The building corner from which horizontal measurements begin is located ___ of the point of beginning.
A. 40′-0″ south and 15′-0″ west
B. 40′-0″ north and 15′-0″ east
C. 40′-0″ south and 15′-0″ east
D. 40′-0″ north and 15′-0″ west

______________ **6.** The front of the Stewart Residence is finished with ___.
A. 8″ vinyl siding and wood shingles
B. 4″ vinyl siding, brick veneer, and wood shingles
C. 8″ vinyl siding, brick veneer, and asphalt shingles
D. none of the above

T F **7.** The native stone chimney is south of the ridge board.

T F **8.** Casement windows in the Kitchen provide a view of Leawood Drive.

T F **9.** One #3 door is required.

______________ **10.** Regarding the floor plans, ___.
A. the Stewart Residence has four bedrooms
B. a separate Breakfast Area adjoins the Kitchen
C. the Den and Future Family Room are on the same level
D. the Master Bedroom measures 12′-9″ × 13′-6″

T F **11.** Two hose bibbs are shown for the Stewart Residence.

T F **12.** Four 1⅜″ × 2′-8″ × 6′-8″ hollow-core, six-panel birch doors are required.

T F **13.** Six telephone outlets are to be installed during construction.

T F **14.** Glass block windows for the Stewart Residence are custom made.

_______________ **15.** Skylights in the entry measure ___.

_______________ **16.** Duplex outlets in the Stewart Residence are placed either 12″ or ___″ above the finished floor.

T F **17.** Floor joists for the main floor run north to south.

_______________ **18.** All roofs of the Stewart Residence have a(n) ___ in 12 slope.

_______________ **19.** Roof overhang at exterior walls measures ___.

_______________ **20.** Regarding typical concrete foundations, ___.

A. three #4 horizontal rebar are continuous
B. 2″ rigid insulation is placed around foundation walls
C. a 16″ × 8″ concrete footing supports the foundation wall
D. all of the above

_______________ **21.** Regarding typical exterior walls, ___.

A. 2 × 6 studs are 16″ OC
B. ½″ diameter ABs are 2′-0″ OC
C. insulation is blown in all exterior walls
D. sill plates are 2 × 8s

_______________ **22.** Regarding doors, ___.

A. all are flush panel
B. #11 is an exterior door
C. a full glass door leads to the deck
D. none of the above

_______________ **23.** Regarding windows, ___.

A. Andersen C15 casement windows are shown in Bedrooms 2 and 3
B. the vertical dimension for Den windows is 5′-0⅜″
C. all windows in the residence are the same type
D. none of the above

_______________ **24.** Four ___ can fixtures are installed in the Den ceiling.

T F **25.** The hall closet has a pull chain lighting fixture.

T F **26.** The deck and handrail are ACA treated.

T F **27.** Perforated PVC drain tile is placed around foundation walls.

T F **28.** Elevation details of the kitchen are drawn to the scale of ¼″ = 1′-0″.

T F **29.** An 8′ utility easement runs along the north side of the lot.

_______________ **30.** All receptacles near vanities are ___ protected.

_______________ **31.** The Laundry Room is entered through the ___.

_______________ **32.** The Garage measures ___ × 26′-2½″.

_______________ **33.** N windows are shown on the ___ Elevation.

T F **34.** The Stewart Residence contains gable and shed roofs.

T F **35.** The exterior wall finish of the Stewart Residence is 8″ vinyl siding and brick veneer.

Appendix

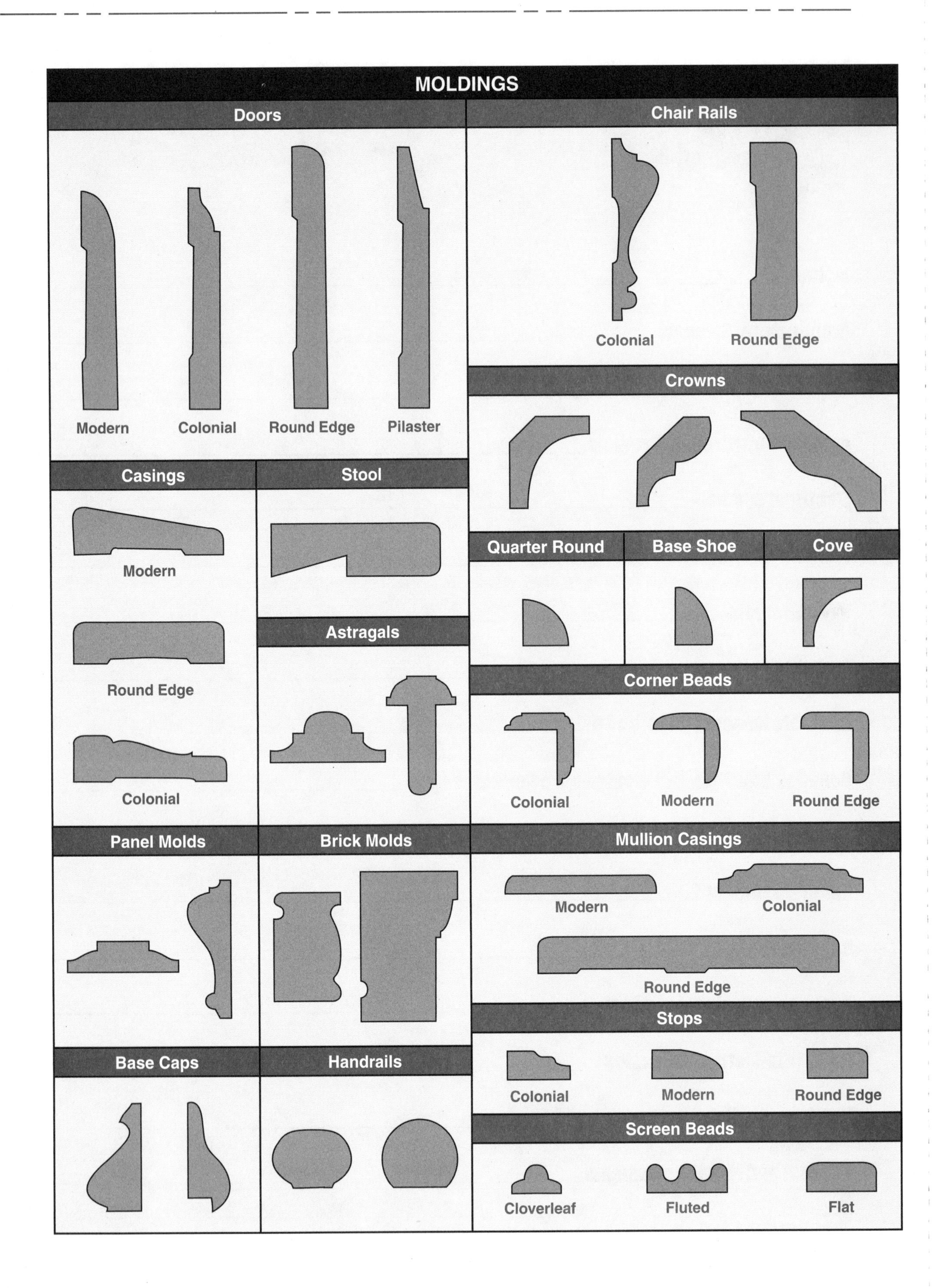
MOLDINGS
Doors
Modern
Colonial
Round Edge
Pilaster
Chair Rails
Colonial
Round Edge
Crowns
Casings
Modern
Round Edge
Colonial
Stool
Astragals
Quarter Round
Base Shoe
Cove
Corner Beads
Colonial
Modern
Round Edge
Panel Molds
Brick Molds
Mullion Casings
Modern
Colonial
Round Edge
Stops
Colonial
Modern
Round Edge
Base Caps
Handrails
Screen Beads
Cloverleaf
Fluted
Flat

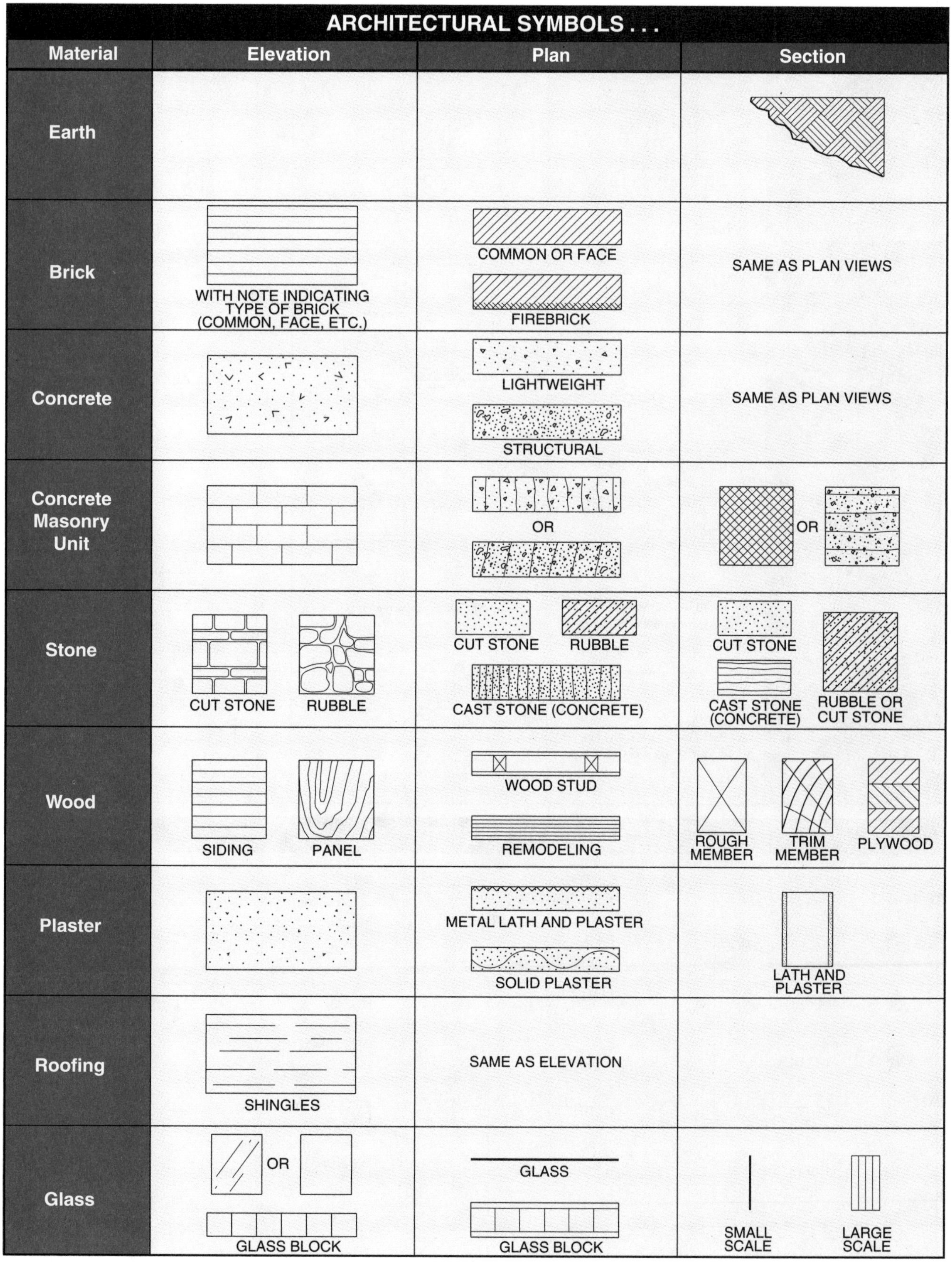

ARCHITECTURAL SYMBOLS . . .			
Material	**Elevation**	**Plan**	**Section**
Earth			
Brick	WITH NOTE INDICATING TYPE OF BRICK (COMMON, FACE, ETC.)	COMMON OR FACE FIREBRICK	SAME AS PLAN VIEWS
Concrete		LIGHTWEIGHT STRUCTURAL	SAME AS PLAN VIEWS
Concrete Masonry Unit		OR	OR
Stone	CUT STONE RUBBLE	CUT STONE RUBBLE CAST STONE (CONCRETE)	CUT STONE CAST STONE (CONCRETE) RUBBLE OR CUT STONE
Wood	SIDING PANEL	WOOD STUD REMODELING	ROUGH MEMBER TRIM MEMBER PLYWOOD
Plaster		METAL LATH AND PLASTER SOLID PLASTER	LATH AND PLASTER
Roofing	SHINGLES	SAME AS ELEVATION	
Glass	OR GLASS BLOCK	GLASS GLASS BLOCK	SMALL SCALE LARGE SCALE

... ARCHITECTURAL SYMBOLS			
Material	**Elevation**	**Plan**	**Section**
Facing Tile	CERAMIC TILE	FLOOR TILE	CERAMIC TILE LARGE SCALE CERAMIC TILE SMALL SCALE
Structural Clay Tile			SAME AS PLAN VIEW
Insulation		LOOSE FILL OR BATTS RIGID SPRAY FOAM	SAME AS PLAN VIEWS
Sheet Metal Flashing		OCCASIONALLY INDICATED BY NOTE	
Metals Other Than Flashing	INDICATED BY NOTE OR DRAWN TO SCALE	SAME AS ELEVATION	SMALL SCALE STEEL CAST IRON ALUMINUM BRONZE OR BRASS
Structural Steel	INDICATED BY NOTE OR DRAWN TO SCALE	OR	REBARS LI SMALL SCALE LARGE SCALE L-ANGLES, S-BEAMS, ETC.

PLOT PLAN SYMBOLS			
N NORTH	FIRE HYDRANT	WALK	E OR ELECTRIC SERVICE
POINT OF BEGINNING (POB)	MAILBOX	IMPROVED ROAD	G OR NATURAL GAS LINE
UTILITY METER OR VALVE	MANHOLE	UNIMPROVED ROAD	W OR WATER LINE
POWER POLE AND GUY	TREE	℄ BUILDING LINE	T OR TELEPHONE LINE
LIGHT STANDARD	BUSH	℘ PROPERTY LINE	NATURAL GRADE
TRAFFIC SIGNAL	HEDGE ROW	PROPERTY LINE	FINISH GRADE
STREET SIGN	FENCE	TOWNSHIP LINE	+ XX.00′ EXISTING ELEVATION

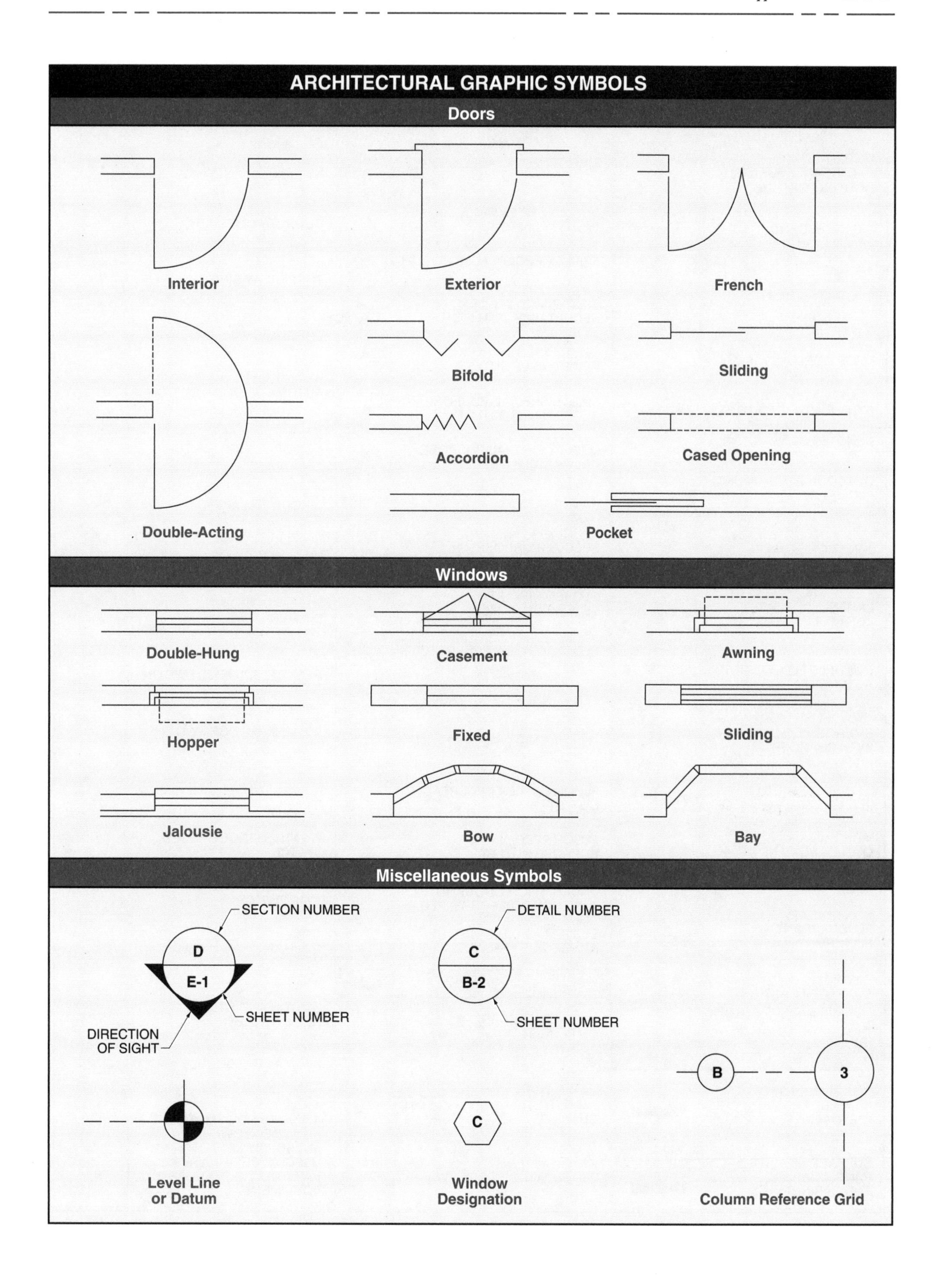
ARCHITECTURAL GRAPHIC SYMBOLS
Doors
Interior
Exterior
French
Bifold
Sliding
Accordion
Cased Opening
Double-Acting
Pocket
Windows
Double-Hung
Casement
Awning
Hopper
Fixed
Sliding
Jalousie
Bow
Bay
Miscellaneous Symbols
SECTION NUMBER
D
E-1
SHEET NUMBER
DIRECTION OF SIGHT
DETAIL NUMBER
C
B-2
SHEET NUMBER
B
3
Level Line or Datum
C
Window Designation
Column Reference Grid

ELECTRICAL SYMBOLS . . .

Lighting Outlets

OUTLET BOX AND INCANDESCENT LIGHTING FIXTURE	CEILING WALL
INCANDESCENT TRACK LIGHTING	
BLANKED OUTLET	B B
DROP CORD	D
EXIT LIGHT AND OUTLET BOX. SHADED AREAS DENOTE FACES.	
OUTDOOR POLE-MOUNTED FIXTURES	
JUNCTION BOX	J J
LAMPHOLDER WITH PULL SWITCH	L_{PS} L_{PS}
MULTIPLE FLOODLIGHT ASSEMBLY	
EMERGENCY BATTERY PACK WITH CHARGER	B
INDIVIDUAL FLUORESCENT FIXTURE	
OUTLET BOX AND FLUORESCENT LIGHTING TRACK FIXTURE	
CONTINUOUS FLUORESCENT FIXTURE	
SURFACE-MOUNTED FLUORESCENT FIXTURE	

Panelboards

FLUSH-MOUNTED PANELBOARD AND CABINET	
SURFACE-MOUNTED PANELBOARD AND CABINET	

Convenience Outlets

SINGLE RECEPTACLE OUTLET	
DUPLEX RECEPTACLE OUTLET–120 V	
TRIPLEX RECEPTACLE OUTLET–240 V	
SPLIT-WIRED DUPLEX RECEPTACLE OUTLET	
SPLIT-WIRED TRIPLEX RECEPTACLE OUTLET	
SINGLE SPECIAL-PURPOSE RECEPTACLE OUTLET	
DUPLEX SPECIAL-PURPOSE RECEPTACLE OUTLET	
RANGE OUTLET	R
SPECIAL-PURPOSE CONNECTION	DW
CLOSED-CIRCUIT TELEVISION CAMERA	
CLOCK HANGER RECEPTACLE	C
FAN HANGER RECEPTACLE	F
FLOOR SINGLE RECEPTACLE OUTLET	
FLOOR DUPLEX RECEPTACLE OUTLET	
FLOOR SPECIAL-PURPOSE OUTLET	
UNDERFLOOR DUCT AND JUNCTION BOX FOR TRIPLE, DOUBLE, OR SINGLE DUCT SYSTEM AS INDICATED BY NUMBER OF PARALLEL LINES	

Busducts and Wireways

SERVICE, FEEDER, OR PLUG-IN BUSWAY	B B B
CABLE THROUGH LADDER OR CHANNEL	C C C
WIREWAY	W W W

Switch Outlets

SINGLE-POLE SWITCH	S
DOUBLE-POLE SWITCH	S_2
THREE-WAY SWITCH	S_3
FOUR-WAY SWITCH	S_4
AUTOMATIC DOOR SWITCH	S_D
KEY-OPERATED SWITCH	S_K
CIRCUIT BREAKER	S_{CB}
WEATHERPROOF CIRCUIT BREAKER	S_{WCB}
DIMMER	S_{DM}
REMOTE CONTROL SWITCH	S_{RC}
WEATHERPROOF SWITCH	S_{WP}
FUSED SWITCH	S_F
WEATHERPROOF FUSED SWITCH	S_{WF}
TIME SWITCH	S_T
CEILING PULL SWITCH	S
SWITCH AND SINGLE RECEPTACLE	S
SWITCH AND DOUBLE RECEPTACLE	S
A STANDARD SYMBOL WITH AN ADDED LOWERCASE SUBSCRIPT LETTER IS USED TO DESIGNATE A VARIATION IN STANDARD EQUIPMENT	a.b a.b $S_{a.b}$

. . . ELECTRICAL SYMBOLS

Commercial and Industrial Systems

- PAGING SYSTEM DEVICE
- FIRE ALARM SYSTEM DEVICE
- COMPUTER DATA SYSTEM DEVICE
- PRIVATE TELEPHONE SYSTEM DEVICE
- SOUND SYSTEM
- FIRE ALARM CONTROL PANEL — FACP

Signaling System Outlets for Residential Systems

- PUSHBUTTON
- BUZZER
- BELL
- BELL AND BUZZER COMBINATION
- COMPUTER DATA OUTLET
- BELL RINGING TRANSFORMER — BT
- ELECTRIC DOOR OPENER — D
- CHIME — CH
- TELEVISION OUTLET — TV
- THERMOSTAT — T

Underground Electrical Distribution or Electrical Lighting Systems

- MANHOLE — M
- HANDHOLE — H
- TRANSFORMER-MANHOLE OR VAULT — TM
- TRANSFORMER PAD — TP
- UNDERGROUND DIRECT BURIAL CABLE
- UNDERGROUND DUCT LINE
- STREET LIGHT STANDARD FED FROM UNDERGROUND CIRCUIT

Above-Ground Electrical Distribution or Lighting Systems

- POLE
- STREET LIGHT AND BRACKET
- PRIMARY CIRCUIT
- SECONDARY CIRCUIT
- DOWN GUY
- HEAD GUY
- SIDEWALK GUY
- SERVICE WEATHERHEAD

Panel Circuits and Miscellaneous

- LIGHTING PANEL
- POWER PANEL
- WIRING – CONCEALED IN CEILING OR WALL
- WIRING – CONCEALED IN FLOOR
- WIRING EXPOSED
- HOME RUN TO PANEL BOARD
  Indicate number of circuits by number of arrows. Any circuit without such designation indicates a two-wire circuit. For a greater number of wires indicate as follows: /// (3 wires) //// (4 wires), etc.
- FEEDERS
  Use heavy lines and designate by number corresponding to listing in feeder schedule
- WIRING TURNED UP
- WIRING TURNED DOWN
- GENERATOR — G
- MOTOR — M
- INSTRUMENT (SPECIFY) — I
- TRANSFORMER — T
- CONTROLLER
- EXTERNALLY-OPERATED DISCONNECT SWITCH
- PULL BOX

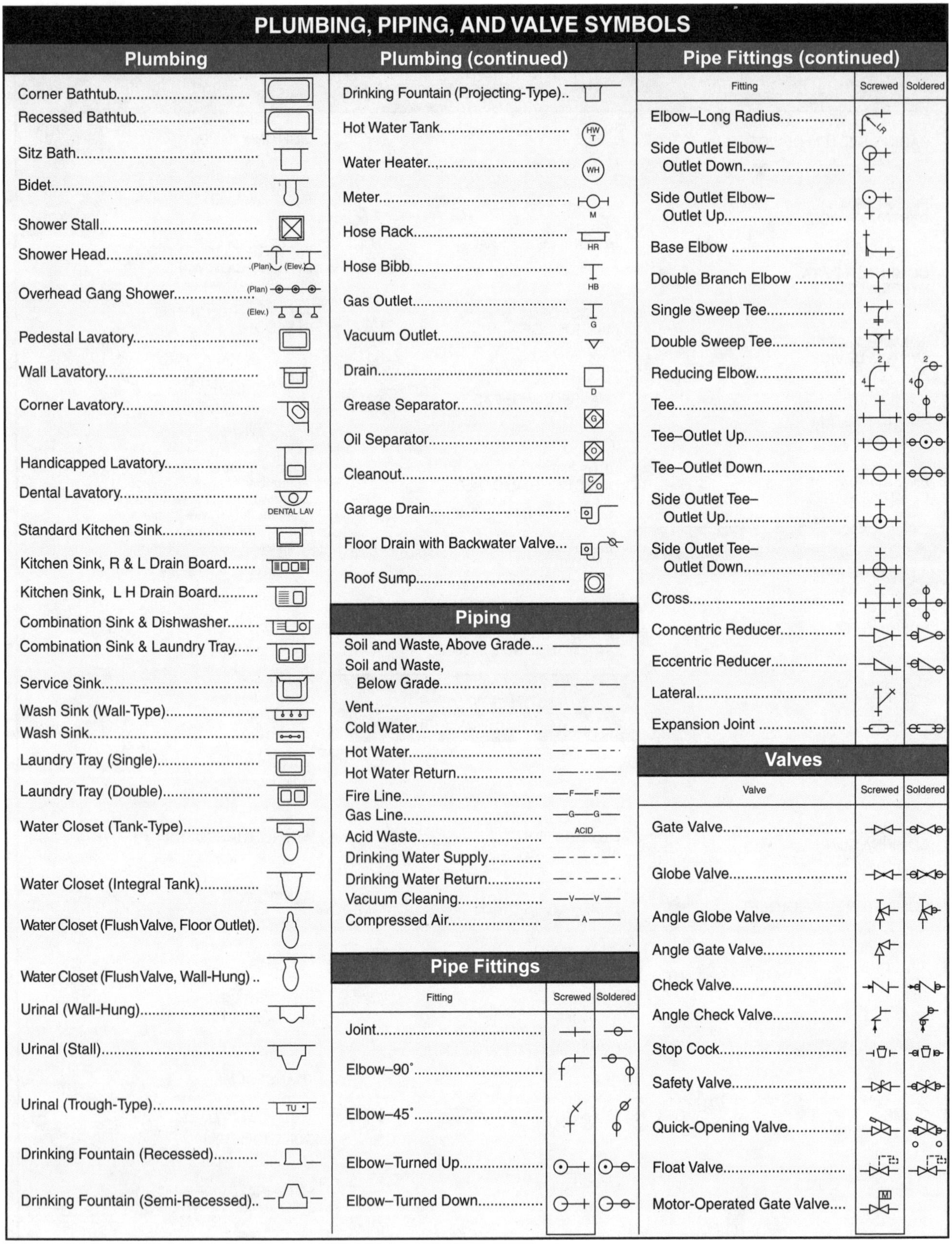

PLUMBING, PIPING, AND VALVE SYMBOLS

Plumbing

Corner Bathtub	
Recessed Bathtub	
Sitz Bath	
Bidet	
Shower Stall	
Shower Head	(Plan) (Elev.)
Overhead Gang Shower	(Plan) (Elev.)
Pedestal Lavatory	
Wall Lavatory	
Corner Lavatory	
Handicapped Lavatory	
Dental Lavatory	DENTAL LAV
Standard Kitchen Sink	
Kitchen Sink, R & L Drain Board	
Kitchen Sink, L H Drain Board	
Combination Sink & Dishwasher	
Combination Sink & Laundry Tray	
Service Sink	
Wash Sink (Wall-Type)	
Wash Sink	
Laundry Tray (Single)	
Laundry Tray (Double)	
Water Closet (Tank-Type)	
Water Closet (Integral Tank)	
Water Closet (Flush Valve, Floor Outlet)	
Water Closet (Flush Valve, Wall-Hung)	
Urinal (Wall-Hung)	
Urinal (Stall)	
Urinal (Trough-Type)	TU
Drinking Fountain (Recessed)	
Drinking Fountain (Semi-Recessed)	

Plumbing (continued)

Drinking Fountain (Projecting-Type)	
Hot Water Tank	HW T
Water Heater	WH
Meter	M
Hose Rack	HR
Hose Bibb	HB
Gas Outlet	G
Vacuum Outlet	
Drain	D
Grease Separator	G
Oil Separator	O
Cleanout	
Garage Drain	
Floor Drain with Backwater Valve	
Roof Sump	

Piping

Piping	Symbol
Soil and Waste, Above Grade	
Soil and Waste, Below Grade	
Vent	
Cold Water	
Hot Water	
Hot Water Return	
Fire Line	—F—F—
Gas Line	—G—G—
Acid Waste	ACID
Drinking Water Supply	
Drinking Water Return	
Vacuum Cleaning	—V—V—
Compressed Air	—A—

Pipe Fittings

Fitting	Screwed	Soldered
Joint		
Elbow–90°		
Elbow–45°		
Elbow–Turned Up		
Elbow–Turned Down		

Pipe Fittings (continued)

Fitting	Screwed	Soldered
Elbow–Long Radius		
Side Outlet Elbow–Outlet Down		
Side Outlet Elbow–Outlet Up		
Base Elbow		
Double Branch Elbow		
Single Sweep Tee		
Double Sweep Tee		
Reducing Elbow	2 4	2 4
Tee		
Tee–Outlet Up		
Tee–Outlet Down		
Side Outlet Tee–Outlet Up		
Side Outlet Tee–Outlet Down		
Cross		
Concentric Reducer		
Eccentric Reducer		
Lateral		
Expansion Joint		

Valves

Valve	Screwed	Soldered
Gate Valve		
Globe Valve		
Angle Globe Valve		
Angle Gate Valve		
Check Valve		
Angle Check Valve		
Stop Cock		
Safety Valve		
Quick-Opening Valve		
Float Valve		
Motor-Operated Gate Valve	M	

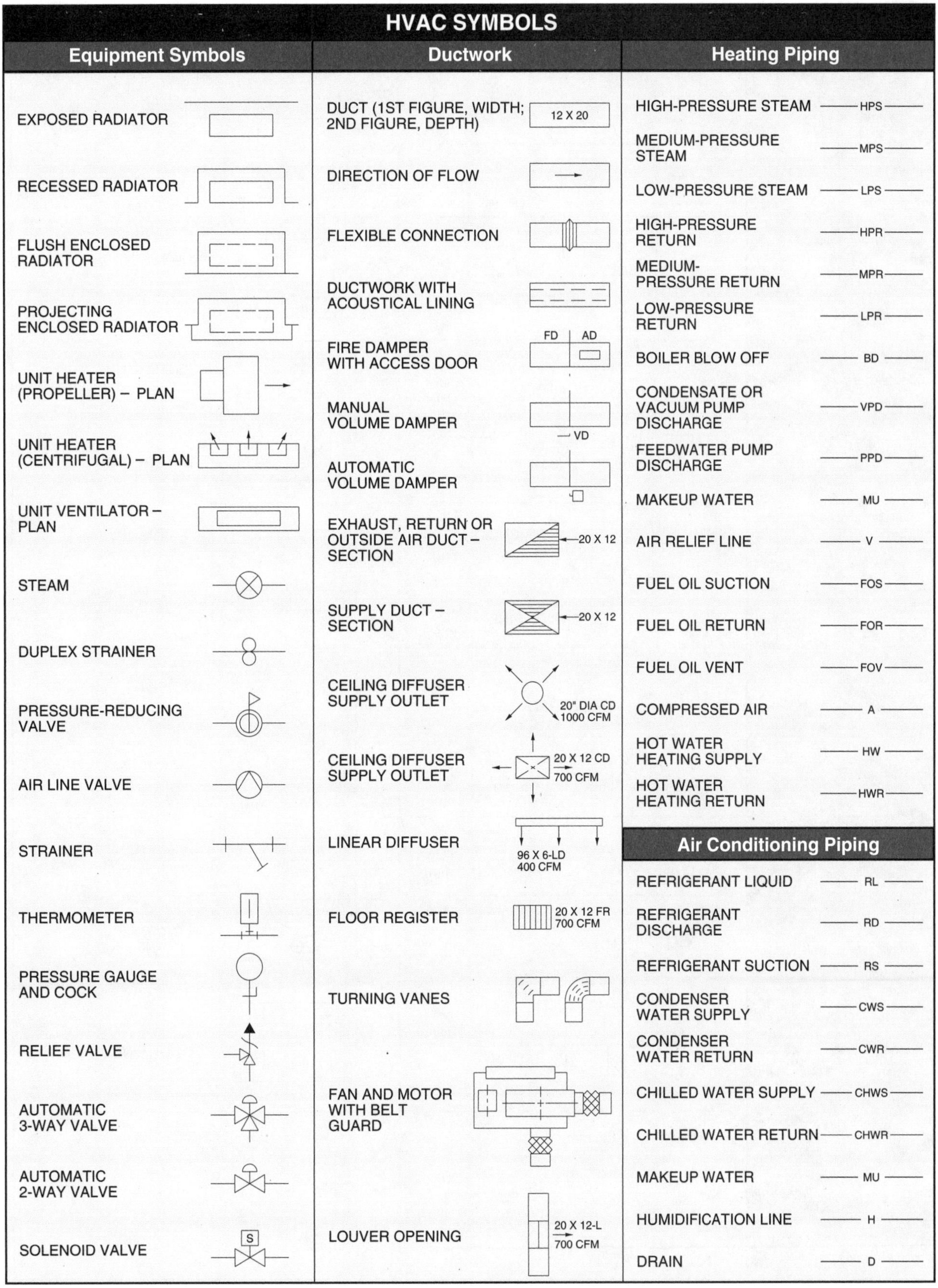

HVAC SYMBOLS

Equipment Symbols

Symbol
EXPOSED RADIATOR
RECESSED RADIATOR
FLUSH ENCLOSED RADIATOR
PROJECTING ENCLOSED RADIATOR
UNIT HEATER (PROPELLER) – PLAN
UNIT HEATER (CENTRIFUGAL) – PLAN
UNIT VENTILATOR – PLAN
STEAM
DUPLEX STRAINER
PRESSURE-REDUCING VALVE
AIR LINE VALVE
STRAINER
THERMOMETER
PRESSURE GAUGE AND COCK
RELIEF VALVE
AUTOMATIC 3-WAY VALVE
AUTOMATIC 2-WAY VALVE
SOLENOID VALVE

Ductwork

DUCT (1ST FIGURE, WIDTH; 2ND FIGURE, DEPTH)	12 X 20
DIRECTION OF FLOW	
FLEXIBLE CONNECTION	
DUCTWORK WITH ACOUSTICAL LINING	
FIRE DAMPER WITH ACCESS DOOR	FD, AD
MANUAL VOLUME DAMPER	VD
AUTOMATIC VOLUME DAMPER	
EXHAUST, RETURN OR OUTSIDE AIR DUCT – SECTION	20 X 12
SUPPLY DUCT – SECTION	20 X 12
CEILING DIFFUSER SUPPLY OUTLET	20" DIA CD 1000 CFM
CEILING DIFFUSER SUPPLY OUTLET	20 X 12 CD 700 CFM
LINEAR DIFFUSER	96 X 6-LD 400 CFM
FLOOR REGISTER	20 X 12 FR 700 CFM
TURNING VANES	
FAN AND MOTOR WITH BELT GUARD	
LOUVER OPENING	20 X 12-L 700 CFM

Heating Piping

HIGH-PRESSURE STEAM	HPS
MEDIUM-PRESSURE STEAM	MPS
LOW-PRESSURE STEAM	LPS
HIGH-PRESSURE RETURN	HPR
MEDIUM-PRESSURE RETURN	MPR
LOW-PRESSURE RETURN	LPR
BOILER BLOW OFF	BD
CONDENSATE OR VACUUM PUMP DISCHARGE	VPD
FEEDWATER PUMP DISCHARGE	PPD
MAKEUP WATER	MU
AIR RELIEF LINE	V
FUEL OIL SUCTION	FOS
FUEL OIL RETURN	FOR
FUEL OIL VENT	FOV
COMPRESSED AIR	A
HOT WATER HEATING SUPPLY	HW
HOT WATER HEATING RETURN	HWR

Air Conditioning Piping

REFRIGERANT LIQUID	RL
REFRIGERANT DISCHARGE	RD
REFRIGERANT SUCTION	RS
CONDENSER WATER SUPPLY	CWS
CONDENSER WATER RETURN	CWR
CHILLED WATER SUPPLY	CHWS
CHILLED WATER RETURN	CHWR
MAKEUP WATER	MU
HUMIDIFICATION LINE	H
DRAIN	D

ALPHABET OF LINES		
Name and Use	**Conventional Representation**	**Example**
Object Line Define shape. Outline and detail objects.	THICK	OBJECT LINE
Hidden Line Show hidden features.	$\frac{1}{8}''$ (3 mm) THIN $\frac{1}{32}''$ (0.75 mm)	HIDDEN LINE
Centerline Locate centerpoints of arcs and circles.	$\frac{1}{16}''$ (1.5 mm) THIN $\frac{1}{8}''$ (3 mm) $\frac{3}{4}''$ (18 mm) TO $1\frac{1}{2}''$ (36 mm)	CENTERLINE CENTERPOINT
Dimension Line Show size or location. **Extension Line** Define size or location.	DIMENSION LINE DIMENSION THIN 2′-6″ EXTENSION LINE	DIMENSION LINE $1\frac{3}{4}$ EXTENSION LINE
Leader Call out specific features.	OPEN ARROWHEAD THIN X CLOSED ARROWHEAD 3X	$1\frac{1}{2}$ DRILL LEADER
Cutting Plane Show internal features.	THICK $\frac{1}{8}''$ (3 mm) $\frac{1}{16}''$ (1.5 mm) A A $\frac{3}{4}''$ (18 mm) TO $1\frac{1}{2}''$ (36 mm)	A A LETTER IDENTIFIES SECTION VIEW CUTTING PLANE LINE
Section Line Identify internal features.	$\frac{1}{16}''$ (1.5 mm) THIN	SECTION LINES
Break Line Show long breaks. **Break Line** Show short breaks.	$\frac{3}{4}''$ (18 mm) TO $1\frac{1}{2}''$ (36 mm) THIN FREEHAND THICK	LONG BREAK LINE SHORT BREAK LINE

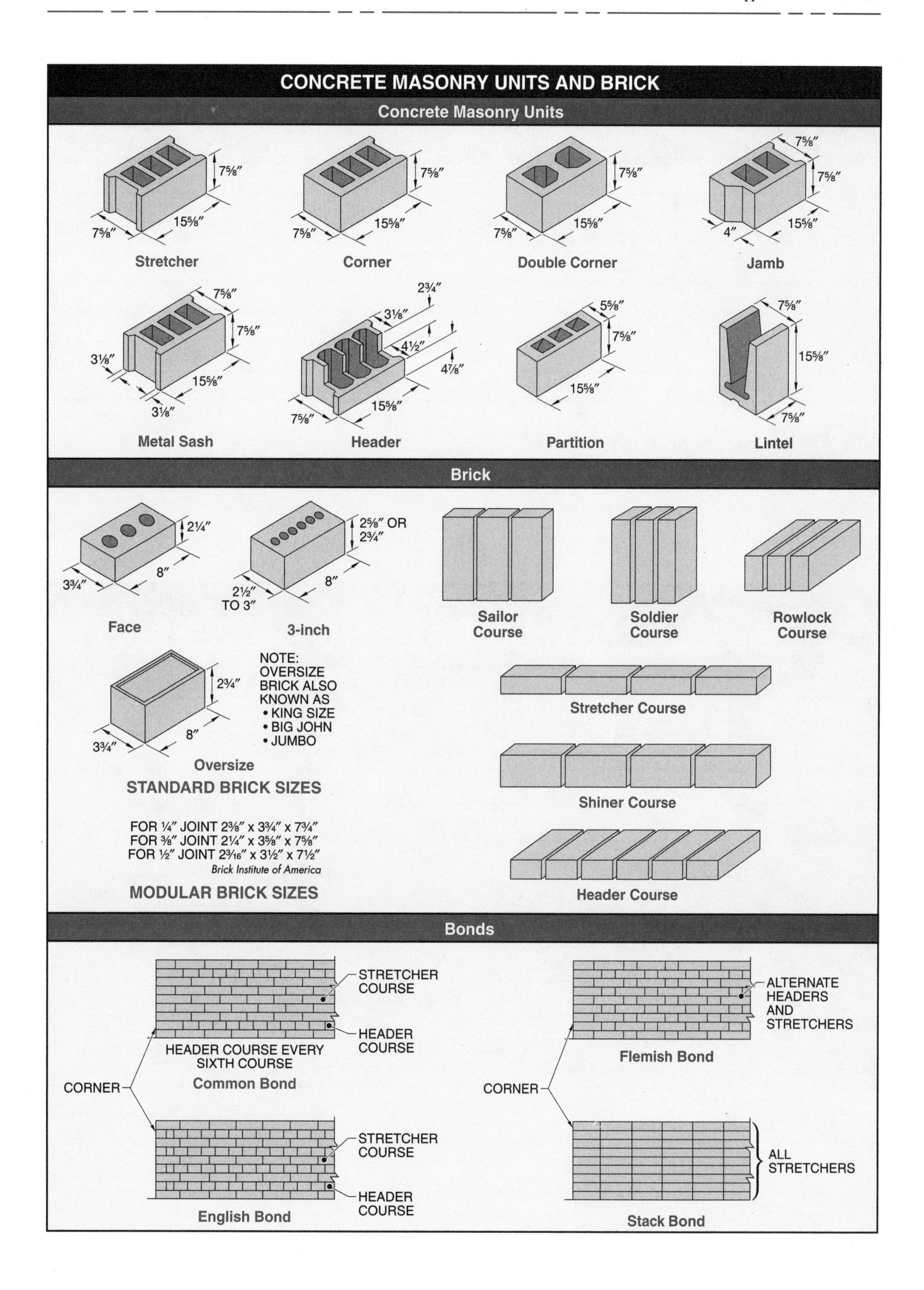
CONCRETE MASONRY UNITS AND BRICK
Concrete Masonry Units
7⅝″
15⅝″
Stretcher
Corner
Double Corner
4″
Jamb
3⅛″
Metal Sash
2¾″
4½″
4⅞″
Header
5⅝″
Partition
Lintel
Brick
2¼″
3¾″
8″
Face
2⅝″ OR
2¾″
2½″
TO 3″
3-inch
Sailor
Course
Soldier
Course
Rowlock
Course
NOTE:
OVERSIZE
BRICK ALSO
KNOWN AS
• KING SIZE
• BIG JOHN
• JUMBO
Oversize
Stretcher Course
Shiner Course
STANDARD BRICK SIZES
FOR ¼″ JOINT 2⅜″ x 3¾″ x 7¾″
FOR ⅜″ JOINT 2¼″ x 3⅝″ x 7⅝″
FOR ½″ JOINT 2³⁄₁₆″ x 3½″ x 7½″
Brick Institute of America
MODULAR BRICK SIZES
Header Course
Bonds
STRETCHER
COURSE
HEADER
COURSE
HEADER COURSE EVERY
SIXTH COURSE
Common Bond
CORNER
English Bond
ALTERNATE
HEADERS
AND
STRETCHERS
Flemish Bond
ALL
STRETCHERS
Stack Bond

COMMON STOCK SIZES OF WELDED WIRE REINFORCEMENT						
New Designation (W-Number)	Old Designation (Wire Gauge)	Diameter*		Steel Area†		Weight‡
				Longitudinal	Transverse	
6 × 6 – W1.4 × W1.4	6 × 6 – 10 × 10	.134	1/8	.028	.028	21
6 × 6 – W2.0 × W2.0	6 × 6 – 8 × 8	.160	5/32	.040	.040	29
6 × 6 – W2.9 × W2.9	6 × 6 – 6 × 6	.192	3/16	.058	.058	42
6 × 6 – W4.0 × W4.0	6 × 6 – 4 × 4	.226	1/4	.080	.080	58
4 × 4 – W1.4 × W1.4	4 × 4 – 10 × 10	.134	1/8	.042	.042	31
4 × 4 – W2.0 × W2.0	4 × 4 – 8 × 8	.160	5/32	.060	.060	43
4 × 4 – W2.9 × W2.9	4 × 4 – 6 × 6	.192	3/16	.087	.087	62
4 × 4 – W4.0 × W4.0	4 × 4 – 4 × 4	.226	1/4	.120	.120	85

* in In.
† in sq in./ft
‡ in lb per 100 sq ft

Wire Reinforcement Institute

STANDARD REBAR SIZES						
Bar Size Designation	Weight per Foot		Diameter		Cross-Sectional Area Squared	
	lb	kg	in.	cm	in.	cm
#3	0.376	0.171	0.375	0.953	0.11	0.71
#4	.0668	0.303	0.500	1.270	0.20	1.29
#5	1.043	0.473	0.625	1.588	0.31	2.00
#6	1.502	0.681	0.750	1.905	0.44	2.84
#7	2.044	0.927	0.875	2.223	0.60	3.87
#8	2.670	1.211	1.000	2.540	0.79	5.10
#9	3.400	1.542	1.128	2.865	1.00	6.45
#10	4.303	1.952	1.270	3.226	1.27	8.19
#11	5.313	2.410	1.410	3.581	1.56	10.07
#14	7.650	3.470	1.693	4.300	2.25	14.52
#18	13.600	6.169	2.257	5.733	4.00	25.81

STANDARD LUMBER SIZES				
Type	Nominal Size*		Actual Size*	
	Thickness	Width	Thickness	Width
Common Boards (Nominal: 1″ × 4″; Actual: ¾″ × 3½″)	1	2	¾	1½
	1	4	¾	3½
	1	6	¾	5½
	1	8	¾	7¼
	1	10	¾	9¼
	1	12	¾	11¼
Dimension (Nominal: 2″ × 4″; Actual: 1½″ × 3½″)	2	2	1½	1½
	2	4	1½	3½
	2	6	1½	5½
	2	8	1½	7¼
	2	10	1½	9¼
	2	12	1½	11¼
Timbers (Nominal: 6″ × 6″; Actual: 5½″ × 5½″)	5	5	4½	4½
	6	6	5½	5½
	6	8	5½	7½
	6	10	5½	9½
	8	8	7½	7½
	8	10	7½	9½

* in in.

DECIMAL EQUIVALENTS OF AN INCH							
Fraction	Decimal	Fraction	Decimal	Fraction	Decimal	Fraction	Decimal
1/64	0.015625	17/64	0.265625	33/64	0.515625	49/64	0.765625
1/32	0.03125	9/32	0.28125	17/32	0.53125	25/32	0.78125
3/64	0.046875	19/64	0.296875	35/64	0.546875	51/64	0.796875
1/16	0.0625	5/16	0.3125	9/16	0.5625	13/16	0.8125
5/64	0.078125	21/64	0.328125	37/64	0.578125	53/64	0.828125
3/32	0.09375	11/32	0.34375	19/32	0.59375	27/32	0.84375
7/64	0.109375	23/64	0.359375	39/64	0.609375	55/64	0.859375
1/8	0.125	3/8	0.375	5/8	0.625	7/8	0.875
9/64	0.140625	25/64	0.390625	41/64	0.640625	57/64	0.890625
5/32	0.15625	13/32	0.40625	21/32	0.65625	29/32	0.90625
11/64	0.171875	27/64	0.421875	43/64	0.671875	59/64	0.921875
3/16	0.1875	7/16	0.4375	11/16	0.6875	15/16	0.9375
13/64	0.203125	29/64	0.453125	45/64	0.703125	61/64	0.953125
7/32	0.21875	15/32	0.46875	23/32	0.71875	31/32	0.96875
15/64	0.234375	31/64	0.484375	47/64	0.734375	63/64	0.984375
1/4	0.250	1/2	0.500	3/4	0.750	1	1.000

DECIMAL EQUIVALENTS OF A FOOT					
Inches	Decimal Foot Equivalent	Inches	Decimal Foot Equivalent	Inches	Decimal Foot Equivalent
1/16	0.0052	4 1/16	0.3385	8 1/16	0.6719
1/8	0.0104	4 1/8	0.3438	8 1/8	0.6771
3/16	0.0156	4 3/16	0.3490	8 3/16	0.6823
1/4	0.0208	4 1/4	0.3542	8 1/4	0.6875
5/16	0.0260	4 5/16	0.3594	8 5/16	0.6927
3/8	0.0313	4 3/8	0.3646	8 3/8	0.6979
7/16	0.0365	4 7/16	0.3698	8 7/16	0.7031
1/2	0.0417	4 1/2	0.3750	8 1/2	0.7083
9/16	0.0469	4 9/16	0.3802	8 9/16	0.7135
5/8	0.0521	4 5/8	0.3854	8 5/8	0.7188
11/16	0.0573	4 11/16	0.3906	8 11/16	0.7240
3/4	0.0625	4 3/4	0.3958	8 3/4	0.7292
13/16	0.0677	4 13/16	0.4010	8 13/16	0.7344
7/8	0.0729	4 7/8	0.4063	8 7/8	0.7396
15/16	0.0781	4 15/16	0.4115	8 15/16	0.7448
1	0.0833	5	0.4167	9	0.7500
1 1/16	0.0885	5 1/16	0.4219	9 1/16	0.7552
1 1/8	0.0938	5 1/8	0.4271	9 1/8	0.7604
1 3/16	0.0990	5 3/16	0.4323	9 3/16	0.7656
1 1/4	0.1042	5 1/4	0.4375	9 1/4	0.7708
1 5/16	0.1094	5 5/16	0.4427	9 5/16	0.7760
1 3/8	0.1146	5 3/8	0.4479	9 3/8	0.7813
1 7/16	0.1198	5 7/16	0.4531	9 7/16	0.7865
1 1/2	0.1250	5 1/2	0.4583	9 1/2	0.7917
1 9/16	0.1302	5 9/16	0.4635	9 9/16	0.7969
1 5/8	0.1354	5 5/8	0.4688	9 5/8	0.8021
1 11/16	0.1406	5 11/16	0.4740	9 11/16	0.8073
1 3/4	0.1458	5 3/4	0.4792	9 3/4	0.8125
1 13/16	0.1510	5 13/16	0.4844	9 13/16	0.8177
1 7/8	0.1563	5 7/8	0.4896	9 7/8	0.8229
1 15/16	0.1615	5 15/16	0.4948	9 15/16	0.8281
2	0.1667	6	0.5000	10	0.8333
2 1/16	0.1719	6 1/16	0.5052	10 1/16	0.8385
2 1/8	0.1771	6 1/8	0.5104	10 1/8	0.8438
2 3/16	0.1823	6 3/16	0.5156	10 3/16	0.8490
2 1/4	0.1875	6 1/4	0.5208	10 1/4	0.8542
2 5/16	0.1927	6 5/16	0.5260	10 5/16	0.8594
2 3/8	0.1979	6 3/8	0.5313	10 3/8	0.8646
2 7/16	0.2031	6 7/16	0.5365	10 7/16	0.8698
2 1/2	0.2083	6 1/2	0.5417	10 1/2	0.8750
2 9/16	0.2135	6 9/16	0.5469	10 9/16	0.8802
2 5/8	0.2188	6 5/8	0.5521	10 5/8	0.8854
2 11/16	0.2240	6 11/16	0.5573	10 11/16	0.8906
2 3/4	0.2292	6 3/4	0.5625	10 3/4	0.8958
2 13/16	0.2344	6 13/16	0.5677	10 13/16	0.9010
2 7/8	0.2396	6 7/8	0.5729	10 7/8	0.9063
2 15/16	0.2448	6 15/16	0.5781	10 15/16	0.9115
3	0.2500	7	0.5833	11	0.9167
3 1/16	0.2552	7 1/16	0.5885	11 1/16	0.9219
3 1/8	0.2604	7 1/8	0.5938	11 1/8	0.9271
3 3/16	0.2656	7 3/16	0.5990	11 3/16	0.9323
3 1/4	0.2708	7 1/4	0.6042	11 1/4	0.9375
3 5/16	0.2760	7 5/16	0.6094	11 5/16	0.9427
3 3/8	0.2813	7 3/8	0.6146	11 3/8	0.9479
3 7/16	0.2865	7 7/16	0.6198	11 7/16	0.9531
3 1/2	0.2917	7 1/2	0.6250	11 1/2	0.9583
3 9/16	0.2969	7 9/16	0.6302	11 9/16	0.9635
3 5/8	0.3021	7 5/8	0.6354	11 5/8	0.9688
3 11/16	0.3073	7 11/16	0.6406	11 11/16	0.9740
3 3/4	0.3125	7 3/4	0.6458	11 3/4	0.9792
3 13/16	0.3177	7 13/16	0.6510	11 13/16	0.9844
3 7/8	0.3229	7 7/8	0.6563	11 7/8	0.9896
3 15/16	0.3281	7 15/16	0.6615	11 15/16	0.9948
4	0.3333	7	0.6667	12	1.0000

ENGLISH-TO-METRIC CONVERSION			
Quantity	**To Convert**	**To**	**Multiply By**
Length	inches	millimeters	25.4
	inches	centimeters	2.54
	feet	centimeters	30.48
	feet	meters	.3048
	yards	centimeters	91.44
	yards	meters	.9144
Area	square inches	square millimeters	645.2
	square inches	square centimeters	6.452
	square feet	square centimeters	929.0
	square feet	square meters	.0929
	square yards	square meters	.8361
Volume	cubic inches	cubic millimeters	1639
	cubic inches	cubic centimeters	16.39
	cubic feet	cubic centimeters	2.832
	cubic feet	cubic meters	.02832
	cubic yards	cubic meters	.7646
Liquid Measure	pints	cubic centimeters	473.2
	pints	liters	.4732
	quarts	cubic centimeters	946.3
	quarts	liters	.9463
	gallons	cubic centimeters	3785
	gallons	liters	3.785
Weight	ounces	grams	28.35
	ounces	kilograms	.02835
	pounds	grams	453.6
	pounds	kilograms	.4536
	short tons (2000 lb)	kilograms	907.2
	short tons (2000 lb)	metric ton (1000 kg)	.9072
Pressure	inches of water colum	kilopascals	.2491
	feet of water column	kilopascals	2.989
	pounds per square inc	kilopascals	6.895
Temperature	degrees Fahrenheit (°F)	degrees Celsius (°C)	$\frac{5}{9}$(°F−32)

METRIC-TO-ENGLISH CONVERSION			
Quantity	**To Convert**	**To**	**Multiply By**
Length	millimeters	inches	.03937
	centimeters	inches	.3937
	meters	feet	3.281
	meters	yards	1.0937
Area	square millimeters	square inches	.00155
	square centimeters	square inches	.1550
	square centimeters	square feet	.0010
	square meters	square feet	10.76
	square meters	square yards	1.196
Volume	cubic centimeters	cubic inches	.06102
	cubic meters	cubic feet	35.31
	cubic meters	cubic yards	1.308
Liquid Measure	liters	pints	2.113
	liters	quarts	1.057
	liters	gallons	.2642
Weight	grams	ounces	.03527
	kilograms	pounds	2.205
	metric ton (1000 kg)	pounds	2205
Pressure	kilopascals	inches of water colu	4.014
	kilopascals	feet of water colum	.3346
	kilopascals	pounds per square inch	.1450
Temperature	degrees Celsius (°C)	degrees Fahrenheit (°F)	($\frac{9}{5}$ °F) + 32

DECIMAL AND METRIC EQUIVALENTS		
Fractions	**Decimal Inches**	**Millimeters**
1/16	.0625	1.58
1/8	.125	3.18
3/16	.1875	4.76
1/4	.250	6.35
5/16	.3125	7.97
3/8	.375	9.52
7/16	.4375	11.11
1/2	.500	12.70
9/16	.5625	14.29
5/8	.625	15.88
11/16	.6875	17.46
3/4	.750	19.05
13/16	.8125	20.64
7/8	.875	22.22
1	1.00	25.40

ABBREVIATIONS . . .

Term	Abbreviation
A	
above	ABV
access	ACS
access panel	AP
acoustic	AC or ACST
acoustical plaster ceiling	APC
acoustical tile	AT. or ACT.
adjacent	ADJ
adjustable	ADJT or ADJ
aggregate	AGG or AGGR
air circulating	ACIRC
air conditioner	AIR COND
air conditioning	A/C or AIR COND
alloy	ALY
alloy steel	ALY STL
alternate	ALTN
alternating current	AC
aluminum	AL
ambient	AMB
American National Standard	AMER NATL STD
American National Standards Institute	ANSI
American Steel Wire Gauge	ASWG
American Wire Gauge	AWG
ampere	A or AMP
anchor	AHR
anchor bolt	AB
appearance	APP
apartment	APT.
approximate	APX or APPROX
architectural	ARCH.
architecture	ARCH.
area	A
area drain	AD
as drawn	AD
asphalt	ASPH
asphalt roof shingles	ASPHRS
asphalt tile	AT.
automatic	AUTO.
auxiliary	AUX
avenue	AVE
azimuth	AZ
B	
barrier	BARR
barrier, moisture vapor-proof	BMVP
barrier, waterproof	BWP
basement	BSMT
bathroom	B
bathtub	BT
batten	BATT
beam	BM
bearing	BRG
bearing plate	BPL or BRG PL
bedroom	BR
below	BLW
bench mark	BM
beveled wood siding	BWS
bituminous	BIT
blocking	BLKG
blueprint	BP
board	BD
board foot	BF or BD FT
boiler	BLR
bookcase	BC
book shelves	BK SH
boulevard	BLVD
boundary	BDRY
brass	BRS
breaker	BRKR
brick	BRK
British thermal unit	BTU
bronze	BRZ
broom closet	BC
building	BLDG or BL
building line	BL
built-in	BLTIN
built up roofing	BUR
C	
cabinet	CAB.
cable	CA
canopy	CAN.
caulking	CK or CLKG
cantilever	CANV
cased opening	CO
casing	CSG
cast iron	CI
cast-iron pipe	CIP
cast steel	CS
cast stone	CST or CS
catch basin	CB
caulked joint	CLKJ
cavity	CAV
ceiling	CLG
cellar	CEL
cement	CEM
cement floor	CF
cement mortar	CEM MORT
center	CTR
centerline	CL
center matched	CM
center-to-center	C TO C
central	CTL
ceramic	CER
ceramic tile	CT
ceramic-tile base	CTB
ceramic-to-metal (seal)	CERMET
chamfer	CHAM or CHMFR
channel	CHAN
check valve	CV
chimney	CHM
chord	CHD
cinder block	CINBL
circle	CIR
circuit	CKT
circuit breaker	CB or CIR BKR
circuit interrupter	CI
circumference	CRCMF
cleanout	CO
clear glass	CL GL
closet	C, CL, CLO, or CLOS
coaxial	COAX.
cold air	CA
cold-rolled	CR
cold-rolled steel	CRS
cold water	CW
collar beam	COL B
color code	CC
combination	COMB.
combustible	COMBL
combustion	COMB.
common	COM
composition	COMP
concrete	CONC
concrete block	CCB or CONC BLK
concrete floor	CCF, CONC FLR, or CONC FL
concrete masonry unit	CMU
concrete pipe	CP
concrete splash block	CSB
condenser	COND
conductor	CNDCT
conduit	CND
construction	CONSTR
construction joint	CJ or CONSTR JT
continuous	CONT
contour	CTR
contract	CONTR or CONT
contractor	CONTR
control joint	CJ or CLJ
conventional	CVNTL
copper	CU
corner	COR
cornice	COR
corrugate	CORR
counter	CNTR
county	CO
cubic	CU
cubic feet	CFT or CU FT
cubic foot per minute	CFM
cubic foot per second	CFS
cubic inch	CU IN.
cubic yard	CU YD
current	CUR
cutoff	CO
cutoff valve	COV
cut out	CO
D	
damper	DMPR
datum	DAT
decibel	DB

. . . ABBREVIATIONS . . .

Term	Abbreviation
degree	DEG
depth	DP
design	DSGN
detail	DTL or DET
diagonal	DIAG
diagram	DIAG
dimension	DIM.
dimmer	DIM. or DMR
dining room	DR or DNG RM
direct current	DC
direction	DIR
disconnect	DISC.
disconnect switch	DS
dishwasher	DW
distribution panel	DPNL
ditto	DO.
door	DR
door stop	DST
door switch	DSW
dormer	DRM
double-acting	DA or DBL ACT
double-hung window	DHW
double-pole double-throw	DPDT
double-pole double-throw switch	DPDT SW
double-pole single-throw	DPST
double-pole single-throw switch	DPST SW
double-pole switch	DP SW
double-strength glass	DSG
down	DN or D
downspout	DS
dozen	DOZ
drain	D or DR
drain tile	DT
drawer	DWR
drawing	DWG
dryer	D
drywall	DW
dwelling	DWEL
E	
each	EA
east	E
elbow	ELB
electric or electrical	ELEC
electrical metallic tubing	EMT
electric operator	ELECT. OPR
electric panel	EP
electromechanical	ELMCH
elevation	ELEV
enamel	ENAM
end-to-end	E to E
entrance	ENTR
equipment	EQPT
equivalent	EQUIV
estimate	EST
example	EX
excavate	EXCA or EXC
exchange	EXCH
exhaust	EXH
exhaust vent	EXHV
existing	EXST
expanded metal	EM
expansion joint	EXP JT
exterior	EXT
exterior grade	EXT GR
F	
face brick	FB
faceplate	FP
Fahrenheit	°F
fiberboard, solid	FBDS
finish	FIN. or FNSH
finish all over	FAO
finish grade	FG
finish one side	FIS
finish two sides	F2S
finished floor	FIN. FLR, FIN. FL, or FNSH FL
firebrick	FBRK or FBCK
fireplace	FPL or FP
fireproof	FP or FPRF
fire resistant	FRES
fixed transom	FTR
fixed window	FX WDW
fixture	FIX. or FXTR
flashing	FIG or FL
floor	FLR or FL
floor drain	FD
flooring	FLR or FLG
fluorescent	FLUR or FLUOR
flush	FL
footing	FTG
foundation	FND or FDN
frame	FR
frostproof hose bibb	FPHB
full scale	FSC
full size	FS
furnace	FURN
furred ceiling	FC
furring	FUR
fuse	FU
fuse block	FB
fusebox	FUBX
fuseholder	FUHLR
fusible	FSBL
G	
gallon	GAL.
gallon per hour	GPH
gallon per minute	GPM
galvanized iron	GI or GALVI
galvanized steel	GS or GALVS
garage	GAR.
gas	G
gate valve	GTV
gauge	GA
general contractor	GEN CONT
girder	G
glass	GL
glass block	GLB or GL BL
glaze	GLZ
glued laminated	GLULAM
grade	GR
grade line	GL
gravel	GVL
grill	G
gross weight	GRWT
ground	GRD
ground (outlet)	G
ground-fault circuit interrupter	GFCI
ground-fault interrupter	GFI
gypsum	GYP
gypsum board	GYP BD
gypsum-plaster ceiling	GPC
gypsum-plaster wall	GPW
gypsum sheathing board	GSB
gypsum wallboard	GWB
H	
hardboard	HBD
hardware	HDW
header	HDR
heat	HT
heated	HTD
heater	HTR
heating	HTG
heating, ventilating, and air conditioning	HVAC
height	HGT
hexagon	HEX.
high density overlay	HDO
high point	HPT
highway	HWY
hinge	HNG
hollow-core	HC
hollow metal door	HMD
honeycomb	HNYCMB
horizontal	HOR or HORZ
horsepower	HP
hose bibb	HB
hot air	HA
hot water	HW
humidity	HMD
I	
illuminate	ILLUM
incandescent	INCAND
inch	IN.
inch per second	IPS
inside diameter	ID
install	INSTL
insulation	INS or INSUL
interior	INT
iron	I
J	
jamb	JB or JMB
joint	JT
joist	J
K	
kiln dried	KD
kitchen	K, KT, or KIT.

. . . ABBREVIATIONS . . .

Term	Abbreviation
L	
laminate	LAM
laminated veneer lumber	LVL
landing	LDG
lateral	LATL
lath	LTH
laundry	LAU
laundry tray	LT
lavatory	LAV
leader	L
left hand	LH
length	L, LG, or LGTH
level	LVL
library	LIB
living room	LR
light	LT
light switch	LT SW
limestone	LMS or LS
linen closet	L CL
lining	LN
linoleum	LINO
linoleum floor	LF or LINO FLR
lintel	LNTL
living room	LR
local	LCL
louver	LVR or LV
low point	LP
lumber	LBR
M	
main	MN
makeup	MKUP
manufactured	MFD
marble	MRB or MR
masonry	MSNRY
masonry opening	MO
material	MTL or MATL
maximum	MAX
median	MDN
medicine cabinet	MC
medium	MDM
medium density overlay	MDO
meridian	MER
metal	MET.
metal anchor	MA
metal door	METD
metal flashing	METF
metal threshold	MT
mineral	MNRL
minimum	MIN
mirror	MIR
miscellaneous	MISC
miter	MIT
mixture	MIX.
modular	MOD
molding	MLD or MLDG
mortar	MOR
N	
National Electrical Code®	NEC®
National Electrical Safety Code	NESC
natural grade	NG
negative	(–) or NEG
noncombustible	NCOMBL
north	N
nosing	NOS
not to scale	NTS
O	
obscure glass	OBSC GL
octagon	OCT
on center	OC
one-pole	SP
opening	OPG or OPNG
open web joist	OJ, OW J, or OW JOIST
opposite	OPP
optional	OPT
ordinance	ORD
oriented strand board	OSB
outlet	OUT.
outside diameter	OD
out-to-out	O TO O
overall	OA
overcurrent	OC
overcurrent relay	OCR
overhead	OH or OVHD
P	
paint	PNT
panel	PNL
pantry	PAN.
parallel	PRL
parallel strand lumber	PSL
partition	PTN
passage	PASS.
penny (nails, etc.)	d
perimeter	PERIM
perpendicular	PERP
per square inch	PSI
phase	PH
piping	PP
plaster	PLAS or PL
plastered open	PO
plate	PL
plate glass	PG, PL GL, or PLGL
platform	PLAT
plumbing	PLBG
plywood	PLYWD
point	PT
point of beginning	POB
polyvinyl chloride	PVC
porch	P
pound	LB
power	PWR
power supply	PWR SPLY
precast	PRCST
prefabricated	PFB or PREFAB
prefinished	PFN
property	PROP.
property line	PL
pull switch	PS
pump	PMP
Q	
quadrant	QDRNT
quarry tile	QT
quarry tile base	QTB
quarry tile floor	QTF
quarter	QTR
quarter-round	¼RD
R	
radiator	RAD or RDTR
raised	RSD
random	RDM
range	R
receptacle	RCPT
recessed	REC
rectangle	RECT
redwood	RWD
reference	REF
reference line	REFL
reflected	REFLD
refrigerator	REF or REFR
register	REG or RGTR
reinforce or reinforcing	RE or REINF
reinforced concrete	RC
reinforcing steel	RST
reinforcing steel bar	REBAR
required	REQD
retaining	RETG
revision	REV
revolution per minute	RPM
revolution per second	RPS
right hand	RH
riser	R
road	RD
roof	RF
roof drain	RD
roofing	RFG
room	RM or R
rough	RGH
rough opening	RO or RGH OPNG
rough-sawn	RS
round	RND
rubber	RBR
rubber tile	RBT or R TILE
rustproof	RSTPF
S	
safety	SAF
sanitary	S
S-beam	S
scale	SC
schedule	SCH or SCHED
screen	SCN, SCR, or SCRN
screen door	SCD
screw	SCR
scuttle	S
section	SEC or SECT.
select	SEL

. . . ABBREVIATIONS

Term	Abbreviation
self-cleaning	SLFCLN
self-closing	SELF CL
service	SERV or SVCE
sewer	SEW.
sheathing	SHTH or SHTHG
sheet	SHT or SH
sheeting	SH
sheet metal	SM
shelf and rod	SH&RD
shelving	SH or SHELV
shingle	SHGL
shower	SH
shower and toilet	SH & T
shower drain	SD
shutter	SHTR
sidelight	SI LT
sill cock	SC
single-phase	1PH
single-pole	SP
single-pole double-throw	SPDT
single-pole double throw switch	SPDT SW
single-pole single-throw	SPST
single-pole single-throw switch	SPST SW
single-pole switch	SP SW
single-strength glass	SSG
single-throw	ST
sink	SK or S
skylight	SLT
sliding door	SLD or SL DR
slope	SLP
soffit	SF
soil pipe	SP
soil stack	SSK
solid core	SC
soundproof	SNDPRF
south	S
specific	SP
specification	SPEC
splash block	SB
square	SQ
square feet	SQ FT
square inch	SQ IN.
square yard	SQ YD
stack	STK
stained	STN
stainless steel	SST
stairs	ST
stairway	STWY
standard	STD
steel	ST or STL
steel sash	SS
stone	STN
storage	STO or STG
street	ST or STR
structural	STRL
Structural Clay Products Research Foundation	SCR
structural clay tile	SCT
structural glass	SG
supply	SPLY
survey	SURV
suspended	SUSP
switch	SW or S
T	
telephone	TEL
television	TV
temperature	TEMP
tempered plate glass	TEM PL GL
terra cotta	TC
terazzo	TZ or TER
thermostat	THERMO
thick	THK
threshold	TH
tile base	TB
tile drain	TD
tile floor	TF
timber	TMBR
toilet	T
tongue-and-groove	T & G
township	T
tread	TR or T
typical	TYP
U	
underground	UGND
unexcavated	UNEXC
unfinished	UNFIN or UNF
unit heater	UH
unless otherwise specified	UOS
untreated	UTRTD
utility	U or UTIL
utility room	UR or U RM
V	
vacuum	VAC
valley	VAL
valve	V
variance	VAR
vent	V
vent hole	VH
ventilate	VEN
ventilating equipment	VE
vent pipe	VP
vent stack	VS
vertical	VERT
vestibule	VEST.
vinyl tile	VT or V TILE
vitreous tile	VIT TILR
void	VD
volt	V
voltage	V
voltage drop	VD
volt amp	VA
volume	VOL
W	
wainscot	WSCT, WAIN., or WA
walk in closet	WIC
wall	W
wallboard	WLB
wall receptacle	WR
warm air	WA
washing machine	WM
water	WTR or W
water closet	WC
water heater	WH
water line	WL
water meter	WM
waterproof	WTRPRF
water-resistant	WR
watt	W
weatherproof	WTHPRF or WP
weather-resistant	WR
weather stripping	WS
weep hole	WH
welded wire fabric	WWF
west	W
white pine	WP
wide	W
wide flange	W or WF
window	WDO
wood	WD
wood frame	WF
wrought iron	WI
Y	
yard	YD
yellow pine	YP
Z	
Zone	Z

Glossary

abbreviation: Letter or series of letters of words denoting a complete word.

acronym: Abbreviated word formed from the first letter of each word that describes the article.

acute triangle: Triangle in which all angles are less than 90°.

addition: Process of combining two or more whole numbers to obtain a total.

air conditioner: Piece of equipment used to control humidity and temperature and remove airborne impurities in a house.

air conditioning: Process used to control humidity, temperature, and/or cleanliness of air within a house.

arc: Portion of the circumference of a circle.

architect's scale: Drafting scale used when producing drawings of buildings and structural parts.

area: Two-dimensional surface measurement.

balloon framing: System of frame construction in which one-piece exterior wall studs extend from the first floor line or sill plate to the double top plate.

baseline: East-West line that defines a group of townships on a grid.

benchmark: Stake driven into the ground or a point in a street or along a curb that is used as the point of beginning.

blower: Piece of equipment that draws air from rooms and moves it through the furnace where it is heated before being conveyed back into the rooms.

boiler: Closed tank connected to an energy source that heats water to a high temperature.

brick bond: Pattern formed by exposed faces of brick.

brick veneer construction: Frame construction with brick exterior facing.

bridging: Metal or wood cross bridging or solid wood blocking installed between joists to stiffen a floor unit and prevent joists from twisting.

built-up beam: Structural member made of laminated wood members, designed to carry heavy loads.

chord: A line from circumference to circumference that does not pass through the centerpoint of a circle.

circle: Plane figure generated around a centerpoint.

circulating pump: Device that moves water through supply water piping in a hot water heating system.

circumference: Outside edge of a circle.

civil engineer's scale: Drafting scale used when creating maps and survey drawings.

compass: Drafting instrument used to draw arcs and circles.

compression tank: Tank that absorbs and relieves pressure caused by water expansion when heated.

computer: Device that receives information from an input device, processes it, and displays the results on a monitor.

contemporary design: Architectural design that reflects current trends and may include more complex rooflines, geometric windows, and curved or angled walls.

contour line: Dashed or solid line on a plot plan that passes through points having the same elevation.

cutting plane line: Line that identifies the exact place where a cutting plane passes through a feature.

decimal fraction: Fraction with a denominator of 10, 100, 1000, 10,000, and so on.

denominator: Part of a fraction that indicates the total number of parts or divisions of a whole number.

detail: Scaled plan, elevation, or section drawn to a larger scale to show special features.

diameter: Distance from circumference to circumference passing through the centerpoint of a circle.

dividend: Number to be divided in the division operation.

dividers: Drafting instrument used to transfer dimensions.

division: Mathematical process of determining the number of times one number is contained in another number.

divisor: Number that a dividend is divided by in the division operation.

door hand: Direction a door swings.

dormer: Projection from a sloping roof that provides additional interior area.

duplex nail: Double-headed nail designed to be pulled out easily to facilitate the stripping of forms.

easement: Strip of privately owned land set aside for placement of public utilities.

electric radiant heating system: Radiant heating system in which heat is generated when electricity meets resistance as it flows through embedded heating cables or baseboard units.

elevation: **1.** A scaled view looking directly at the walls. **2.** A vertical measurement above or below the point of beginning.

ellipse: Plane figure generated by the sum of the distances from two fixed points.

equilateral triangle: Acute triangle that contains three equal sides and three 60° angles.

even number: Number that can be divided by two without a remainder or decimal occurring.

exterior elevation: Scaled view that shows the shape and size of the outside walls of the house and the roof.

exterior insulation and finish system (EIFS): System used to provide exterior building protection through application of exterior insulation, insulation board, reinforcing mesh, a base coat of acrylic copolymers and portland cement, and a finish coat of acrylic resins.

finish grade: Slope of the land after final grading.

fire cut: Angled cut in the end of a wood joist that allows a burnt joist to fall out without disturbing the solid brick wall.

floor plan: Scaled view of the various floors in a house looking directly down from a horizontal cutting plane taken 5′-0″ above each finished floor.

flush door: Door with flat surfaces with the stiles and rails within the door.

footing: Support base for a foundation wall.

forced warm air heating system: Residential heating system that uses a blower to draw air (return air) from rooms through return air grilles and ductwork.

function key: Key on a keyboard that performs a special task.

furnace: Piece of equipment that heats the air circulating in a forced warm air heating system.

general note: Notation that refers to an entire set of prints.

geothermal heat: Heat that is derived from heat contained within the earth.

hollow-core door: Flush door with wood surface veneers, a solid wood frame, and a mesh core.

horizontal line: Line that is level or parallel to the horizon.

hot water radiant heating (hydronic) system: Radiant heating system in which water is heated in a boiler at a central location and then distributed to the desired rooms through tubing.

improper fraction: Fraction with the numerator larger than the denominator. An improper fraction must be converted to a mixed number.

interior elevation: Scaled view that shows the shapes, sizes, and finishes of interior walls and partitions of a house.

irregular polygon: Polygon that has unequal sides and unequal angles.

isometric drawing: Pictorial drawing with horizontal lines drawn 30° above (or below) the horizon.

isosceles triangle: A triangle that contains two equal sides and two equal angles.

job-built form: Foundation wall form constructed piece by piece on top of a footing.

joist: Horizontal framing member that supports a floor.

keyboard: Input device containing standard alpha and numeric keys as well as function keys.

keyway: Groove formed in fresh concrete that interlocks with concrete from a foundation wall.

kilogram: Basic unit of weight in the SI metric system.

light: Pane of glass.

location dimension: Dimension that locates a particular feature in relation to another feature.

longitudinal section: Section created by passing a cutting plane through the long dimension of a house.

masonry construction: System of construction in which masonry units such as brick, concrete masonry units, stone, or structural clay tile are formed into walls to carry the load of floor and roof joists.

mechanical engineer's scale: Drafting scale used when drawing machines and machine parts.

meridian: North-South line that defines a group of townships on a grid.

meter: Basic unit of measurement in the SI metric system.

minuend: Total number of units prior to subtraction.

mixed decimal number: Decimal number consisting of a whole number and a decimal number separated by a decimal point.

mixed number: Fraction containing a whole number and a fraction.

model code: National building code developed through conferences between building officials and industry representatives around the country.

modular brick: Brick classified by its actual size and designed so every third horizontal joint falls on a multiple of 4″ (modular measure).

monitor: High-resolution color display screen that allows drawings, data, and text to be displayed.

multiplicand: Number being multiplied in the multiplication operation.

multiplication: Mathematical process of adding a number to itself as many times as there are units indicated by the other number.

multiplier: Number in the multiplication operation that indicates the number of times the addition should occur.

natural grade: Slope of the land before rough grading.

nominal size: Size of a piece of wood before it is planed to finished size.

numerator: Part of a fraction that indicates the number of parts included in the fraction.

oblique drawing: Pictorial drawing with one surface drawn in true shape and receding lines projecting back from the face.

obtuse triangle: Triangle that includes one angle greater than 90°.

odd number: Number that cannot be divided by two without a remainder or decimal occurring.

operating controls: Controls that cycle equipment ON and OFF as required.

orthographic projection: Drawing in which each face of an object is projected onto flat planes, generally at 90° to one another.

panel door: Door with individual panels between the stiles and rails.

panel form: Prefabricated concrete wall form consisting of a metal frame and metal or wood panel facing, or wood studs and plates with wood panel facing.

panned ceiling: Ceiling consisting of two ceiling levels connected by sloped surfaces.

perspective drawing: Pictorial drawing with all receding lines converging to vanishing points.

pictorial drawing: Three-dimensional representation of an object.

piping: Tubing used to distribute hot water from the boiler to terminal devices and return water back to the boiler to be reheated.

pitch: Angle a roof slopes from the roof ridge to the outside walls of the building.

place: Position that a digit occupies; it represents the value of the digit.

plane figure: Geometric shape with a flat surface.

platform framing: System of frame construction in which each story of a building is framed as a unit, with studs being one story in height.

plot plan: Scaled drawing that shows the shape and size of the building lot; location, shape, and overall size of a house on the lot; and the finish floor elevation.

point of beginning (POB): Location point from which horizontal dimensions and vertical elevations are made.

polygon: Many-sided plane figure that is bounded by straight lines.

post-and-beam framing: System of frame construction in which posts and beams provide the primary structural support.

potable water: Water free from impurities in amounts that could cause disease or harmful effects.

power controls: Controls located in the electrical conductors leading to the furnace and blower.

prime number: Odd number that can only be divided by 1 or itself without a remainder or decimal occurring.

print: Reproduction of a working drawing.

product: Result of multiplication.

proper decimal number: Decimal number that does not have a whole number, such as .9.

quadrant: One-fourth of a circle.

quadrilateral: Four-sided plane figure.

quotient: The result of the division operation.

raceway: Enclosed channel designed to protec ri-cal conductors.

radiant heating system: Heating system that transfers heat to the living space via hot water tubing or electrical cables embedded in the floor or ceiling.

radius: One-half the diameter of a circle.

rebar: Deformed steel bar used to reinforce concrete structural members.

rectangle: Quadrilateral that contains four 90° angles with opposite sides equal.

regular polygon: Polygon that has equal sides and equal angles.

result: Difference between the minuend and subtrahend in the subtraction operation.

rhomboid: Quadrilateral that has opposite sides equal, opposite angles equal, but does not contain any 90° angles.

rhombus: Quadrilateral that contains four equal sides with opposite angles equal and no 90° angles.

right triangle: Triangle that contains one 90° angle.

riser: Vertical portion of a stair step.

safety controls: Controls that prevent injury to personnel or damage to equipment in the event of equipment malfunction.

scale: Drafting tool used to measure lines and reduce or enlarge them proportionally.

scuttle: Access opening in a ceiling or roof with a removable or movable cover.

section: **1.** Scaled view created by passing a cutting plane through a portion of a building. **2.** Division of a township and is one mile long on each side.

sector: Pie-shaped segment of a circle.

semicircle: One-half of a circle.

size dimension: Dimension that indicates the size of a particular area or feature.

sketching: Process of drawing without instruments.

slab-at-grade foundation: Ground-supported foundation system that combines short concrete foundation walls or a thickened concrete edge with a concrete floor slab.

slanted line: Inclined line that is neither horizontal nor vertical.

slope: Relationship of unit rise to unit run of a roof.

solar energy: Energy available from the sun in the form of sunlight.

solid-core door: Flush door with wood surface veneer and an inner core made of solid wood blocks, engineered wood products, or high-density foam.

specifications: Written supplements to working drawings that provide additional building information.

square: Quadrilateral that contains four equal sides and four 90° angles.

standard brick: Brick classified by its nominal size.

structural insulated panel: Structural member consisting of a thick layer of rigid foam insulation pressed between two OSB or plywood panels.

subtraction: Mathematical process of taking one number away from another number.

subtrahend: Number of units to be removed from the minuend in the subtraction operation.

sum: Total that results from addition.

survey plat: Map showing a division of land, such as a portion of a quarter section of a township subdivided into streets and lots.

symbol: Pictorial representation of a structural or material component used on prints.

terminal device: Device that extracts heat from hot water to heat the air in the desired rooms.

title block: Area on a working drawing or print that is used to provide written information about the drawing or print.

township: Square area that is six miles long on each side, or 36 square miles.

traditional design: Architectural design that reflects long-standing design elements.

transverse section: Section created by passing a cutting plane through the short dimension of a house.

trapezium: Quadrilateral that has no sides parallel.

trapezoid: Quadrilateral that has two sides parallel.

tread: Horizontal portion of a stair step.

triangle: **1.** Drafting tool used to draw vertical and inclined lines. **2.** Three-sided plane figure that contains 180°.

T-square: Drafting tool used to draw horizontal lines and as a reference base for positioning triangles.

typical note: Notation that refers to all similar items on the prints.

unit controls: Controls installed on a furnace by the manufacturer or installer to maintain safe and efficient operation of the furnace.

unit rise: Vertical increase in height per foot of run.

unit run: Unit of the total run based on 12″.

vertical line: A line that is plumb or upright.

voice/data/video (VDV) system: Low-voltage electrical system designed for use with various types of information technology equipment, including telephone systems, computer systems and networks, television video systems for cable and/or satellite dish systems, and security and fire alarm systems.

volume: Three-dimensional capacity of a space.

walkout basement: Basement with standard-sized windows to provide light and standard-sized doors for entry and exit.

whole number: Number that does not have a fractional or decimal part.

working drawing: Drawing that contains the graphic information necessary to complete a construction job.

wythe: Single continuous masonry wall, one unit thick.

Index

USING THE PRINTREADING FOR RESIDENTIAL CONSTRUCTION—PART 1 CD-ROM

Before removing the CD-ROM from the protective sleeve, please note that the book cannot be returned for refund or credit if the CD-ROM sleeve seal is broken.

System Requirements

The *Printreading for Residential Construction–Part 1* CD-ROM is designed to work best on a computer meeting the following hardware/software requirements:

- Intel® Pentium® processor
- Microsoft® Windows® 95, 98, 98 SE, Me, NT®, 2000, or XP® operating system
- 64 MB of free available system RAM (128 MB recommended)
- 90 MB of available disk space
- 800 × 600 16-bit (thousands of colors) color display or better
- Sound output capability and speakers
- CD-ROM drive

Opening Files

Insert the CD-ROM into the computer CD-ROM drive. Within a few seconds, the home screen will be displayed allowing access to all features of the CD-ROM. Information about the usage of the CD-ROM can be accessed by clicking on USING THIS CD-ROM. The Chapter Quick Quizzes™, Illustrated Glossary, Stewart Residence Prints, Stewart Residence 3-D Model, Master Math™ Problems, Media Clips, and Reference Material can be accessed by clicking on the appropriate button on the home screen. Clicking on the American Tech web site button (www.go2atp.com) accesses information on related educational products. Unauthorized reproduction of the material on this CD-ROM is strictly prohibited.

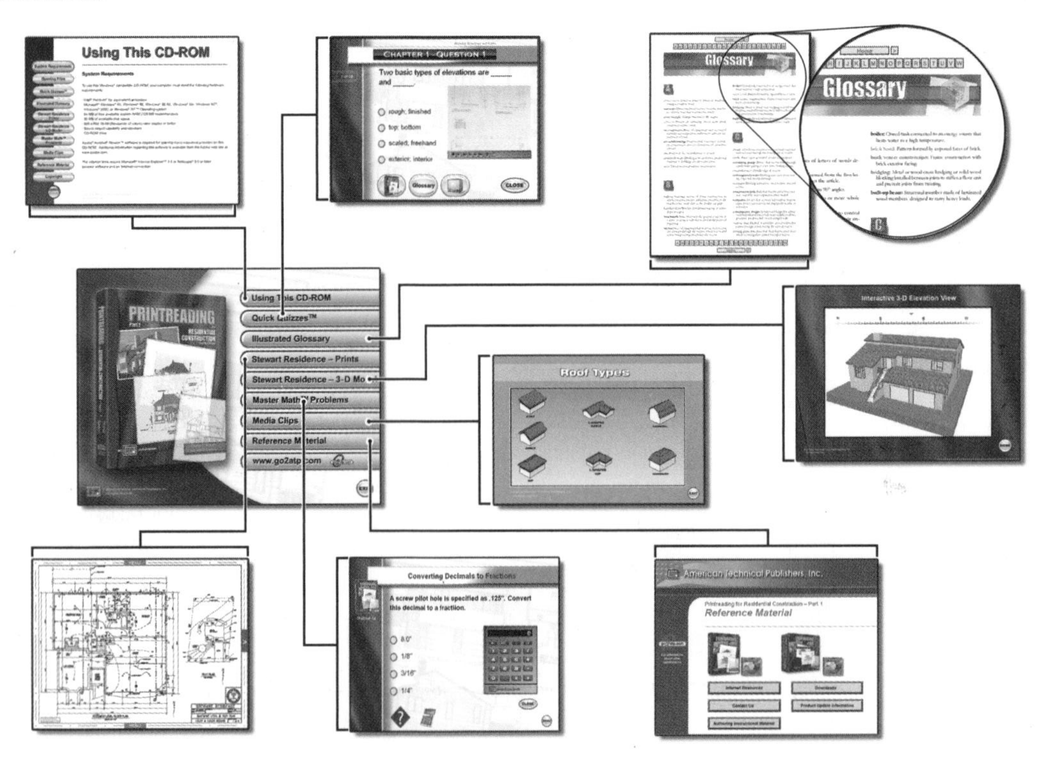

Pentium is a registered trademark of Intel Corporation. Microsoft, Windows, Windows NT, Windows XP, and Internet Explorer are either registered trademarks or trademarks of Microsoft Corporation in the United States and/or other countries. Adobe, Acrobat, and Reader are trademarks of Adobe Systems Incorporated. Netscape is a registered trademark of Netscape Communications Corporation. Quick Quizzes and Master Math are trademarks of American Technical Publishers, Inc.